Mathematical Methods of Many-Body Quantum Field Theory

CHAPMAN & HALL/CRC
Research Notes in Mathematics Series

Submission of proposals for consideration
Suggestions for publication, in the form of outlines and representative samples, are invited by the Editorial Board for assessment. Intending authors should approach one of the main editors or another member of the Editorial Board, citing the relevant AMS subject classifications. Alternatively, outlines may be sent directly to the publisher's offices. Refereeing is by members of the board and other mathematical authorities in the topic concerned, throughout the world.

Preparation of accepted manuscripts
On acceptance of a proposal, the publisher will supply full instructions for the preparation of manuscripts in a form suitable for direct photo-lithographic reproduction. Specially printed grid sheets can be provided. Word processor output, subject to the publisher's approval, is also acceptable.

Illustrations should be prepared by the authors, ready for direct reproduction without further improvement. The use of hand-drawn symbols should be avoided wherever possible, in order to obtain maximum clarity of the text.

The publisher will be pleased to give guidance necessary during the preparation of a typescript and will be happy to answer any queries.

Important note
In order to avoid later retyping, intending authors are strongly urged not to begin final preparation of a typescript before receiving the publisher's guidelines. In this way we hope to preserve the uniform appearance of the series.

CRC Press UK
Chapman & Hall/CRC Statistics and Mathematics
23 Blades Court
Deodar Road
London SW15 2NU
Tel: 020 8875 4370

Detlef Lehmann

Mathematical Methods of Many-Body Quantum Field Theory

CRC Press
Taylor & Francis Group
Boca Raton London New York

CRC Press is an imprint of the
Taylor & Francis Group, an **informa** business
A CHAPMAN & HALL BOOK

CRC Press
Taylor & Francis Group
6000 Broken Sound Parkway NW, Suite 300
Boca Raton, FL 33487-2742

First issued in paperback 2019

ISBN-13: 978-1-58488-490-3 (hbk)
ISBN-13: 978-0-367-39390-8 (pbk)

Library of Congress Cataloging-in-Publication Data

Lehmann, Detlef.
    Mathematical methods of many-body quantum field theory / Detlef Lehmann.
        p. cm. -- (Chapman & Hall/CRC research notes in mathematics series ; 436)
    ISBN 1-58488-490-8 (alk. paper)
    1. Many-body problem. 2. Quantum field theory--Mathematical models. I. Title. II. Series.

QC174.17.P7L44 2004
530.14'3--dc22
                                                      2004056042

Library of Congress Card Number 2004056042

Visit the Taylor & Francis Web site at
http://www.taylorandfrancis.com

and the CRC Press Web site at
http://www.crcpress.com

Preface

In this book we develop the mathematical tools for the description of quantum many-body systems and apply them to the many-electron system. These are the formalism of second quantization, field theoretical perturbation theory, functional integral methods, bosonic and fermionic, and estimation and resummation techniques for Feynman diagrams. The physical effects discussed in this context are mainly BCS superconductivity, s-wave and higher l-wave, and we take a short look to the fractional quantum Hall effect. A central question of this book is, to what extent the approximations, which are done in the BCS theory of superconductivity, or more generally, in the theory of the weak coupling many-electron system, can be mathematically rigorously justified. Thus the style is mathematical in the sense of working with precise definitions and statements at all times, but, as we hope, close to the physics point of view in that we tried to emphasize actually how to compute things, not just proving that they exist and are well defined.

This book came into being as a combination of lecture notes, handed out to students attending the course Mathematical Physics III at TU Berlin, and the Habilitationsschrift of the author, entitled 'Perturbation Theory for the Many-Electron System with Short-Range Interaction and Its Resummation'. As such, we think that the text may be useful for the following groups of readers. First those who want to pursue the project of trying to mathematically rigorously explore the issue of exactly when the standard approximation schemes in this field actually do work, and when they can be shown to break down. Those researchers may find many results of this kind together with a great deal of worked-out detail which should also be useful for approaching similar problems. And second students, who are having trouble figuring out exactly what is going on in one or the other computation while reading the established physics literature, may find some useful supplemental explanations.

I owe special thanks to my former supervisors, J. Feldman, H. Knoerrer and E. Trubowitz, who taught me field theoretical methods and to my former and current employers, ETH Zürich, IAS Princeton, UBC Vancouver and TU Berlin, for excellent working conditions and their, in part, very generous financial support over the last years.

Berlin, February 2004
Detlef Lehmann

Contents

Dependence of Chapters

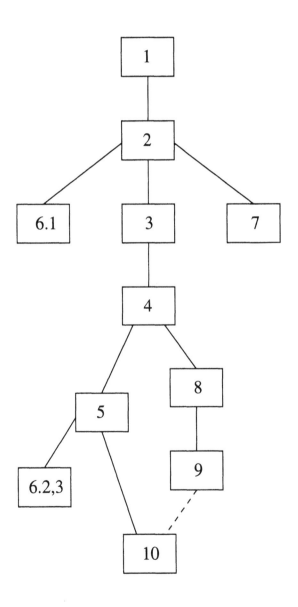

Chapter 1

Introduction

The computation of field theoretical correlation functions is a very difficult problem. These functions encode the physical properties of the model under consideration and therefore it is important to know how these functions behave. As it is the case for many mathematical objects which describe some not too idealized systems, also these functions, in most cases, cannot be computed explicitly. Thus the question arises how these functions can be controlled.

A quantum many-body system is given by a Hamiltonian $H(\lambda) = H_0 + \lambda H_{\text{int}}$. Here, usually, the kinetic energy part H_0 is exactly diagonalizable and H_{int} describes the particle-particle, the many-body interaction. There are many situations where it makes sense to consider a small coupling λ. In such a situation it is reasonable to start with perturbation theory. That is, one writes down the Taylor series around $\lambda = 0$ which is the expansion into Feynman diagrams. Typically, some of these diagrams diverge if the cutoffs of the theory are removed. This does not mean that something is wrong with the model, but merely means first of all that the function which has been expanded is not analytic if the cutoffs are removed. The following example may be instructive. Let

$$G_\delta(\lambda) := \int_0^\infty dx \int_0^1 dk \frac{1}{2\sqrt{k+\lambda x+\delta}} e^{-x} \tag{1.1}$$

where $\delta > 0$ is some cutoff and the coupling λ is small and positive. One may think of $\delta = T$, the temperature, or $\delta = 1/L$, L^d being the volume of the system, and G_δ corresponds to some correlation function. By explicit computation

$$G_0(\lambda) = \lim_{\delta \to 0} G_\delta(\lambda) = \int_0^\infty dx \left(\sqrt{1+\lambda x} - \sqrt{\lambda x}\right) e^{-x} = 1 + O(\lambda) - O(\sqrt{\lambda}) \tag{1.2}$$

Thus, the $\delta \to 0$ limit is well defined but it is not analytic. This fact has to show up in the Taylor expansion. It reads

$$G_\delta(\lambda) = \sum_{j=0}^n \binom{-\frac{1}{2}}{j} \int_0^\infty dx \int_0^1 dk \frac{x^j e^{-x}}{2(k+\delta)^{j+\frac{1}{2}}} \lambda^j + r_{n+1} \tag{UR}$$

Apparently, all integrals over k diverge for $j \geq 1$ in the limit $\delta \to 0$. Now, the whole problem in field theoretic perturbation theory is to find a rearrangement which reorders the expansion (UR) ('UR' for 'unrenormalized') into a new expansion

$$G_0(\lambda) = \sum_{\ell=0}^{n} \binom{\frac{1}{2}}{\ell} \int_0^\infty dx\, x^\ell \, e^{-x} \lambda^\ell \; - c\sqrt{\lambda} + R_{n+1} \qquad (R)$$

('R' for 'renormalized') which, in this explicitly solvable example, can be obtained from (1.2) by expanding the $\sqrt{1+\lambda x}$ term. In (R), all coefficients are finite and, for small λ, the lowest order terms are a good approximation since $|R_{n+1}| \le n!\,\lambda^{n+1}$, although the whole series in (R), obtained by letting $n \to \infty$, still has radius of convergence zero. That is, the expansion (R) is asymptotic, the lowest order terms give us information about the behavior of the correlation function, but the expansion (UR) is not, its lowest order terms do not give us any information. The problem is of course that in a typical field theoretic situation we do not know the exact answer (1.2) and then it is not clear how to obtain (R) from (UR). Roughly speaking, this book is about the passage from (UR) to (R) for the many-electron system with short-range interaction which serves as a typical quantum many-body system. Thereby we will develop the standard perturbation theory formalism, derive the fermionic and bosonic functional integral representations, consider approximations like BCS theory, estimate Feynman diagrams and set up the renormalization group framework. In the last chapter we discuss a somewhat novel method which is devoted to the resummation of the nonanalytic parts of a field theoretical perturbation series.

In the first three chapters (2-4) we present the standard perturbation theory formalism, the expansion into Feynman diagrams. We start in chapter 2 with second quantization. In relativistic quantum mechanics this concept is important to describe the creation and destruction of particles. In nonrelativistic many-body theory this is simply a rewriting of the Hamiltonian, a very useful one of course. The perturbation expansion for $\exp\{-\beta(H_0 + \lambda V)\}$ is presented and Wick's theorem is proven. In chapter 4 we introduce anticommuting Grassmann variables and derive the Grassmann integral representations for the correlation functions. Grassmann integrals are a very suitable tool to handle the combinatorics and the rearrangement of fermionic perturbation series.

In the fifth chapter, we use these formulae to write down the bosonic functional integral representations for the correlation functions. These are typically of the form $\int F(\phi)\, e^{-V_{\rm eff}(\phi)} d\phi / \int e^{-V_{\rm eff}(\phi)} d\phi$. Here F depends on the particular correlation function under consideration but the effective potential $V_{\rm eff}$ is fixed once the model is fixed. Usually it is given by a quadratic part minus the logarithm of a functional determinant. In particular, we consider the case of an attractive delta-interaction and we give a rigorous proof that the global minimum of the full effective potential in that case is in fact given by the BCS configuration. This is obtained by estimating the functional determinant as a whole without any expansions and is thus a completely nonperturbative result.

In chapter 6, we discuss BCS theory, the Bardeen-Cooper-Schrieffer theory of superconductivity. Basically the BCS approximation consists of two steps. The interacting part of the full Hamiltonian, which is quartic in the annihilation and creation operators, comes, because of conservation of momentum, with three independent momentum sums. The first step of the approximation consists in putting the total momentum of two incoming electrons equal to zero. The result is a Hamiltonian, which is still quartic in the annihilation and creation operators, but which has only two independent momentum sums. Sometimes this model is called the 'reduced BCS model' but one may also call it the 'quartic BCS model'. The model, which has been solved by Bardeen, Cooper and Schrieffer in 1958 [6] is a quadratic model. It is obtained from the quartic BCS model by substituting the product of two annihilation or creation operators by a number, which is chosen to be the expectation value of these operators with respect to the quadratic Hamiltonian, to be determined selfconsistently. This mean field approximation is the second step of the BCS approximation.

In section 6.2 we show that the quartic BCS model is already explicitly solvable, it is not necessary to make the quadratic mean field approximation. This result follows from the observation that in going from three to two independent momentum sums one changes the volume dependence of the model in such a way that in the bosonic functional integral representation the integration variables are forced to take values at the global minimum of the effective potential in the infinite volume limit. That is, the saddle point approximation becomes exact. Even for the quartic BCS model the effective potential is a complicated function of many variables but with the results of chapter 5 we are able to determine the global minimum which results in explicit expressions for the correlation functions. For an s-wave interaction the results coincide with those of the quadratic mean field formalism, but for higher ℓ-wave interactions this is no longer necessarily the case.

Chapter 7 provides a nice application of the second quantization formalism to the fractional quantum Hall effect. We show that, in a certain long range limit, the interacting many-body Hamiltonian in the lowest Landau level can be exactly diagonalized. However, the long range approximation which is used there has to be considered as unphysical. Nevertheless we think it is worth discussing this approximate model since it has an, in finite volume, explicitly given eigenvalue spectrum which, in the infinite volume limit, most likely has a gap for rational fillings and no gap for irrational fillings. This is interesting since a similar behavior one would like to prove for the original model.

Chapters 8 and 9 are devoted to the rigorous control of perturbation theory in the weak coupling case. These are the most technical chapters. Chapter 8 contains bounds on individual Feynman diagrams whereas chapter 9 estimates sums of diagrams. First it is shown that the value of a diagram depends on its subgraph structure. This is basic for an understanding of renormalization. Then it is shown that, for the many-electron system with short range interaction, an n'th order diagram without two- and four-legged subgraphs allows a $const^n$ bound which is the best possible case. Roughly speaking, one can expect that a sum of diagrams, where each diagram allows a $const^n$ bound, is at least asymptotic. That is, the lowest order terms of such a series would be a good approximation in the weak coupling case and this is all one would like to have. Then it is shown that n'th order diagrams with four-legged subgraphs but without two-legged subgraphs are still finite but they produce $n!$'s. This is bad since, roughly speaking, a sum of such diagrams cannot expected to be asymptotic. That is, the computation of the lowest order terms of such an expansion does not give any information on the behavior of the whole sum. For that reason diagrams without two- and four-legged subgraphs are called 'convergent' diagrams but this does not refer to diagrams with four-legged but without two-legged subgraphs, although the latter ones are also finite. Finally diagrams with two-legged subdiagrams are in general infinite when cutoffs are removed (volume to infinity, temperature to zero).

In the ninth chapter we consider the sum of convergent diagrams. As already mentioned, such a sum can be expected to be asymptotic. More precisely, for a bosonic model one can expect an asymptotic series and for a fermionic model, one may even expect a series with a positive radius of convergence. In fact this is what we prove. We choose a fermionic model which has the same power counting as the many-electron system and show that the sum of convergent diagrams has a positive radius of convergence. The same result has been proven for the many-electron system in two dimensions and can be found in the research literature [18]. For those who wonder at this point how objects like the 'sum of all diagrams without two- and four-legged subgraphs' are treated technically we shortly remark that these sums are generated inductively by integrating out scales in a fermionic functional integral and then at each step Grassmann monomials with two and four ψ's are taken out by hand.

Diagrams with two-legged subdiagrams have to be renormalized. Conceptually, renormalization is nothing else than a rearrangement of the perturbation series. However, due to technical reasons, it may be implemented in different ways. One way of doing this is by the use of counterterms. In this approach one changes the model under consideration. Instead of a model with kinetic energy, say, $e_k = k^2/(2m) - \mu$, μ the chemical potential, one starts with a model with kinetic energy $e_k + \delta e$. The counterterm δe depends on the coupling and may also depend on k. Typically, for problems with an infrared singularity, like the many-electron system, where the singularity is on the Fermi surface $e_k = 0$, the counterterm is a finite quantity. It can be chosen in such a way, that the perturbation series for the altered model with kinetic energy $e(k) + \delta e$ does no longer contain any divergent diagrams. In fact, for the many-electron system with short-range interaction, it can be proven

[18, 20, 16] that, in two dimensions, the renormalized sum of all diagrams without four-legged subgraphs is analytic for sufficiently small coupling. This is true for the model with kinetic energy $e_k = k^2/(2m) - \mu$ which has a round Fermi surface $F = \{k \,|\, e_k = 0\}$ but also holds for models with a more general e_k which may have an anisotropic Fermi surface. Then, the last and the most complicated step in the perturbative analysis consists in adding in the four-legged diagrams. These diagrams determine the physics of the model.

At low temperatures the many-electron system may undergo a phase transition to the superconducting state by the formation of Cooper pairs. Two electrons, with opposite momenta k and $-k$, with an effective interaction which has an attractive part, may form a bound state. Since at small temperatures only those momenta close to the Fermi surface are relevant, the formation of Cooper pairs can be suppressed, if one substitutes (by hand) the energy momentum relation $e_k = k^2/(2m) - \mu$ by a more general expression with an anisotropic Fermi surface. That is, if momentum k is on the Fermi surface, then momentum $-k$ is not on F for almost all k. For such an e_k one can prove that four-legged subdiagrams no longer produce any factorials, an n'th order diagram without two-legged but not necessarily without four-legged subgraphs is bounded by $const^n$. As a result, Feldman, Knörrer and Trubowitz could prove that, in two space dimensions, the renormalized perturbation series for such a model has in fact a small positive radius of convergence and that the momentum distribution $\langle a_{k\sigma}^+ a_{k\sigma} \rangle$ has a jump discontinuity across the Fermi surface of size $1 - \delta_\lambda$ where $\delta_\lambda > 0$ can be chosen arbitrarily small if the coupling λ is made small. Because of the latter property this theorem is referred to as the Fermi liquid theorem.

The complete rigorous proof of this fact is a larger technical enterprise [20]. It is distributed over a series of 10 papers with a total volume of 680 pages. J. Feldman has setup a webpage under www.math.ubc.ca/~feldman/fl.html where all the relevant material can be found. The introductory paper 'A Two Dimensional Fermi Liquid, Part 1: Overview' gives the precise statement of results and illustrates, in several model computations, the main ingredients and difficulties of the proof.

As FKT remark in that paper, this theorem is still not the complete story. Since two-legged subdiagrams have been renormalized by the addition of a counterterm, the model has been changed. Because e_k has been chosen anisotropic, also the counterterm δe_k is a nontrivial function of k, not just a constant. Thus, one is led to an invertability problem: For given e_k, is there a $\tilde{e}_k$ such that $\tilde{e}_k + \delta \tilde{e}_k = e_k$? If this question is addressed on a rigorous level, it also becomes very difficult. See [28, 55] for the current status. The articles of [28] and [20] add up to one thousand pages.

Another way to get rid of anomalously large or divergent diagrams is to resum them, if this is possible somehow. Typically this leads to integral equations for the correlation functions. The good thing in having integral equations is that the renormalization is done more or less automatically. The correlation functions are obtained from a system of integral equations whose solution can have all kinds of nonanalytic terms (which are responsible for the divergence of the coefficients in the naive perturbation expansion). If one works with counterterms one more or less has to know the answer in advance in order to choose the right counterterms. However, the bad thing with integral equations is that usually it is impossible to get a closed system of equations without making an uncontrolled approximation. If one tries to get an integral equation for a two-point function, one gets an expression with two- and four-point functions. Then, dealing with the four-point function, one obtains an expression with two-, four- and six-point functions and so on. Thus, in order to get a closed system of equations, at some place one is forced to approximate a, say, six-point function by a sum of products of two- and four-point functions.

In the last chapter we present a somewhat novel formalism which allows the resummation of two- and four-legged subdiagrams in a systematic and relatively elegant way which leads to integral equations for the correlation functions. Although this method too does not lead to a complete rigorous control of the correlation functions, we hope that the reader feels like the author who found it quite instructive to see renormalization from this point of view.

Chapter 2

Second Quantization

In this chapter we introduce the many-body Hamiltonian for the N-electron system and rewrite it in terms of annihilation and creation operators. This rewriting is called second quantization. We introduce the canonical and the grand canonical ensemble which is the framework in which quantum statistical mechanics has to be formulated. By considering the ideal Fermi gas, we try to motivate that the grand canonical ensemble may be more practical for computations than the canonical ensemble.

2.1   Coordinate and Momentum Space

Consider one electron in d dimensions in a finite box of size $[0, L]^d$. Its kinetic energy is given by

$$h_0 = \frac{\hbar^2}{2m} \Delta \tag{2.1}$$

and its Schrödinger equation $h_0 \varphi = \varepsilon \varphi$ is solved by plane waves $\varphi(\mathbf{x}) = e^{i\mathbf{k}\mathbf{x}}$. Since we are in a finite box, we have to impose some boundary conditions. Probably the most natural ones are Dirichlet boundary conditions $\varphi(\mathbf{x}) = 0$ on the boundary of $[0, L]^d$ but it is more convenient to choose periodic boundary conditions, $\varphi(\mathbf{x}) = \varphi(\mathbf{x} + L\mathbf{e}_j)$ for all $1 \leq j \leq d$. Hence $e^{ik_j L}$ must be equal to 1 which gives $\mathbf{k} = (k_1, ..., k_d) = \frac{2\pi}{L}(m_1, ..., m_d)$ with $m_j \in \mathbb{Z}$. Thus, a continuous but bounded coordinate space gives a discrete but unbounded momentum space. Similarly, a discrete but unbounded coordinate space gives a continuous but bounded momentum space and a discrete and bounded coordinate space, with a finite number of points, gives a discrete and bounded momentum space with the same number of points.

To write down the Hamiltonian for the many-electron system in second quantized form, we will introduce annihilation and creation operators in coordinate space, $\psi(\mathbf{x})$ and $\psi^+(\mathbf{x})$. Strictly speaking, for a continuous coordinate space, these are operator-valued distributions. To keep the formalism simple, we found it convenient to introduce a small lattice spacing $1/M > 0$ in coordinate space which makes everything finite dimensional. We then derive suitable expressions for the correlation functions in the next chapters and at the very end, the limits lattice spacing to zero and volume to infinity are considered.

Thus, let coordinate space be

$$\Gamma = \left\{ \mathbf{x} = \tfrac{1}{M}(n_1, \cdots, n_d) \mid 0 \leq n_i \leq ML - 1 \right\}$$
$$= \left(\tfrac{1}{M}\mathbb{Z}\right)^d / (L\mathbb{Z})^d \tag{2.2}$$

Momentum space is given by

$$\mathcal{M} := \Gamma^\sharp = \left\{ \mathbf{k} = \tfrac{2\pi}{L}(m_1, \cdots, m_d) \mid 0 \leq m_i \leq ML - 1 \right\}$$
$$= \left(\tfrac{2\pi}{L}\mathbb{Z}\right)^d / (2\pi M\mathbb{Z})^d \tag{2.3}$$

such that $0 \leq k_j \leq 2\pi M$ or $-\pi M \leq k_j \leq \pi M$ since $-k_j = 2\pi M - k_j$. Removing the cutoffs, one gets

$$\frac{1}{L^d} \sum_{\mathbf{m}} = \frac{1}{(2\pi)^d} \left(\frac{2\pi}{L}\right)^d \sum_{\mathbf{m}} \overset{L \to \infty}{\longrightarrow} \int_{[-\pi M, \pi M]^d} \frac{d^d\mathbf{k}}{(2\pi)^d}, \tag{2.4}$$

$$\frac{1}{M^d} \sum_{\mathbf{n}} \overset{M \to \infty}{\longrightarrow} \int_{[-L/2, L/2]^d} d^d\mathbf{x} \tag{2.5}$$

A complete orthonormal system of $L^2(\Gamma) = \mathbb{C}^{N^d}$, $N = ML$, is given by the plane waves

$$\varphi_{\mathbf{k}}(\mathbf{x}) \equiv \varphi_{\mathbf{m}}(\mathbf{n}) = \frac{1}{(ML)^{\frac{d}{2}}} e^{i\frac{2\pi}{ML}\sum_{i=0}^{d} m_i n_i} = \frac{1}{N^{\frac{d}{2}}} e^{2\pi i \frac{\mathbf{m}\mathbf{n}}{N}} \tag{2.6}$$

The unitary matrix of discrete Fourier transform is given by $F = (F_{\mathbf{mn}})$ where

$$F_{\mathbf{mn}} = \frac{1}{N^{\frac{d}{2}}} e^{-2\pi i \frac{\mathbf{mn}}{N}} \tag{2.7}$$

One has

$$F^* = F^{-1} = \bar{F} = \begin{pmatrix} & | & \\ \cdots & \varphi_{\mathbf{k}}(\mathbf{x}) & \cdots \\ & | & \end{pmatrix} \tag{2.8}$$

The discretized version of

$$\hat{f}(\mathbf{k}) = \int d^d\mathbf{x}\, e^{-i\mathbf{k}\mathbf{x}} f(\mathbf{x}), \qquad f(\mathbf{x}) = \int \frac{d^d\mathbf{k}}{(2\pi)^d} e^{i\mathbf{k}\mathbf{x}} \hat{f}(\mathbf{k}) \tag{2.9}$$

reads in terms of F

$$\hat{f}(\mathbf{k}) = \frac{1}{M^d} \sum_{\mathbf{x}} e^{-i\mathbf{k}\mathbf{x}} f(\mathbf{x}) = \left(\tfrac{L}{M}\right)^{\frac{d}{2}} \sum_{\mathbf{x}} F_{\mathbf{kx}} f(\mathbf{x}) = \left(\tfrac{L}{M}\right)^{\frac{d}{2}} (Ff)(\mathbf{k})$$

$$f(\mathbf{x}) = \frac{1}{L^d} \sum_{\mathbf{k}} e^{i\mathbf{k}\mathbf{x}} \hat{f}(\mathbf{k}) = \left(\tfrac{M}{L}\right)^{\frac{d}{2}} \sum_{\mathbf{k}} F^*_{\mathbf{xk}} \hat{f}(\mathbf{k}) = \left(\tfrac{M}{L}\right)^{\frac{d}{2}} (F^*\hat{f})(\mathbf{x}) \tag{2.10}$$

Derivatives are given by difference operators

$$\frac{\partial}{\partial x_i} f(\mathbf{x}) = M \left(f(\frac{\mathbf{n}+\mathbf{e}_i}{M}) - f(\frac{\mathbf{n}}{M}) \right)$$

$$= \frac{1}{M^d} \sum_{\mathbf{y}} M M^d (\delta_{\mathbf{x}+\frac{\mathbf{e}_i}{M},\mathbf{y}} - \delta_{\mathbf{x},\mathbf{y}}) f(\mathbf{y}) \qquad (2.11)$$

$$(\Delta f)(\mathbf{x}) = \frac{1}{M^d} \sum_{\mathbf{y}} M^2 M^d \sum_{i=1}^{d} (\delta_{\mathbf{x}+\frac{\mathbf{e}_i}{M},\mathbf{y}} + \delta_{\mathbf{x}-\frac{\mathbf{e}_i}{M},\mathbf{y}} - 2\delta_{\mathbf{x},\mathbf{y}}) f(\mathbf{y}) \quad (2.12)$$

which are diagonalized by F,

$$\frac{\partial}{\partial x_j} \varphi_{\mathbf{k}}(\mathbf{x}) = M (e^{2\pi i \frac{m_j}{N}} - 1) \varphi_{\mathbf{k}}(\mathbf{x}) \qquad (2.13)$$

which gives

$$\left[F \frac{1}{i} \frac{\partial}{\partial x_j} F^* \right]_{\mathbf{m},\mathbf{m}'} = M (e^{2\pi i \frac{m_j}{N}} - 1) \delta_{\mathbf{m},\mathbf{m}'}$$

$$\overset{M \to \infty}{\to} \frac{2\pi}{L} m_j \, \delta_{\mathbf{m},\mathbf{m}'} = k_j \, \delta_{\mathbf{k},\mathbf{k}'} \qquad (2.14)$$

$$[F(-\Delta) F^*]_{\mathbf{m},\mathbf{m}'} = \sum_{i=1}^{d} M^2 \left(2 - 2\cos(\frac{2\pi m_i}{ML}) \right) = \sum_{i=1}^{d} 4M^2 \sin^2 \frac{\pi m_i}{ML} \, \delta_{\mathbf{m},\mathbf{m}'}$$

$$\overset{M \to \infty}{\to} \left(\frac{2\pi}{L} \mathbf{m} \right)^2 \delta_{\mathbf{m},\mathbf{m}'} = \mathbf{k}^2 \, \delta_{\mathbf{k},\mathbf{k}'} \qquad (2.15)$$

In the following we will write k_j for the Fourier transform of $\frac{1}{i}\frac{\partial}{\partial x_j}$ instead of writing the exact discretized expressions.

2.2   The Many-Electron System

The N-particle Hamiltonian $H_N : \mathcal{F}_N \to \mathcal{F}_N$ is given by

$$H_N = -\frac{1}{2m} \sum_{i=1}^{N} \Delta_{\mathbf{x}_i} + \frac{1}{2} \sum_{\substack{i,j=1 \\ i \neq j}} V(\mathbf{x}_i - \mathbf{x}_j) \qquad (2.16)$$

which acts on the antisymmetric N-particle Fock space

$$\mathcal{F}_N = \left\{ F_N \in L^2 \left[(\Gamma \times \{\uparrow,\downarrow\})^N \right] = (\mathbb{C}^{2|\Gamma|})^N \, | \, \forall \pi \in S_n : \qquad (2.17) \right.$$

$$\left. F_N(\mathbf{x}_{\pi 1}\sigma_{\pi 1}, \cdots, \mathbf{x}_{\pi N}\sigma_{\pi N}) = \mathrm{sign}\pi \, F_N(\mathbf{x}_1\sigma_1, \cdots, \mathbf{x}_N\sigma_N) \right\}$$

Since we assume a small but positive temperature $T = 1/\beta > 0$, we have to do quantum statistical mechanics. Conceptually, the most natural setting would be the

Canonical Ensemble: An observable has to be represented by some operator $A_N : \mathcal{F}_N \to \mathcal{F}_N$ and measurements correspond to the expectation values

$$\langle A_N \rangle_{\mathcal{F}_N} = \frac{Tr_{\mathcal{F}_N} A_N e^{-\beta H_N}}{Tr_{\mathcal{F}_N} e^{-\beta H_N}} \tag{2.18}$$

Example (The Ideal Fermi Gas): The ideal Fermi gas is given by

$$H_{0,N} = -\frac{1}{2m} \sum_{i=1}^{N} \Delta_{\mathbf{x}_i} \tag{2.19}$$

We compute the canonical partition function

$$Q_N := Tr_{\mathcal{F}_N} e^{-\beta H_{0,N}} \tag{2.20}$$

To this end introduce an orthonormal basis of $\mathcal{F}_1$ of eigenvectors of $-\Delta$ which is given by the plane waves

$$\left\{ \phi_{\mathbf{k}\sigma}(\mathbf{x}\tau) := \delta_{\sigma\tau} \frac{1}{L^{\frac{d}{2}}} e^{i\mathbf{k}\mathbf{x}} \,\middle|\, (\mathbf{k}, \sigma) \in \mathcal{M} \times \{\uparrow, \downarrow\} \right\} \tag{2.21}$$

where the set of momenta $\mathcal{M}$ is given by (2.3). The scalar product is

$$(\phi_{\mathbf{k}\sigma}, \phi_{\mathbf{k}',\sigma',})_{\mathcal{F}_1} := \frac{1}{M^d} \sum_{\mathbf{x}\tau} \phi_{\mathbf{k}\sigma}(\mathbf{x}\tau) \bar{\phi}_{\mathbf{k}'\sigma'}(\mathbf{x}\tau) = \delta_{\sigma,\sigma'} \delta_{\mathbf{k},\mathbf{k}'} \tag{2.22}$$

and we have

$$-\Delta\phi_{\mathbf{k}\sigma} = \varepsilon(\mathbf{k})\phi_{\mathbf{k}\sigma}, \quad \varepsilon(\mathbf{k}) = \sum_{i=1}^{d} 2M^2(1 - \cos[k_i/M]) \overset{M\to\infty}{\to} \mathbf{k}^2 \tag{2.23}$$

An orthogonal basis of $\mathcal{F}_n$ is given by wedge products or Slater determinants

$$\phi_{\mathbf{k}_1\sigma_1} \wedge \cdots \wedge \phi_{\mathbf{k}_n\sigma_n}(\mathbf{x}_1\tau_1, \cdots, \mathbf{x}_n\tau_n)$$

$$:= \frac{1}{n!} \sum_{\pi \in S_n} \mathrm{sign}\pi \, \phi_{\mathbf{k}_1\sigma_1}(\mathbf{x}_{\pi 1}\tau_{\pi 1}) \cdots \phi_{\mathbf{k}_n\sigma_n}(\mathbf{x}_{\pi n}\tau_{\pi n})$$

$$= \frac{1}{n!} \det\left[\phi_{\mathbf{k}_i\sigma_i}(\mathbf{x}_j\tau_j)\right]_{1\le i,j\le n} \tag{2.24}$$

The orthogonality relation reads

$$\left(\phi_{\mathbf{k}_1\sigma_1} \wedge \cdots \wedge \phi_{\mathbf{k}_n\sigma_n} \,,\, \phi_{\mathbf{k}_1'\sigma_1'} \wedge \cdots \wedge \phi_{\mathbf{k}_n'\sigma_n'} \right)_{\mathcal{F}_N}$$

$$= \frac{1}{M^{nd}} \sum_{\mathbf{x}_1\tau_1\cdots\mathbf{x}_n\tau_n} \phi_{\mathbf{k}_1\sigma_1} \wedge \cdots \wedge \phi_{\mathbf{k}_n\sigma_n}(\mathbf{x}_1\tau_1, \cdots, \mathbf{x}_n\tau_n) \times$$

$$\overline{\phi_{\mathbf{k}_1'\sigma_1'} \wedge \cdots \wedge \phi_{\mathbf{k}_n'\sigma_n'}}(\mathbf{x}_1\tau_1, \cdots, \mathbf{x}_n\tau_n)$$

$$= \frac{1}{M^{nd}} \sum_{\mathbf{x}_1\tau_1\cdots\mathbf{x}_n\tau_n} \frac{1}{n!^2} \sum_{\pi\in S_n} \operatorname{sign}\pi \, \phi_{\mathbf{k}_1\sigma_1}(\mathbf{x}_{\pi 1}\tau_{\pi 1}) \cdots \phi_{\mathbf{k}_n\sigma_n}(\mathbf{x}_{\pi n}\tau_{\pi n}) \times$$

$$\det\left[\phi_{\mathbf{k}_i'\sigma_i'}(\mathbf{x}_j\tau_j) \right]$$

$$= \frac{1}{M^{nd}} \sum_{\mathbf{x}_1\tau_1\cdots\mathbf{x}_n\tau_n} \frac{1}{n!} \phi_{\mathbf{k}_1\sigma_1}(\mathbf{x}_1\tau_1) \cdots \phi_{\mathbf{k}_n\sigma_n}(\mathbf{x}_n\tau_n) \det\left[\phi_{\mathbf{k}_i'\sigma_i'}(\mathbf{x}_j\tau_j) \right]$$

$$= \frac{1}{M^{nd}} \sum_{\mathbf{x}_1\tau_1\cdots\mathbf{x}_n\tau_n} \frac{1}{n!} \sum_{\pi\in S_n} \operatorname{sign}\pi \, \phi_{\mathbf{k}_1\sigma_1}(\mathbf{x}_1\tau_1) \cdots \phi_{\mathbf{k}_n\sigma_n}(\mathbf{x}_n\tau_n) \times$$

$$\phi_{\mathbf{k}_{\pi 1}\sigma_{\pi 1}}(\mathbf{x}_1\tau_1) \cdots \phi_{\mathbf{k}_{\pi n}\sigma_{\pi n}}(\mathbf{x}_n\tau_n)$$

$$= \frac{1}{n!} \det\left[(\phi_{\mathbf{k}_i\sigma_i}, \phi_{\mathbf{k}_j'\sigma_j'})_{\mathcal{F}_1} \right]$$

$$= \frac{1}{n!} \begin{cases} \pm 1 & \text{if } \{\mathbf{k}_1\sigma_1, \cdots, \mathbf{k}_n\sigma_n\} = \{\mathbf{k}_1'\sigma_1', \cdots, \mathbf{k}_n'\sigma_n'\} \\ 0 & \text{else} \end{cases} \qquad (2.25)$$

Thus an orthonormal basis of $\mathcal{F}_N$ is given by

$$\left\{ \sqrt{N!}\, \phi_{\mathbf{k}_1\sigma_1} \wedge \cdots \wedge \phi_{\mathbf{k}_N\sigma_N} \,\Big|\, \mathbf{k}_1\sigma_1 \prec \cdots \prec \mathbf{k}_N\sigma_N, \, (\mathbf{k}_i, \sigma_i) \in \mathcal{M} \times \{\uparrow,\downarrow\} \right\}$$

where $\prec$ is any ordering on $\mathcal{M} \times \{\uparrow,\downarrow\}$. Another way of writing this is

$$\left\{ \sqrt{N!} \bigwedge_{\mathbf{k}\sigma} (\phi_{\mathbf{k}\sigma})^{n_{\mathbf{k}\sigma}} \,\Big|\, n_{\mathbf{k}\sigma} \in \{0,1\}, \sum_{\mathbf{k}\sigma} n_{\mathbf{k}\sigma} = N \right\} \qquad (2.26)$$

Since

$$-\frac{1}{2m} \sum_{i=1}^{N} \Delta_{\mathbf{x}_i} \bigwedge_{\mathbf{k}\sigma} (\phi_{\mathbf{k}\sigma})^{n_{\mathbf{k}\sigma}} = \sum_{\mathbf{k}\sigma} n_{\mathbf{k}\sigma}\varepsilon(\mathbf{k}) \bigwedge_{\mathbf{k}\sigma} (\phi_{\mathbf{k}\sigma})^{n_{\mathbf{k}\sigma}} \qquad (2.27)$$

one ends up with

$$Q_N = Tr\, e^{-\beta H_{0,N}} = \sum_{\substack{\{n_{\mathbf{k}\sigma}\} \\ \sum n_{\mathbf{k}\sigma}=N}} e^{-\beta \sum_{\mathbf{k}\sigma} \varepsilon(\mathbf{k}) n_{\mathbf{k}\sigma}} \qquad (2.28)$$

for the canonical partition function of the ideal Fermi gas. ∎

Because of the constraint $\sum n_{\mathbf{k}\sigma} = N$ formula (2.28) cannot be simplified any further. However, if we consider a generating function for the Q_N's which involves a sum over N, we arrive at a more compact expression:

$$Z(z) := \sum_{N=0}^{\infty} z^N Q_N = \sum_{N=0}^{\infty} \sum_{\substack{\{n_{\mathbf{k}\sigma}\} \\ \sum n_{\mathbf{k}\sigma} = N}} \prod_{\mathbf{k}\sigma} \left(z\, e^{-\beta \varepsilon(\mathbf{k})} \right)^{n_{\mathbf{k}\sigma}}$$

$$= \sum_{\{n_{\mathbf{k}\sigma}\}} \prod_{\mathbf{k}\sigma} \left(z\, e^{-\beta \varepsilon(\mathbf{k})} \right)^{n_{\mathbf{k}\sigma}} = \prod_{\mathbf{k}\sigma} \left[1 + z\, e^{-\beta \varepsilon(\mathbf{k})} \right] \tag{2.29}$$

Thus, from a computational point of view, it is not too practical to have the constraint of a given number of particles, $\sum n_{\mathbf{k}\sigma} = N$. Therefore, instead of using the canonical ensemble one usually computes in the

Grand Canonical Ensemble: Let $\mathcal{F} = \oplus_{N=0}^{\infty} \mathcal{F}_N$, $H = \oplus_{N=0}^{\infty} H_N$. An observable has to be represented by some operator $A = \oplus_{N=0}^{\infty} A_N : \mathcal{F} \to \mathcal{F}$ and measurements correspond to the expectation values

$$\langle A \rangle_{\mathcal{F}} = \frac{Tr_{\mathcal{F}}\, A\, e^{-\beta(H - \mu \mathbf{N})}}{Tr_{\mathcal{F}}\, e^{-\beta(H - \mu \mathbf{N})}} \tag{2.30}$$

where the chemical potential μ has to be determined by the condition $\langle \mathbf{N} \rangle = N$, N being the given number of particles and $\mathbf{N}$ the number operator, $\mathbf{N}(1, F_1, F_2, \cdots) := (0, F_1, 2F_2, \cdots)$.

Example (The Ideal Fermi Gas): We compute the chemical potential $\mu = \mu(\beta, N, L) = \mu(\beta, \rho)$ for the ideal Fermi gas with density $\rho = N/L^d$ and we calculate the energy for the ideal Fermi gas.

To this end introduce the 'fugacity' z which is related to μ according to $z = e^{\beta \mu}$. One has

$$N = \langle \mathbf{N} \rangle_{\mathcal{F}} = \frac{\sum_{N=0}^{\infty} N z^N Q_N}{\sum_{N=0}^{\infty} z^N Q_N} = \frac{z \frac{d}{dz} Z(z)}{Z(z)} = z \frac{d}{dz} \log Z(z)$$

$$= z \frac{d}{dz} \sum_{\mathbf{k}\sigma} \log \left[1 + z\, e^{-\beta \varepsilon(\mathbf{k})} \right] = 2 \sum_{\mathbf{k}} \frac{z\, e^{-\beta \varepsilon(\mathbf{k})}}{1 + z\, e^{-\beta \varepsilon(\mathbf{k})}}$$

$$\approx 2L^d \int \frac{d^d \mathbf{k}}{(2\pi)^d} \frac{z\, e^{-\beta \varepsilon(\mathbf{k})}}{1 + z\, e^{-\beta \varepsilon(\mathbf{k})}} \tag{2.31}$$

where we have used (2.4) in the last line and the integral goes over $[-\pi M, \pi M]^d$. Recalling that $z = e^{\beta \mu}$ and introducing

$$e_{\mathbf{k}} := \varepsilon(\mathbf{k}) - \mu \tag{2.32}$$

we obtain in the zero temperature limit

$$N = 2L^d \int \frac{d^d \mathbf{k}}{(2\pi)^d} \frac{e^{-\beta e_{\mathbf{k}}}}{1 + e^{-\beta e_{\mathbf{k}}}} \xrightarrow{\beta \to \infty} 2L^d \int \frac{d^d \mathbf{k}}{(2\pi)^d} \chi(e_{\mathbf{k}} < 0) \tag{2.33}$$

which determines μ as a function of the density $\rho = N/L^d$. The expectation value of the energy is obtained from $Z(z)$ according to

$$\langle H_0 \rangle_{\mathcal{F}} = -\tfrac{d}{d\beta} \log Z + \mu N = -\tfrac{d}{d\beta} 2 \sum_{\mathbf{k}} \log\left[1 + e^{-\beta e_{\mathbf{k}}}\right] + \mu 2 \sum_{\mathbf{k}} \frac{e^{-\beta e_{\mathbf{k}}}}{1 + e^{-\beta e_{\mathbf{k}}}}$$

$$= 2 \sum_{\mathbf{k}} \varepsilon(\mathbf{k}) \frac{e^{-\beta e_{\mathbf{k}}}}{1 + e^{-\beta e_{\mathbf{k}}}} \approx 2L^d \int \frac{d^d \mathbf{k}}{(2\pi)^d} \varepsilon(\mathbf{k}) \chi\big(\varepsilon(\mathbf{k}) < \mu\big) \qquad (2.34)$$

and the last approximation holds for large volume and small temperature. ∎

2.3 Annihilation and Creation Operators

2.3.1 Coordinate Space

Let $\alpha \in \{\uparrow, \downarrow\}$ be a spin index and let

$$\delta_{\mathbf{x}\alpha}(\mathbf{x}'\alpha') := \delta_{\alpha,\alpha'} \, M^d \delta_{\mathbf{x},\mathbf{x}'} \qquad (2.35)$$

For $F_N \in \mathcal{F}_N$ the wedge product $\delta_{\mathbf{x}\alpha} \wedge F_N \in \mathcal{F}_{N+1}$ is defined by

$$(\delta_{\mathbf{x}\alpha} \wedge F_N)(\mathbf{x}_1\alpha_1, \cdots, \mathbf{x}_{N+1}\alpha_{N+1}) :=$$

$$\tfrac{1}{N+1} \sum_{i=1}^{N+1} (-1)^{i-1} \delta_{\mathbf{x}\alpha}(\mathbf{x}_i\alpha_i) \, F_N(\mathbf{x}_1\alpha_1, \cdots, \widehat{\mathbf{x}_i\alpha_i}, \cdots, \mathbf{x}_{N+1}\alpha_{N+1}) \qquad (2.36)$$

Then the creation operator at $\mathbf{x}$ is defined by $\psi^+(\mathbf{x}\alpha) : \mathcal{F}_N \to \mathcal{F}_{N+1}$,

$$\psi^+(\mathbf{x}\alpha) F_N := \sqrt{N+1}\, \delta_{\mathbf{x}\alpha} \wedge F_N \qquad (2.37)$$

and the annihilation operator $\psi(\mathbf{x}\alpha) : \mathcal{F}_{N+1} \to \mathcal{F}_N$ is defined by the adjoint of ψ^+, $\psi(\mathbf{x}\alpha) = [\psi^+(\mathbf{x}\alpha)]^*$.

Lemma 2.3.1 (i) *The adjoint operator* $\psi(\mathbf{x}\alpha) : \mathcal{F}_{N+1} \to \mathcal{F}_N$ *is given by*

$$(\psi(\mathbf{x}\alpha) F_{N+1})(\mathbf{x}_1\alpha_1, \cdots, \mathbf{x}_N\alpha_N) = \sqrt{N+1}\, F_{N+1}(\mathbf{x}\alpha, \mathbf{x}_1\alpha_1, \cdots, \mathbf{x}_N\alpha_N) \qquad (2.38)$$

(ii) *The following canonical anticommutation relations hold:*

$$\{\psi(\mathbf{x}\alpha), \psi(\mathbf{y}\beta)\} = \{\psi^+(\mathbf{x}\alpha), \psi^+(\mathbf{y}\beta)\} = 0,$$

$$\{\psi(\mathbf{x}\alpha), \psi^+(\mathbf{y}\beta)\} = \delta_{\alpha\beta}\, M^d \delta_{\mathbf{x}\mathbf{y}}. \qquad (2.39)$$

Proof: (i) We abbreviate $\xi = (\mathbf{x}\alpha)$ and $\eta = (\mathbf{y}\beta)$. One has

$$\left(F_{N+1}, \psi^+(\xi)G_N\right)_{\mathcal{F}_{N+1}} =$$

$$\sum_{\xi_1 \cdots \xi_{N+1}} \bar{F}_{N+1}(\xi_1, \cdots, \xi_{N+1}) \sqrt{N+1}\, (\delta_\xi \wedge G_N)(\xi_1, \cdots, \xi_{N+1})$$

$$= \frac{1}{\sqrt{N+1}} \sum_{\xi_1 \cdots \xi_{N+1}} \bar{F}_{N+1}(\xi_1, \cdots, \xi_{N+1}) \times$$

$$\sum_{i=1}^{N+1} (-1)^{i-1} \delta_\xi(\xi_i) G_N(\xi_1, \cdots, \hat{\xi}_i, \cdots \xi_{N+1})$$

$$= \frac{1}{\sqrt{N+1}} \sum_{i=1}^{N+1} (-1)^{i-1} \sum_{\xi_1 \cdots, \hat{\xi}_i, \cdots \xi_{N+1}} \bar{F}_{N+1}(\xi_1, \cdots, \xi, \cdots, \xi_{N+1}) \times$$

$$G_N(\xi_1, \cdots, \hat{\xi}_i, \cdots \xi_{N+1})$$

$$= \frac{1}{\sqrt{N+1}} \sum_{i=1}^{N+1} \sum_{\eta_1 \cdots \eta_N} \bar{F}_{N+1}(\xi, \eta_1, \cdots, \eta_N) G_N(\eta_1, \cdots, \eta_N)$$

$$= \sqrt{N+1}\, \left(F_{N+1}(\xi, \cdot), G_N(\cdot)\right)_{\mathcal{F}_N} = \left(\psi(\xi)F_{N+1}, G_N\right)_{\mathcal{F}_N}. \tag{2.40}$$

(ii) We compute $\{\psi, \psi^+\}$. One has

$$\left(\psi(\xi)\psi^+(\eta)F_N\right)(\xi_1, \cdots, \xi_N) = \sqrt{N+1}\, (\psi(\xi)\, \delta_\eta \wedge F_N)(\xi_1, \cdots, \xi_N)$$

$$= (N+1)\, (\delta_\eta \wedge F_N)(\xi, \xi_1, \cdots, \xi_N) \tag{2.41}$$

$$= \delta_\eta(\xi)F_N(\xi_1, \cdots, \xi_N) + \sum_{i=1}^{N} (-1)^{(i+1)-1} \delta_\eta(\xi_i) F_N(\xi, \xi_1, \cdots, \hat{\xi}_i, \cdots, \xi_N)$$

Since

$$\left(\psi^+(\eta)\psi(\xi)F_N\right)(\xi_1, \cdots, \xi_N) = \sqrt{N}\, (\psi^+(\eta)F_N(\xi, \cdot))(\xi_1, \cdots, \xi_N)$$

$$= N\frac{1}{N} \sum_{i=1}^{N} (-1)^{i-1} \delta_\eta(\xi_i) F_N(\xi, \xi_1, \cdots, \hat{\xi}_i, \cdots, \xi_N) \tag{2.42}$$

the lemma follows. ∎

In the following theorem we show that the familiar expressions for the Hamiltonian in terms of annihilation and creation operators is just another representation for an N-particle Hamiltonian of quantum mechanics. So although these representations are sometimes referred to as 'second quantization', there is conceptually nothing new. We use the notation $\Gamma_s = \Gamma \times \{\uparrow, \downarrow\}$ ('s' for 'spin') and write $L^2(\Gamma_s) = \mathbb{C}^{|\Gamma_s|}$.

Theorem 2.3.2 (i) *Let* $h = (h_{\mathbf{x}\alpha,\mathbf{y}\beta}) : L^2(\Gamma_s) \to L^2(\Gamma_s)$ *(one particle Hamiltonian) and let*

$$H_{0,N} = \sum_{i=1}^{N} h_i \ : \ \mathcal{F}_N \to \mathcal{F}_N \tag{2.43}$$

where

$$(h_i F_N)(\mathbf{x}_1\alpha_1, \cdots, \mathbf{x}_N\alpha_N) =$$
$$\frac{1}{M^d} \sum_{\mathbf{y}\beta} h(\mathbf{x}_i\alpha_i, \mathbf{y}\beta)\, F_N(\mathbf{x}_1\alpha_1, \cdots, \mathbf{y}\beta, \cdots, \mathbf{x}_N\alpha_N) \tag{2.44}$$

Then

$$\frac{1}{M^{2d}} \sum_{\substack{\mathbf{x}\mathbf{y} \\ \alpha\beta}} \psi^+(\mathbf{x}\alpha)h(\mathbf{x}\alpha, \mathbf{y}\beta)\psi(\mathbf{y}\beta)\Big|_{\mathcal{F}_N} = H_{0,N} \tag{2.45}$$

(ii) *Let* $v : \Gamma \to \mathbb{R}$ *and let* $V_N : \mathcal{F}_N \to \mathcal{F}_N$ *be the multiplication operator*

$$(V_N F_N)(\mathbf{x}_1\alpha_1, \cdots, \mathbf{x}_N\alpha_N) = \frac{1}{2} \sum_{\substack{i,j=1 \\ i \neq j}}^{N} v(\mathbf{x}_i - \mathbf{x}_j) F_N(\mathbf{x}_1\alpha_1, \cdots, \mathbf{x}_N\alpha_N) \tag{2.46}$$

Then

$$\frac{1}{2}\frac{1}{M^{2d}} \sum_{\substack{\mathbf{x}\mathbf{y} \\ \alpha\beta}} \psi^+(\mathbf{x}\alpha)\psi^+(\mathbf{y}\beta)\, v(\mathbf{x}-\mathbf{y})\, \psi(\mathbf{y}\beta)\psi(\mathbf{x}\alpha)\Big|_{\mathcal{F}_N} = V_N \tag{2.47}$$

Proof: We abbreviate again $\xi = (\mathbf{x}\alpha)$ and $\eta = (\mathbf{y}\beta)$. One has

$$\left(\psi^+(\xi)h(\xi, \eta)\psi(\eta)F_N\right)(\xi_1, \cdots, \xi_N) =$$
$$= h(\xi, \eta)\frac{1}{\sqrt{N}} \sum_{i=1}^{N}(-1)^{i-1}\delta_\xi(\xi_i)\,(\psi(\eta)F_N)\,(\xi_1, \cdots \hat{\xi}_i \cdots, \xi_N)$$
$$= h(\xi, \eta) \sum_{i=1}^{N}\delta_\xi(\xi_i)F_N(\xi_1, \cdots, \eta, \cdots, \xi_N) \tag{2.48}$$

which gives

$$\frac{1}{M^{2d}} \sum_{\xi,\eta}\left(\psi^+(\xi)h(\xi, \eta)\psi(\eta)F_N\right)(\xi_1, \cdots, \xi_N)$$
$$= \sum_{i=1}^{N} \frac{1}{M^{2d}} \sum_{\xi,\eta}\delta_\xi(\xi_i)h(\xi, \eta)F_N(\xi_1, \cdots, \eta, \cdots, \xi_N)$$
$$= \sum_{i=1}^{N} \frac{1}{M^{d}} \sum_{\eta}h(\xi_i, \eta)F_N(\xi_1, \cdots, \eta, \cdots, \xi_N)$$

$$= \sum_{i=1}^{N}(h_i F_N)(\xi_1, \cdots, \xi_N) \tag{2.49}$$

This proves part (i). To obtain (ii) observe that because of (2.48) one has

$$\left(\psi^+(\xi)\psi(\xi)F_N\right)(\xi_1, \cdots, \xi_N) = \sum_{i=1}^{N} \delta_\xi(\xi_i)\, F_N(\xi_1, \cdots, \xi_N) \tag{2.50}$$

Since

$$\psi^\pm(\mathbf{x}\alpha)\psi^+(\mathbf{y}\beta)\, v(\mathbf{x} - \mathbf{y})\, \psi(\mathbf{y}\beta)\psi(\mathbf{x}\alpha)$$
$$= \psi^+(\mathbf{x}\alpha)\psi(\mathbf{x}\alpha)\, v(\mathbf{x} - \mathbf{y})\, \psi^+(\mathbf{y}\beta)\psi(\mathbf{y}\beta)$$
$$- \delta_{\mathbf{x}\alpha}(\mathbf{y}\beta)\, \psi^+(\mathbf{x}\alpha)v(\mathbf{x} - \mathbf{y})\psi(\mathbf{y}\beta)$$

one gets, using (2.48) again,

$$\left(\psi^+(\xi)\psi^+(\eta)v(\mathbf{x} - \mathbf{y})\, \psi(\eta)\psi(\xi)F_N\right)(\xi_1, \cdots, \xi_N)$$

$$= \sum_{i,j=1}^{N} \delta_\xi(\xi_i)\, \delta_\eta(\xi_j)\, v(\mathbf{x} - \mathbf{y})\, F_N(\xi_1, \cdots, \xi_N)$$

$$- \sum_{i=1}^{N} \delta_\xi(\eta)\delta_\xi(\xi_i)\, v(\mathbf{x} - \mathbf{y})\, F_N(\xi_1, \cdots, \xi_N)$$

$$= \sum_{\substack{i,j=1 \\ i \neq j}}^{N} \delta_\xi(\xi_i)\, \delta_\eta(\xi_j)\, v(\mathbf{x} - \mathbf{y})\, F_N(\xi_1, \cdots, \xi_N) \tag{2.51}$$

$$+ \sum_{i=1}^{N} [\delta_\xi(\xi_i)\delta_\eta(\xi_i) - \delta_\xi(\eta)\delta_\xi(\xi_i)]\, v(\mathbf{x} - \mathbf{y})\, F_N(\xi_1, \cdots, \xi_N)$$

Since the terms in the last line cancel part (ii) is proven. ∎

2.3.2 Momentum Space

Recall that the plane waves $\phi_{\mathbf{k}\sigma}(\mathbf{x}\tau) = \delta_{\sigma,\tau} L^{-\frac{d}{2}} e^{i\mathbf{k}\mathbf{x}}$ are an orthonormal basis of $\mathcal{F}_1$. We define

$$a_{\mathbf{k}\sigma} = \frac{1}{M^d} \sum_{\mathbf{x}\tau} L^{\frac{d}{2}} \bar{\phi}_{\mathbf{k}\sigma}(\mathbf{x}\tau)\psi_{\mathbf{x}\tau} = \frac{1}{M^d} \sum_{\mathbf{x}} e^{-i\mathbf{k}\mathbf{x}}\psi_{\mathbf{x}\sigma} \tag{2.52}$$

$$\Rightarrow a_{\mathbf{k}\sigma}^+ = \frac{1}{M^d} \sum_{\mathbf{x}\tau} L^{\frac{d}{2}} \phi_{\mathbf{k}\sigma}(\mathbf{x}\tau)\psi_{\mathbf{x}\tau}^+ = \frac{1}{M^d} \sum_{\mathbf{x}} e^{i\mathbf{k}\mathbf{x}}\psi_{\mathbf{x}\sigma}^+ \tag{2.53}$$

The following corollary follows immediately from the properties of ψ and ψ^+.

Corollary 2.3.3 (i) *One has*

$$(a_{\mathbf{k}\sigma} F_{N+1})(\mathbf{x}_1\sigma_1,\cdots,\mathbf{x}_N\sigma_N) \tag{2.54}$$

$$= \sqrt{N+1}\tfrac{1}{M^d}\sum_{\mathbf{x}} e^{-i\mathbf{k}\mathbf{x}} F_{N+1}(\mathbf{x}\sigma,\mathbf{x}_1\sigma_1,\cdots,\mathbf{x}_N\sigma_N)$$

$$(a_{\mathbf{k}\sigma}^+ F_N)(\mathbf{x}_1\sigma_1,\cdots,\mathbf{x}_{N+1}\sigma_{N+1}) \tag{2.55}$$

$$= \tfrac{1}{\sqrt{N+1}}\sum_{j=1}^{N+1}(-1)^{j-1}e^{i\mathbf{k}\mathbf{x}_j} F_N(\mathbf{x}_1\sigma_1,\cdots,\widehat{\mathbf{x}_j\sigma_j},\cdots,\mathbf{x}_N\sigma_N)$$

(ii) *The following canonical anticommutation relations hold*

$$\{a_{\mathbf{k}\sigma}, a_{\mathbf{k}'\sigma'}\} = \{a_{\mathbf{k}\sigma}^+, a_{\mathbf{k}'\sigma'}^+\} = 0$$
$$\{a_{\mathbf{k}\sigma}, a_{\mathbf{k}'\sigma'}^+\} = \delta_{\sigma,\sigma'} L^d \delta_{\mathbf{k},\mathbf{k}'} \tag{2.56}$$

Theorem 2.3.4 *Let* $H = \oplus_{N=0}^{\infty} H_N$ *where*

$$H_N = -\tfrac{1}{2m}\sum_{i=1}^{N}\Delta_{\mathbf{x}_i} + \sum_{\substack{i,j=1 \\ i\neq j}} V(\mathbf{x}_i - \mathbf{x}_j) - \mu N$$

Then there is the representation

$$H = \tfrac{1}{L^d}\sum_{\mathbf{k}\sigma} e_{\mathbf{k}}\, a_{\mathbf{k}\sigma}^+ a_{\mathbf{k}\sigma} + \tfrac{1}{L^{3d}}\sum_{\sigma,\tau}\sum_{\mathbf{k},\mathbf{p},\mathbf{q}} V(\mathbf{k}-\mathbf{p})\, a_{\mathbf{k}\sigma}^+ a_{\mathbf{q}-\mathbf{k},\tau}^+ a_{\mathbf{q}-\mathbf{p},\tau} a_{\mathbf{p}\sigma} \tag{2.57}$$

where

$$e_{\mathbf{k}} = \tfrac{1}{2m}\sum_{i=1}^{d} 2M^2(1 - \cos[k_i/M]) - \mu \overset{M\to\infty}{\longrightarrow} \tfrac{\mathbf{k}^2}{2m} - \mu \tag{2.58}$$

Proof: Let $H_{0,N} = \tfrac{1}{2m}\sum_{i=1}^{N}(-\Delta_{\mathbf{x}_i} - \mu)$. According to Theorem 1.1.2, part (i), we have

$$H_0 = \tfrac{1}{M^d}\sum_{\mathbf{x},\sigma} \psi_{\mathbf{x}\sigma}^+(-\Delta - \mu)\psi_{\mathbf{x}\sigma}$$

$$= \tfrac{1}{M^d}\sum_{\mathbf{x},\sigma}\tfrac{1}{L^d}\sum_{\mathbf{k}} e^{-i\mathbf{k}\mathbf{x}} a_{\mathbf{k}\sigma}^+(-\Delta - \mu)\tfrac{1}{L^d}\sum_{\mathbf{p}} e^{i\mathbf{p}\mathbf{x}} a_{\mathbf{p}\sigma}$$

$$= \tfrac{1}{L^{2d}}\sum_{\mathbf{k},\mathbf{p}} L^d \delta_{\mathbf{k},\mathbf{p}}(\varepsilon(\mathbf{p}) - \mu) a_{\mathbf{k}\sigma}^+ a_{\mathbf{p}\sigma}$$

$$= \tfrac{1}{L^d}\sum_{\mathbf{k}}(\varepsilon(\mathbf{k}) - \mu) a_{\mathbf{k}\sigma}^+ a_{\mathbf{k}\sigma} \tag{2.59}$$

Similarly we obtain for the interacting part I according to the second part of Theorem 1.1.2

$$I = \frac{1}{M^{2d}} \sum_{\mathbf{x},\mathbf{y}} \sum_{\sigma,\tau} \psi^+_{\mathbf{x}\sigma} \psi^+_{\mathbf{y}\tau} V(\mathbf{x} - \mathbf{y}) \psi_{\mathbf{y}\tau} \psi_{\mathbf{x}\sigma}$$

$$= \frac{1}{M^{2d}} \sum_{\mathbf{x},\mathbf{y}} \sum_{\sigma,\tau} \frac{1}{L^{4d}} \sum_{\mathbf{k}_1 \mathbf{k}_2 \mathbf{k}_3 \mathbf{k}_4} e^{-i k_1 \mathbf{x} - i k_2 \mathbf{y} + i k_3 \mathbf{y} + i k_4 \mathbf{x}} \times$$

$$a^+_{\mathbf{k}_1 \sigma} a^+_{\mathbf{k}_2 \tau} V(\mathbf{x} - \mathbf{y}) a_{\mathbf{k}_3 \tau} a_{\mathbf{k}_4 \sigma} \qquad (2.60)$$

The exponential equals $e^{-i(\mathbf{k}_1 - \mathbf{k}_4)(\mathbf{x}-\mathbf{y})} e^{-i(\mathbf{k}_1 + \mathbf{k}_2 - \mathbf{k}_3 - \mathbf{k}_4)\mathbf{y}}$. Since $\frac{1}{M^d} \sum_{\mathbf{y}} e^{i\mathbf{p}\mathbf{y}} = L^d \delta_{\mathbf{p},0}$, we arrive at

$$I = \frac{1}{L^{4d}} \sum_{\sigma,\tau} \sum_{\mathbf{k}_1 \mathbf{k}_2 \mathbf{k}_3 \mathbf{k}_4} L^d \delta_{\mathbf{k}_1 + \mathbf{k}_2, \mathbf{k}_3 + \mathbf{k}_4} \hat{V}(\mathbf{k}_1 - \mathbf{k}_4) a^+_{\mathbf{k}_1 \sigma} a^+_{\mathbf{k}_2 \tau} a_{\mathbf{k}_3 \tau} a_{\mathbf{k}_4 \sigma} \qquad (2.61)$$

If we choose $\mathbf{k}_1 =: \mathbf{k}$, $\mathbf{k}_4 =: \mathbf{p}$ and $\mathbf{q} := \mathbf{k}_1 + \mathbf{k}_2$ we obtain (2.57). ∎

Chapter 3

Perturbation Theory

In this chapter we derive the perturbation series for the partition function. Since it is given by a trace of an exponential of an operator $H_0 + \lambda V$, we first expand the exponential with respect to λ. The result is a power series with operator valued coefficients. Then we compute the trace of the n'th order coefficient. This can be done with Wick's theorem which lies at the bottom of any diagrammatic expansion. It states that the expectation value, with respect to $\exp[-\beta H_0]$, of a product of $2n$ annihilation and creation operators is given by a sum of terms where each term is a product of expectation values of only two annihilation and creation operators. Depending on whether the model is fermionic or bosonic or complex or scalar, this sum is given by a determinant, a pfaffian, a permanent or simply the sum over all pairings which is the bosonic analog of a pfaffian. For the many-electron system, one obtains an $n \times n$ determinant. The final result is summarized in Theorem 3.2.4 below.

3.1 The Perturbation Series for $e^{H_0 + \lambda V}$

The goal of this section is to prove the following

Theorem 3.1.1 *Let* $H_0, V \in \mathbb{C}^{N \times N}$ *and let* $H_\lambda = H_0 + \lambda V$. *Then*

$$e^{(t_2 - t_1)H_\lambda} = e^{t_2 H_0} \, \mathrm{T} e^{\lambda \int_{t_1}^{t_2} ds \, V(s)} \, e^{-t_1 H_0} \qquad (3.1)$$

where

$$\mathrm{T} e^{\lambda \int_{t_1}^{t_2} ds \, V(s)} := \sum_{n=0}^{\infty} \frac{\lambda^n}{n!} \int_{[t_1, t_2]^n} ds_1 \cdots ds_n \, \mathrm{T} \left[V(s_1) \cdots V(s_n) \right] \qquad (3.2)$$

and

$$\mathrm{T} \left[V(s_1) \cdots V(s_n) \right] = V(s_{\pi 1}) \cdots V(s_{\pi n}) \qquad (3.3)$$

if π *is a permutation such that* $s_{\pi 1} \geq \cdots \geq s_{\pi n}$. *Furthermore* $V(s) = e^{-s H_0} V e^{s H_0}$.

For the proof we need the following

Lemma 3.1.2 *Let H_λ be as above. Then:*

$$\tfrac{d}{d\lambda} e^{(t_2-t_1)H_\lambda} = \int_{t_1}^{t_2} ds\, e^{(t_2-s)H_\lambda} V e^{(s-t_1)H_\lambda} \qquad (3.4)$$

Proof: We have

$$
\begin{aligned}
e^{(t_2-t_1)H_\lambda} - e^{(t_2-t_1)H_{\lambda'}} &= -e^{(t_2-s)H_\lambda} e^{(s-t_1)H_{\lambda'}} \big|_{s=t_1}^{s=t_2} \\
&= -\int_{t_1}^{t_2} ds\, \tfrac{d}{ds}\left(e^{(t_2-s)H_\lambda} e^{(s-t_1)H_{\lambda'}} \right) \\
&= -\int_{t_1}^{t_2} ds\, e^{(t_2-s)H_\lambda}(-H_\lambda + H_{\lambda'}) e^{(s-t_1)H_{\lambda'}} \\
&= \int_{t_1}^{t_2} ds\, e^{(t_2-s)H_\lambda}(\lambda - \lambda') V e^{(s-t_1)H_{\lambda'}} \qquad (3.5)
\end{aligned}
$$

which gives

$$
\begin{aligned}
\tfrac{d}{d\lambda} e^{(t_2-t_1)H_\lambda} &= \lim_{\lambda' \to \lambda} \frac{e^{(t_2-t_1)H_\lambda} - e^{(t_2-t_1)H_{\lambda'}}}{\lambda - \lambda'} \\
&= \int_{t_1}^{t_2} ds\, e^{(t_2-s)H_\lambda} V e^{(s-t_1)H_\lambda} \quad\blacksquare
\end{aligned}
$$

Proof of Theorem: Since for matrices H_0, V the exponential $e^{H_0 + \lambda V}$ is analytic in λ, one has

$$e^{(t_2-t_1)H_\lambda} = \sum_{n=0}^{\infty} \tfrac{\lambda^n}{n!} \left(\tfrac{d}{d\lambda}\right)^n_{|\lambda=0} e^{(t_2-t_1)H_\lambda} \qquad (3.6)$$

Thus we have to prove

$$
\left(\tfrac{d}{d\lambda}\right)^n_{|\lambda=0} e^{(t_2-t_1)H_\lambda} =
$$
$$
e^{t_2 H_0} \int_{[t_1,t_2]^n} ds_1 \cdots ds_n\, \mathrm{T}[V(s_1)\cdots V(s_n)]\, e^{-t_1 H_0} \qquad (3.7)
$$

We claim that for arbitrary λ

$$
\left(\tfrac{d}{d\lambda}\right)^n e^{(t_2-t_1)H_\lambda} =
$$
$$
e^{t_2 H_\lambda} \int_{[t_1,t_2]^n} ds_1 \cdots ds_n\, \mathrm{T}[V_\lambda(s_1)\cdots V_\lambda(s_n)]\, e^{-t_1 H_\lambda} \qquad (3.8)
$$

where $V_\lambda(s) = e^{-sH_\lambda} V e^{sH_\lambda}$. Apparently (3.8) for $\lambda = 0$ gives (3.7). (3.8) is proven by induction on n. For $n = 1$, (3.8) reduces to the lemma above.

Suppose (3.8) is correct for $n - 1$. Since $T[V_\lambda(s_1) \cdots V_\lambda(s_n)]$ is a symmetric function, one has

$$\int_{[t_1,t_2]^n} ds_1 \cdots ds_n \, T[V_\lambda(s_1) \cdots V_\lambda(s_n)]$$

$$= n! \int_{[t_1,t_2]^n} ds_1 \cdots ds_n \, \chi(s_1 \geq s_2 \geq \cdots \geq s_n) T[V_\lambda(s_1) \cdots V_\lambda(s_n)]$$

$$= n! \int_{t_1}^{t_2} ds_1 \int_{t_1}^{s_1} ds_2 \cdots \int_{t_1}^{s_{n-1}} ds_n \, V(s_1) \cdots V(s_n) \quad (3.9)$$

and the induction hypothesis reads

$$\left(\tfrac{d}{d\lambda}\right)^{n-1} e^{(t_2-t_1)H_\lambda} =$$

$$(n-1)! \int_{t_1}^{t_2} ds_1 \cdots \int_{t_1}^{s_{n-2}} ds_{n-1} \, e^{(t_2-s_1)H_\lambda} V e^{(s_1-s_2)H_\lambda} \ldots V e^{(s_{n-1}-t_1)H_\lambda}$$

Thus

$$\left(\tfrac{d}{d\lambda}\right)^n e^{(t_2-t_1)H_\lambda} = (n-1)! \int_{t_1}^{t_2} ds_1 \cdots \int_{t_1}^{s_{n-2}} ds_{n-1} \, \tfrac{d}{d\lambda}\{ \quad \} \quad (3.10)$$

where, if we put $s_0 := t_2$ and $s_n := t_1$

$$\tfrac{d}{d\lambda}\{ \quad \} = \tfrac{d}{d\lambda}\left\{ e^{(t_2-s_1)H_\lambda} V e^{(s_1-s_2)H_\lambda} \ldots V e^{(s_{n-1}-t_1)H_\lambda} \right\}$$

$$= \sum_{i=0}^{n-1} e^{(t_2-s_1)H_\lambda} V \cdots \tfrac{d}{d\lambda} e^{(s_i-s_{i+1})H_\lambda} \ldots V e^{(s_{n-1}-t_1)H_\lambda}$$

$$= \sum_{i=0}^{n-1} e^{(t_2-s_1)H_\lambda} V \cdots \int_{s_{i+1}}^{s_i} ds \, e^{(s_i-s)H_\lambda} V e^{(s-s_{i+1})H_\lambda} \cdots$$

$$\cdots V e^{(s_{n-1}-t_1)H_\lambda} \quad (3.11)$$

Substituting this in (3.10), we obtain the following integrals

$$\int_{t_1}^{t_2} ds_1 \cdots \int_{t_1}^{s_{i-1}} ds_i \int_{s_{i+1}}^{s_i} ds \int_{t_1}^{s_i} ds_{i+1} \cdots \int_{t_1}^{s_{n-2}} ds_{n-1}$$

$$= \int_{[t_1,t_2]^n} ds_1 \cdots ds_i \, ds \, ds_{i+1} \cdots ds_n \, \chi(\cdots)$$

where the characteristic function is

$$\chi(\cdots) = \chi(s_1 \geq \cdots \geq s_{i-1} \geq s_i \geq s \geq s_{i+1} \geq \cdots \geq s_{n-1})$$

which results in (renaming the integration variables)

$$\left(\tfrac{d}{d\lambda}\right)^n e^{(t_2-t_1)H_\lambda} =$$

$$n(n-1)! \int_{[t_1,t_2]^n} \chi(s_1 \geq \cdots \geq s_n) \, e^{(t_2-s_1)H_\lambda} V \ldots V e^{(s_n-t_1)H_\lambda} \quad \blacksquare$$

3.2 The Perturbation Series for the Partition Function

In this section we prove Theorem 3.2.4 below in which the perturbation series for the partition function is written down. Let $H = H(\lambda) = H_0 + \lambda H_{\text{int}}$: $\mathcal{F} \to \mathcal{F}$ be the Hamiltonian where

$$H_0 = \tfrac{1}{M^{2d}} \sum_{\mathbf{x}\sigma, \mathbf{y}\tau} \psi^+_{\mathbf{x}\sigma} h(\mathbf{x}\sigma, \mathbf{y}\tau) \psi_{\mathbf{y}\tau} \tag{3.12}$$

$$H_{\text{int}} = \tfrac{1}{M^{2d}} \sum_{\mathbf{x}\sigma, \mathbf{y}\tau} \psi^+_{\mathbf{x}\sigma} \psi^+_{\mathbf{y}\tau} V(\mathbf{x}-\mathbf{y}) \psi_{\mathbf{y}\tau} \psi_{\mathbf{x}\sigma} \tag{3.13}$$

According to Theorem 3.1.1 of the last section we have

$$Tr\, e^{-\beta H(\lambda)} = \tag{3.14}$$

$$\sum_{n=0}^{\infty} \frac{(-\lambda)^n}{n!} \int_{[0,\beta]^n} ds_1 \cdots ds_n\, Tr\left[e^{-\beta H_0} T H_{\text{int}}(s_1) \cdots H_{\text{int}}(s_n) \right]$$

where

$$H_{\text{int}}(s) = e^{sH_0} H_{\text{int}} e^{-sH_0} \tag{3.15}$$

$$= \tfrac{1}{M^{2d}} \sum_{\mathbf{x}\sigma, \mathbf{y}\tau} \psi^+_{\mathbf{x}\sigma}(s) \psi^+_{\mathbf{y}\tau}(s) V(\mathbf{x}-\mathbf{y}) \psi_{\mathbf{y}\tau}(s) \psi_{\mathbf{x}\sigma}(s)$$

$$\tag{3.16}$$

if we define

$$\psi_{\mathbf{x}\sigma}(s) := e^{sH_0} \psi_{\mathbf{x}\sigma} e^{-sH_0} \tag{3.17}$$

$$\psi^+_{\mathbf{x}\sigma}(s) := e^{sH_0} \psi^+_{\mathbf{x}\sigma} e^{-sH_0} \tag{3.18}$$

Observe that for $s \neq 0$ $\psi^+_{\mathbf{x}\sigma}(s)$ is no longer the adjoint of $\psi_{\mathbf{x}\sigma}(s)$. In (3.14) the 'temperature' ordering operator T acts on the operators $H_{\text{int}}(s)$. Instead of this, we may let T act directly on the annihilation and creation operators $\psi(s), \psi^+(s)$. We introduce the following notation: For

$$\xi = (s, \mathbf{x}, \sigma, b) \in [0, \beta] \times \Gamma \times \{\uparrow, \downarrow\} \times \{0, 1\} =: \bar{\Gamma} \tag{3.19}$$

let

$$\psi(\xi) := \begin{cases} \psi^+_{\mathbf{x}\sigma}(s) & \text{if } b = 1 \\ \psi_{\mathbf{x}\sigma}(s) & \text{if } b = 0 \end{cases} \tag{3.20}$$

Then we define

$$T[\psi(\xi_1) \cdots \psi(\xi_n)] := \text{sign}\pi\, \psi(\xi_{\pi 1}) \cdots \psi(\xi_{\pi n}) \tag{3.21}$$

if π is a permutation such that $s_{\pi 1} \geq \cdots \geq s_{\pi n}$ and, if $s_{\pi j} = s_{\pi(j+1)}$, one has $b_{\pi j} \geq b_{\pi(j+1)}$. The last requirement is necessary since in (3.15) there are $\psi(s)$ and $\psi^+(s)$ for equal values of s and these operators do not anticommute. We also define

$$h(b) = \begin{cases} h & \text{if } b = 0 \\ -h^T & \text{if } b = 1 \end{cases} \tag{3.22}$$

Lemma 3.2.1 *The following relations hold:*

(i) *Let $\psi^{(+)}(s) = e^{sH_0}\psi^{(+)}e^{-sH_0}$. Then*

$$\psi_{\mathbf{x}\sigma}(s) = \sum_{\mathbf{y}\tau} e^{-\frac{s}{M^d}h}(\mathbf{x}\sigma, \mathbf{y}\tau)\,\psi_{\mathbf{y}\tau} \tag{3.23}$$

$$\psi^+_{\mathbf{x}\sigma}(s) = \sum_{\mathbf{y}\tau} e^{\frac{s}{M^d}h}(\mathbf{y}\tau, \mathbf{x}\sigma)\,\psi^+_{\mathbf{y}\tau} \tag{3.24}$$

(ii) *The following anticommutation relations hold*

$$\{\psi_{\mathbf{x}\sigma}(s), \psi_{\mathbf{y}\tau}(t)\} = \{\psi^+_{\mathbf{x}\sigma}(s), \psi^+_{\mathbf{y}\tau}(t)\} = 0 \tag{3.25}$$

$$\{\psi^+_{\mathbf{x}\sigma}(s), \psi_{\mathbf{y}\tau}(t)\} = M^d\, e^{\frac{(s-t)}{M^d}h}(\mathbf{y}\tau, \mathbf{x}\sigma) \tag{3.26}$$

Using the notation (3.22), this can be written more compactly as

$$\{\psi(\xi), \psi(\xi')\} = \delta_{b,1-b'}\,M^d\, e^{\frac{-(s-s')}{M^d}h(b)}(\mathbf{x}\sigma, \mathbf{x}'\sigma') \tag{3.27}$$

In particular, for $h = -\frac{1}{2m}\Delta - \mu$,

$$\{\psi^+_{\mathbf{x}\sigma}(s), \psi_{\mathbf{y}\tau}(t)\} = \delta_{\sigma,\tau}\,\frac{1}{L^d}\sum_{\mathbf{k}} e^{i\mathbf{k}(\mathbf{x}-\mathbf{y})}e^{(s-t)e_{\mathbf{k}}} \tag{3.28}$$

Proof: We have

$$\tfrac{d}{ds}\psi_{\mathbf{x}\sigma}(s)^{(+)} = e^{sH_0}(H_0\psi^{(+)}_{\mathbf{x}\sigma} - \psi^{(+)}_{\mathbf{x}\sigma}H_0)e^{-sH_0} \tag{3.29}$$

Furthermore

$$H_0\psi_{\mathbf{x}\sigma} = \tfrac{1}{M^{2d}}\sum_{\mathbf{x}'\sigma',\mathbf{y}\tau} \psi^+_{\mathbf{x}'\sigma'}h(\mathbf{x}'\sigma', \mathbf{y}\tau)\psi_{\mathbf{y}\tau}\psi_{\mathbf{x}\sigma}$$

$$= -\tfrac{1}{M^{2d}}\sum_{\mathbf{x}'\sigma',\mathbf{y}\tau} \psi^+_{\mathbf{x}'\sigma'}\psi_{\mathbf{x}\sigma}h(\mathbf{x}'\sigma', \mathbf{y}\tau)\psi_{\mathbf{y}\tau}$$

$$= -\tfrac{1}{M^{2d}}\sum_{\mathbf{x}'\sigma',\mathbf{y}\tau} \{\psi^+_{\mathbf{x}'\sigma'}, \psi_{\mathbf{x}\sigma}\}h(\mathbf{x}'\sigma', \mathbf{y}\tau)\psi_{\mathbf{y}\tau}$$

$$\qquad + \psi_{\mathbf{x}\sigma}\tfrac{1}{M^{2d}}\sum_{\mathbf{x}'\sigma',\mathbf{y}\tau} \psi^+_{\mathbf{x}'\sigma'}h(\mathbf{x}'\sigma', \mathbf{y}\tau)\psi_{\mathbf{y}\tau}$$

$$= -\tfrac{1}{M^d}\sum_{\mathbf{y}\tau} h(\mathbf{x}\sigma, \mathbf{y}\tau)\psi_{\mathbf{y}\tau} + \psi_{\mathbf{x}\sigma}H_0 \tag{3.30}$$

and .

$$H_0\psi^+_{\mathbf{x}\sigma} = \tfrac{1}{M^{2d}} \sum_{\mathbf{x}'\sigma',\mathbf{y}\tau} \psi^+_{\mathbf{x}'\sigma'} h(\mathbf{x}'\sigma',\mathbf{y}\tau)\psi_{\mathbf{y}\tau}\psi^+_{\mathbf{x}\sigma}$$

$$= \tfrac{1}{M^{2d}} \sum_{\mathbf{x}'\sigma',\mathbf{y}\tau} \psi^+_{\mathbf{x}'\sigma'} h(\mathbf{x}'\sigma',\mathbf{y}\tau)\{\psi_{\mathbf{y}\tau},\psi^+_{\mathbf{x}\sigma}\}$$

$$- \tfrac{1}{M^{2d}} \sum_{\mathbf{x}'\sigma',\mathbf{y}\tau} \psi^+_{\mathbf{x}'\sigma'} h(\mathbf{x}'\sigma',\mathbf{y}\tau)\psi^+_{\mathbf{x}\sigma}\psi_{\mathbf{y}\tau}$$

$$= \tfrac{1}{M^{d}} \sum_{\mathbf{x}'\sigma'} \psi^+_{\mathbf{x}'\sigma'} h(\mathbf{x}'\sigma',\mathbf{x}\sigma)$$

$$+ \psi^+_{\mathbf{x}\sigma} \tfrac{1}{M^{2d}} \sum_{\mathbf{x}'\sigma',\mathbf{y}\tau} \psi^+_{\mathbf{x}'\sigma'} h(\mathbf{x}'\sigma',\mathbf{y}\tau)\psi_{\mathbf{y}\tau}$$

$$= \tfrac{1}{M^{d}} \sum_{\mathbf{x}'\sigma'} \psi^+_{\mathbf{x}'\sigma'} h(\mathbf{x}'\sigma',\mathbf{x}\sigma) + \psi^+_{\mathbf{x}\sigma} H_0 \tag{3.31}$$

Thus we get the differential equations

$$\tfrac{d}{ds}\psi_{\mathbf{x}\sigma}(s) = -\tfrac{1}{M^{d}} \sum_{\mathbf{y}\tau} h(\mathbf{x}\sigma,\mathbf{y}\tau)\psi_{\mathbf{y}\tau}(s) \tag{3.32}$$

$$\tfrac{d}{ds}\psi^+_{\mathbf{x}\sigma}(s) = \tfrac{1}{M^{d}} \sum_{\mathbf{y}\tau} h(\mathbf{y}\tau,\mathbf{x}\sigma)\psi^+_{\mathbf{y}\tau}(s) \tag{3.33}$$

which proves part (i) of the lemma. Equations (3.25) of part (ii) are obvious. To obtain (3.26), we use part (i):

$$\{\psi^+_{\mathbf{x}\sigma}(s),\psi_{\mathbf{x}'\sigma'}(t)\} = \sum_{\mathbf{y}\tau,\mathbf{y}'\tau'} e^{\frac{s}{M^{d}}h}(\mathbf{y}\tau,\mathbf{x}\sigma) e^{-\frac{t}{M^{d}}h}(\mathbf{x}'\sigma',\mathbf{y}'\tau')\{\psi^+_{\mathbf{y}\tau},\psi_{\mathbf{y}'\tau'}\}$$

$$= M^{d} \sum_{\mathbf{y}\tau} e^{\frac{s}{M^{d}}h}(\mathbf{y}\tau,\mathbf{x}\sigma) e^{-\frac{t}{M^{d}}h}(\mathbf{x}'\sigma',\mathbf{y}\tau)$$

$$= M^{d} e^{\frac{s-t}{M^{d}}h}(\mathbf{x}'\sigma',\mathbf{x}\sigma) \tag{3.34}$$

Suppose now that h is the discrete Laplacian. That is $h = [h(\mathbf{x}\sigma,\mathbf{y}\tau)]_{\mathbf{x}\sigma,\mathbf{y}\tau}$ where

$$h(\mathbf{x}\sigma,\mathbf{y}\tau) = \tfrac{1}{2m}\delta_{\sigma.\tau}M^{d}M^{2}\sum_{i=1}^{d}(\delta_{\mathbf{x}+\frac{\mathbf{e}_i}{M},\mathbf{y}} + \delta_{\mathbf{x}-\frac{\mathbf{e}_i}{M},\mathbf{y}} - 2\delta_{\mathbf{x},\mathbf{y}}) - \mu M^{d}\delta_{\mathbf{x},\mathbf{y}} \tag{3.35}$$

Let

$$F = \frac{1}{(ML)^{\frac{d}{2}}}(e^{-i\mathbf{k}\mathbf{x}})_{\mathbf{k},\mathbf{x}} \tag{3.36}$$

be the matrix of discrete Fourier transform. Since

$$FhF^* = [(e_{\mathbf{k}}\delta_{\mathbf{k},\mathbf{p}})_{\mathbf{k},\mathbf{p}}] \tag{3.37}$$

we have, since $h^T = h$,

$$\{\psi^+_{\mathbf{x}\sigma}(s), \psi_{\mathbf{y}\tau}(t)\} = M^d \, e^{\frac{(s-t)}{M^d}h}(\mathbf{x}\sigma, \mathbf{y}\tau)$$

$$= \delta_{\sigma,\tau} \left[F^* M^d \, e^{\frac{(s-t)}{M^d}FhF^*} F \right]_{\mathbf{x},\mathbf{y}}$$

$$= \delta_{\sigma,\tau} \frac{1}{L^d} \sum_{\mathbf{k}} e^{i\mathbf{k}(\mathbf{x}-\mathbf{y})} e^{(s-t)e_{\mathbf{k}}} \qquad (3.38)$$

which proves the lemma. ∎

Lemma 3.2.2 (Wick) *For some operator* $A : \mathcal{F} \to \mathcal{F}$ *let*

$$\langle A \rangle_0 := \frac{Tr \, e^{-\beta H_0} A}{Tr \, e^{-\beta H_0}} \qquad (3.39)$$

Recall notation (3.20). Then one has $\langle \psi(\xi_1) \cdots \psi(\xi_n) \rangle_0 = 0$ *if* n *is odd and for even* n

$$\langle \psi(\xi_1) \cdots \psi(\xi_n) \rangle_0 = \sum_{j=2}^{n} (-1)^j \langle \psi(\xi_1)\psi(\xi_j) \rangle_0 \, \langle \psi(\xi_2) \cdots \widehat{\psi(\xi_j)} \cdots \psi(\xi_n) \rangle_0 \quad (3.40)$$

where $\widehat{\psi}$ *means omission of that factor. Furthermore*

$$\langle \psi_{\mathbf{x}_1\sigma_1}(s_1, b_1)\psi_{\mathbf{x}_2\sigma_2}(s_2, b_2) \rangle_0 = \qquad\qquad (3.41)$$

$$\sum_{\mathbf{y}\tau} [Id + e^{-\frac{\beta}{M^d}h(b_1)}]^{-1}(\mathbf{x}_1\sigma_1, \mathbf{y}\tau)\{\psi_{\mathbf{y}\tau}(s_1, b_1), \psi_{\mathbf{x}_2\sigma_2}(s_2, b_2)\}$$

if we use the notation (3.22).

Proof: Because of part (i) of the previous lemma we have

$$\psi_{\mathbf{x}\sigma}(s + \beta) = \sum_{\mathbf{y}\tau} e^{-\frac{\beta}{M^d}h}(\mathbf{x}\sigma, \mathbf{y}\tau)\psi_{\mathbf{y}\tau}(s) \qquad (3.42)$$

$$\psi^+_{\mathbf{x}\sigma}(s + \beta) = \sum_{\mathbf{y}\tau} e^{\frac{\beta}{M^d}h^T}(\mathbf{x}\sigma, \mathbf{y}\tau)\psi^+_{\mathbf{y}\tau}(s) \qquad (3.43)$$

or in matrix notation, regarding $\psi(s) = [\psi_{\mathbf{x}\sigma}(s)]_{\mathbf{x}\sigma \in \Gamma \times \{\uparrow,\downarrow\}}$ as a column vector,

$$\psi(s + \beta) = e^{-\frac{\beta}{M^d}h}\psi(s), \quad \psi^+(s + \beta) = e^{\frac{\beta}{M^d}h^T}\psi^+(s) \qquad (3.44)$$

To write this more compact, we use the definition (3.22) which gives

$$\psi(s + \beta, b) = e^{-\frac{\beta}{M^d}h_b}\psi(s, b) \qquad (3.45)$$

26

Since

$$Tr[e^{-\beta H_0}\psi(s,b)] = Tr[\psi(s,b)\,e^{-\beta H_0}] = Tr[e^{-\beta H_0}e^{\beta H_0}\psi(s,b)\,e^{-\beta H_0}]$$
$$= Tr[e^{-\beta H_0}\psi(s+\beta,b)] = e^{-\frac{\beta}{M^d}h_b}Tr[e^{-\beta H_0}\psi(s,b)] \qquad (3.46)$$

we get

$$(Id - e^{-\frac{\beta}{M^d}h_b})Tr[e^{-\beta H_0}\psi(s,b)] = 0 \qquad (3.47)$$

which implies

$$Tr[e^{-\beta H_0}\psi(s,b)] = 0 \qquad (3.48)$$

since all eigenvalues of $Id - e^{-\frac{\beta}{M^d}h_b}$, namely $1 - e^{-\beta(-1)^b e_{\mathbf{k}}}$, are nonzero (we may assume $e_{\mathbf{k}} \neq 0$ for all $\mathbf{k} \in \left(\frac{2\pi}{L}\mathbb{Z}\right)^d$).

If we have n operators $\psi(\xi_1) \equiv \psi_1, \cdots, \psi(\xi_n) \equiv \psi_n$, we write

$$\psi_1\psi_2\cdots\psi_n = \{\psi_1,\psi_2\}\psi_3\cdots\psi_n - \psi_2\psi_1\psi_3\cdots\psi_n$$
$$= \{\psi_1,\psi_2\}\psi_3\cdots\psi_n - \psi_2\{\psi_1,\psi_3\}\psi_4\cdots\psi_n + \psi_2\psi_3\psi_1\psi_4\cdots\psi_n$$
$$= \sum_{j=2}^{n}\{\psi_1,\psi_j\}(-1)^j\psi_2\cdots\widehat{\psi_j}\cdots\psi_n - (-1)^n\psi_2\cdots\psi_n\psi_1 \qquad (3.49)$$

which gives

$$\langle\psi_1\psi_2\cdots\psi_n\rangle_0 + (-1)^n\langle\psi_2\cdots\psi_n\psi_1\rangle_0$$
$$= \sum_{j=2}^{n}\{\psi_1,\psi_j\}(-1)^j\langle\psi_2\cdots\widehat{\psi_j}\cdots\psi_n\rangle_0 \qquad (3.50)$$

Since

$$\langle\psi_2\cdots\psi_n\psi_1\rangle_0 \equiv \langle\psi_2\cdots\psi_n\psi(s_1)\rangle_0$$
$$= \langle\psi(s_1+\beta)\psi_2\cdots\psi_n\rangle_0$$
$$= e^{-\frac{\beta}{M^d}h(b_1)}\Big|_1\langle\psi(s_1)\psi_2\cdots\psi_n\rangle_0 \qquad (3.51)$$

where the subscript $|_1$ at the exponential means matrix multiplication with respect to $\mathbf{x}_1\sigma_1$ from the ψ_1 operator, we get

$$[Id + (-1)^n e^{-\frac{\beta}{M^d}h(b_1)}]\Big|_1\langle\psi_1\psi_2\cdots\psi_n\rangle_0$$
$$= \sum_{j=2}^{n}\{\psi_1,\psi_j\}(-1)^j\langle\psi_2\cdots\widehat{\psi_j}\cdots\psi_n\rangle_0 \qquad (3.52)$$

or

$$\langle\psi_1\psi_2\cdots\psi_n\rangle_0 =$$
$$\sum_{j=2}^{n}[Id + (-1)^n e^{-\frac{\beta}{M^d}h(b_1)}]\Big|_1^{-1}\{\psi_1,\psi_j\}(-1)^j\langle\psi_2\cdots\widehat{\psi_j}\cdots\psi_n\rangle_0 \qquad (3.53)$$

In particular, for $n = 2$,

$$\langle \psi_1 \psi_2 \rangle_0 = [Id + e^{-\frac{\beta}{M^d} h(b_1)}]^{-1}_{|_1} \{\psi_1, \psi_2\} \tag{3.54}$$

which proves the lemma. ∎

The following theorem computes the trace on the right hand side of (3.14).

Theorem 3.2.3 (Wick) (i) *Let T be the temperature ordering operator defined in 3.21. Then*

$$\langle T\psi(\xi_1) \cdots \psi(\xi_n) \rangle_0 = \tag{3.55}$$

$$\sum_{j=2}^{n} (-1)^j \langle T\psi(\xi_1)\psi(\xi_j) \rangle_0 \, \langle T\psi(\xi_2) \cdots \widehat{\psi(\xi_j)} \cdots \psi(\xi_n) \rangle_0$$

and

$$\langle T\psi(\xi)\psi(\xi') \rangle_0 = \tag{3.56}$$

$$\delta_{b,1-b'} M^d \left\{ \frac{e^{-\frac{(s-s')}{M^d} h(b)}}{Id + e^{-\frac{\beta}{M^d} h(b)}} (\mathbf{x}\sigma, \mathbf{x}'\sigma')[\chi(s > s') + \chi(b > b')\chi(s = s')] \right.$$

$$\left. - \frac{e^{-\frac{(s'-s)}{M^d} h(b')}}{Id + e^{-\frac{\beta}{M^d} h(b')}} (\mathbf{x}'\sigma', \mathbf{x}\sigma)[\chi(s' > s) + \chi(b' > b)\chi(s = s')] \right\}$$

(ii) *Let $x = (\mathbf{x}, x_0)$ and $\psi_{x\sigma}^{(+)} = \psi_{\mathbf{x}\sigma}^{(+)}(x_0)$. Then*

$$\langle T\psi_{x_1\sigma_1}^+ \psi_{y_1\tau_1} \cdots \psi_{x_n\sigma_n}^+ \psi_{y_n\tau_n} \rangle_0 = \det \left[\langle T\psi_{x_i\sigma_i}^+ \psi_{y_j\tau_j} \rangle \right]_{1 \le i,j \le n} \tag{3.57}$$

(iii) *Let $h = -\frac{1}{2m}\Delta - \mu$ and $x = (\mathbf{x}, x_0)$. Then $\langle T\psi_{\mathbf{x}\sigma}^+(x_0)\psi_{\mathbf{x}'\sigma'}(x_0') \rangle_0 = \delta_{\sigma,\sigma'} C(\mathbf{x}' - \mathbf{x}, x_0' - x_0)$ where*

$$C(x) = \frac{1}{L^d} \sum_{\mathbf{k}} e^{i\mathbf{k}\mathbf{x}} e^{-x_0 e_\mathbf{k}} \left[\frac{\chi(x_0 \le 0)}{1 + e^{\beta e_\mathbf{k}}} - \frac{\chi(x_0 > 0)}{1 + e^{-\beta e_\mathbf{k}}} \right], \quad x_0 \in (-\beta, \beta) \tag{3.58}$$

$$= \lim_{\epsilon \nearrow 0} \frac{1}{\beta L^d} \sum_{\mathbf{k}, k_0} e^{i(\mathbf{k}\mathbf{x} - k_0 x_0)} \frac{e^{-ik_0\epsilon}}{ik_0 - e_\mathbf{k}}, \quad x_0 \in \mathbb{R} \tag{3.59}$$

Here the k_0 sum ranges over $\frac{\pi}{\beta}(2\mathbb{Z} + 1)$ such that C as a function of x_0 is 2β-periodic and satisfies $C(x_0 + \beta, \mathbf{x}) = -C(x_0, \mathbf{x})$.

Remarks: 1) In the zero temperature limit (3.58) reads

$$C(x) = \frac{1}{L^d} \sum_{\mathbf{k}} e^{i\mathbf{k}x} e^{-x_0 e_{\mathbf{k}}} \left[\chi(x_0 \leq 0)\chi(e_{\mathbf{k}} < 0) - \chi(x_0 > 0)\chi(e_{\mathbf{k}} > 0) \right] \quad (3.60)$$

2) The ϵ prescription in (3.59) is only necessary for $x_0 = x_0'$. In that case we only get the part inside the Fermi surface given by $\chi(e_{\mathbf{k}} < 0)$ in (3.60).

Proof: (i) Equation (3.55) is an immediate consequence of (3.40) since if $s_1 \geq s_2 \geq \cdots \geq s_n$ then we also have $s_1 \geq s_j$ and $s_2 \geq \cdots \hat{s}_j \cdots \geq s_n$. The temperature-ordered expectation of two fields is evaluated with (3.41) and (3.27). By definition of the T-operator

$$\langle T\psi(\xi)\psi(\xi')\rangle_0 = \langle \psi(\xi)\psi(\xi')\rangle_0 [\chi(s > s') + \chi(b > b')\chi(s = s')]$$
$$- \langle \psi(\xi')\psi(\xi)\rangle_0 [\chi(s < s') + \chi(b < b')\chi(s = s')]$$

By (3.41) and (3.27),

$$\langle \psi(\xi)\psi(\xi')\rangle_0 = \sum_{\mathbf{y}\tau} [Id + e^{-\frac{\beta}{M^d}h(b)}]^{-1}(\mathbf{x}\sigma, \mathbf{y}\tau)\{\psi(\mathbf{y}, \tau, s, b), \psi(\xi')\}$$

$$= \sum_{\mathbf{y}\tau} [Id + e^{-\frac{\beta}{M^d}h(b)}]^{-1}(\mathbf{x}\sigma, \mathbf{y}\tau)\delta_{b,1-b'} M^d e^{\frac{-(s-s')}{M^d}h(b)}(\mathbf{y}\tau, \mathbf{x}'\sigma')$$

which proves the first part of the theorem.

(ii) We use part (i) and induction. For $n = 1$, the formula is correct. Suppose it holds for n. Then, expanding the determinant with respect to the first row,

$$\det \left[\langle T\psi_{x_i\sigma_i}^+ \psi_{y_j\tau_j}\rangle_0 \right]_{1 \leq i,j \leq n+1} =$$

$$= \sum_{j=1}^{n+1} (-1)^{1+j} \langle T\psi_{x_1\sigma_1}^+ \psi_{y_j\tau_j}\rangle_0 \det \left[\langle T\psi_{x_i\sigma_i}^+ \psi_{y_k\tau_k}\rangle_0 \right]_{\substack{1 \leq i,k \leq n+1 \\ i \neq 1, k \neq j}}$$

$$= \sum_{j=1}^{n+1} (-1)^{1+j} \langle T\psi_{x_1\sigma_1}^+ \psi_{y_j\tau_j}\rangle_0 \langle T\psi_{x_2\sigma_2}^+ \psi_{y_1\tau_1} \cdots \psi_{x_j\sigma_j}^+ \psi_{y_{j-1}\tau_{j-1}} \circ$$

$$\circ \psi_{x_{j+1}\sigma_{j+1}}^+ \psi_{y_{j+1}\tau_{j+1}} \cdots \psi_{x_{n+1}\sigma_{n+1}}^+ \psi_{y_{n+1}\tau_{n+1}}\rangle_0$$

$$= \sum_{j=1}^{n+1} (-1)^{1+j} \langle T\psi_{x_1\sigma_1}^+ \psi_{y_j\tau_j}\rangle_0 (-1)^{j-1} \langle T\psi_{y_1\tau_1} \psi_{x_2\sigma_2}^+ \cdots \psi_{y_{j-1}\tau_{j-1}} \circ$$

$$\circ \psi_{x_j\sigma_j}^+ \widehat{\psi_{y_j\tau_j}} \psi_{x_{j+1}\sigma_{j+1}}^+ \psi_{y_{j+1}\tau_{j+1}} \cdots \psi_{x_{n+1}\sigma_{n+1}}^+ \psi_{y_{n+1}\tau_{n+1}}\rangle_0$$

$$= \langle T\psi_{x_1\sigma_1}^+ \psi_{y_1\tau_1} \cdots \psi_{x_{n+1}\sigma_{n+1}}^+ \psi_{y_{n+1}\tau_{n+1}}\rangle_0 \quad (3.61)$$

where we used part (i) in the last line and the induction hypothesis in the second line.

(iii) From (3.56) we get

$$\langle T\psi^+_{\mathbf{x}\sigma}(x_0)\psi_{\mathbf{x}'\sigma'}(x'_0)\rangle_0 = \delta_{\sigma,\sigma'}M^d\left\{\frac{e^{\frac{-(x'_0-x_0)}{M^d}h}}{Id+e^{\frac{\beta}{M^d}h}}(\mathbf{x}'\sigma',\mathbf{x}\sigma)\chi(x'_0 \le x_0)\right.$$

$$\left. -\frac{e^{-\frac{(x'_0-x_0)}{M^d}h}}{Id+e^{-\frac{\beta}{M^d}h}}(\mathbf{x}'\sigma',\mathbf{x}\sigma)\chi(x'_0 > x_0)\right\}$$

which gives

$$C(\mathbf{x}x_0,\mathbf{x}'x'_0) = \frac{1}{L^d}\sum_{\mathbf{k}}e^{i\mathbf{k}(\mathbf{x}-\mathbf{x}')}e^{-e_{\mathbf{k}}(x_0-x'_0)}\left[\frac{\chi(x_0-x'_0\le 0)}{1+e^{\beta e_{\mathbf{k}}}}-\frac{\chi(x_0-x'_0>0)}{1+e^{-\beta e_{\mathbf{k}}}}\right]$$

This proves (3.58). The x_0 variables in

$$C(x,x') = C(x-x') = C(x_0-x'_0,\mathbf{x}-\mathbf{x}') = \langle T\psi^+_{\mathbf{x}'\sigma}(x'_0)\psi_{\mathbf{x}\sigma}(x_0)\rangle_0$$

range over $x_0,x'_0 \in (0,\beta)$. In particular, $x_0-x'_0 \in (-\beta,\beta)$. On this interval, C is an antisymmetric function. Namely, let $x_0-x'_0 \in (-\beta,\beta)$ such that also $x_0-x'_0+\beta \in (-\beta,\beta)$. That is, $x_0 < x'_0$. Then

$$C(x_0-x'_0+\beta,\mathbf{x}-\mathbf{x}') = \langle T\psi^+_{\mathbf{x}'\sigma}(x'_0)\psi_{\mathbf{x}\sigma}(x_0+\beta)\rangle_0$$

$$= -Tr\,e^{-\beta H_0}\psi_{\mathbf{x}\sigma}(x_0+\beta)\psi^+_{\mathbf{x}'\sigma}(x'_0)/Tr\,e^{-\beta H_0}$$

$$= -Tr\,\psi_{\mathbf{x}\sigma}(x_0)e^{-\beta H_0}\psi^+_{\mathbf{x}'\sigma}(x'_0)/Tr\,e^{-\beta H_0}$$

$$= -Tr\,e^{-\beta H_0}\psi^+_{\mathbf{x}'\sigma}(x'_0)\psi_{\mathbf{x}\sigma}(x_0)/Tr\,e^{-\beta H_0}$$

$$= -\langle T\psi^+_{\mathbf{x}'\sigma}(x'_0)\psi_{\mathbf{x}\sigma}(x_0+\beta)\rangle_0$$

$$= -C(x_0-x'_0,\mathbf{x}-\mathbf{x}') \tag{3.62}$$

Since C is only defined on $(-\beta,\beta)$, we may expand it into a 2β-periodic Fourier series. This gives us 'frequencies' $k_0 \in \frac{2\pi}{2\beta}\mathbb{Z}$. Because of (3.62), we only get the odd frequencies $k_0 \in \frac{2\pi}{2\beta}(2\mathbb{Z}+1)$. Equation (3.59) is then equivalent to

$$\int_{-\beta}^{\beta}e^{ik_0x_0}C(x_0,\mathbf{k})\,dx_0 = \frac{2\beta}{\beta}C(k_0,\mathbf{k}) = 2C(k_0,\mathbf{k})$$

where

$$C(x_0,\mathbf{k}) = e^{-x_0 e_{\mathbf{k}}}\left[\frac{\chi(x_0\le 0)}{1+e^{\beta e_{\mathbf{k}}}}-\frac{\chi(x_0>0)}{1+e^{-\beta e_{\mathbf{k}}}}\right], \quad C(k_0,\mathbf{k}) = \frac{1}{ik_0-e_{\mathbf{k}}}$$

This is checked by direct computation. ∎

With the above theorem, we are able to write down the perturbation series for the partition function. If we substitute the trace on the right hand side of (3.14) by the determinant of (3.57), we arrive at the following series:

Theorem 3.2.4 Let $x = (x_0, \mathbf{x})$, $U(x) = \delta(x_0) V(\mathbf{x})$ and let C be given by (3.58). Then there is the following power series expansion

$$\frac{Tr\, e^{-\beta H_\lambda}}{Tr\, e^{-\beta H_0}} = \sum_{n=0}^{\infty} \frac{(-\lambda)^n}{n!} D_n \qquad (3.63)$$

where the n'th order coefficient D_n is given by

$$D_n = \int_{[0,\beta]^{2n}} \prod_{j=1}^{2n} dx_{0,j} \; \frac{1}{M^{2dn}} \sum_{\mathbf{x}_1\sigma_1,\cdots,\mathbf{x}_{2n}\sigma_{2n}} \prod_{i=1}^{n} U(x_{2i-1} - x_{2i}) \det \left[\delta_{\sigma_i,\sigma_j} C(x_i, x_j) \right]$$

and $\det \left[\delta_{\sigma_i,\sigma_j} C(x_i, x_j) \right] = \det \left[\delta_{\sigma_i,\sigma_j} C(x_i - x_j) \right]_{1 \le i,j \le 2n}$.

If we expand the determinant in D_n and interchange the sum over permutations with the space and temperature sums or integrals, we obtain the expansion into Feynman diagrams. That is,

$$D_n = \sum_{\pi \in S_n} \mathrm{sign}\pi\, G_n(\pi) \qquad (3.64)$$

where the value of the diagram generated by the permutation π is given by

$$G_n(\pi) = \qquad (3.65)$$

$$\int_{[0,\beta]^{2n}} \prod_{j=1}^{2n} dx_{0,j} \; \frac{1}{M^{2dn}} \sum_{\mathbf{x}_1\sigma_1,\cdots,\mathbf{x}_{2n}\sigma_{2n}} \prod_{i=1}^{n} U(x_{2i-1} - x_{2i}) \prod_{i=1}^{2n} \delta_{\sigma_i,\sigma_{\pi i}} C(x_i - x_{\pi i})$$

The diagrammatic interpretation we postpone to section 4.2.

3.3   The Perturbation Series for the Correlation Functions

The perturbation series for the correlation functions can be obtained by differentiating the perturbation series for the partition function. To this end we introduce parameters $s_{\mathbf{k}\sigma}$ and $\lambda_{\mathbf{q}}$ and the Hamiltonian

$$H = H(\{s_{\mathbf{k}\sigma}, \lambda_{\mathbf{q}}\}) \qquad (3.66)$$

$$= \frac{1}{L^d} \sum_{\mathbf{k}\sigma} (e_{\mathbf{k}} + s_{\mathbf{k}\sigma}) a_{\mathbf{k}\sigma}^+ a_{\mathbf{k}\sigma} + \frac{1}{L^{3d}} \sum_{\sigma,\tau} \sum_{\mathbf{k},\mathbf{p},\mathbf{q}} \lambda_{\mathbf{q}} V(\mathbf{k} - \mathbf{p}) a_{\mathbf{k}\sigma}^+ a_{\mathbf{q}-\mathbf{k},\tau}^+ a_{\mathbf{q}-\mathbf{p},\tau} a_{\mathbf{p}\sigma}$$

$$H_\lambda = H(s_{\mathbf{k}\sigma} = 0, \lambda_{\mathbf{q}} = \lambda) \qquad (3.67)$$

We consider the two point function

$$\langle a_{\mathbf{k}\sigma}^+ a_{\mathbf{k}\sigma} \rangle = \frac{Tr\, e^{-\beta H_\lambda} a_{\mathbf{k}\sigma}^+ a_{\mathbf{k}\sigma}}{Tr\, e^{-\beta H_\lambda}} \tag{3.68}$$

and the four point function

$$\Lambda(\mathbf{q}) = \tfrac{1}{L^{3d}} \sum_{\sigma,\tau} \sum_{\mathbf{k},\mathbf{p}} V(\mathbf{k}-\mathbf{p}) \langle a_{\mathbf{k}\sigma}^+ a_{\mathbf{q}-\mathbf{k},\tau}^+ a_{\mathbf{q}-\mathbf{p},\tau} a_{\mathbf{p}\sigma} \rangle \tag{3.69}$$

The expectation on the right hand side of (3.69) is also with respect to $e^{-\beta H_\lambda}$, as in (3.68). From these functions we can compute the expectation of the energy according to

$$\langle H_\lambda \rangle = \tfrac{1}{L^d} \sum_{\mathbf{k}\sigma} e_{\mathbf{k}} \langle a_{\mathbf{k}\sigma}^+ a_{\mathbf{k}\sigma} \rangle + \sum_{\mathbf{q}} \Lambda(\mathbf{q}) \tag{3.70}$$

Since we defined the creation and annihilation operators such that $\{a_{\mathbf{k}\sigma}, a_{\mathbf{p}\sigma}^+\} = L^d \delta_{\mathbf{k},\mathbf{p}}$ we will find that $\tfrac{1}{L^d} \langle a_{\mathbf{k}\sigma}^+ a_{\mathbf{k}\sigma} \rangle$ and $\Lambda(\mathbf{q})$ are intensive quantities such that $\langle H_\lambda \rangle$ is proportional to the volume as it should be (for constant density and short-range interaction).

Theorem 3.3.1 *Let*

$$Z(\{s_{\mathbf{k}\sigma}, \lambda_{\mathbf{q}}\}) = \frac{Tr\, e^{-\beta H(\{s_{\mathbf{k}\sigma}, \lambda_{\mathbf{q}}\})}}{Tr\, e^{-\beta H_0}} \tag{3.71}$$

Then

$$\tfrac{1}{L^d} \langle a_{\mathbf{k}\sigma}^+ a_{\mathbf{k}\sigma} \rangle = -\tfrac{1}{\beta} \tfrac{\partial}{\partial s_{\mathbf{k}\sigma}}\big|_{s=0} \log Z(\{s_{\mathbf{k}\sigma}\}) \tag{3.72}$$

$$\Lambda(\mathbf{q}) = -\tfrac{1}{\beta} \tfrac{\partial}{\partial \lambda_{\mathbf{q}}}\big|_{\lambda_{\mathbf{q}}=\lambda} \log Z(\{\lambda_{\mathbf{q}}\}) \tag{3.73}$$

Proof: According to Lemma 3.1.2 one has for some differentiable matrix $A = A(s)$

$$\tfrac{d}{ds} e^{A(s)} = \int_0^1 e^{(1-v)A(s)} \dot{A}(s)\, e^{vA(s)} dv \tag{3.74}$$

which gives, since the trace is linear and cyclic,

$$\begin{aligned}
\tfrac{d}{ds} Tr\, e^{A(s)} &= \int_0^1 Tr[e^{(1-v)A(s)} \dot{A}(s)\, e^{vA(s)}] dv \\
&= \int_0^1 Tr[e^{vA(s)} e^{(1-v)A(s)} \dot{A}(s)] dv \\
&= Tr[e^{A(s)} \dot{A}(s)]
\end{aligned} \tag{3.75}$$

Using (3.75) with A being the Hamiltonian (3.66) proves the theorem. ∎

Chapter 4

Gaussian Integration and Grassmann Integrals

In this chapter we introduce anticommuting variables and derive the Grassmann integral representations for the partition function and the correlation functions. Grassmann integrals are a suitable algebraic tool for the rearrangement of fermionic perturbation series. We demonstrate this in the first section by considering a model with a quadratic perturbation which is explicitely solvable. In that case the perturbation series can be resumed completely. By doing this in two ways, first a direct computation without anticommuting variables and then a calculation using Grassmann integrals, we hope to be able to convince the reader of the utility of that formalism.

In the last two sections Gaussian integrals are discussed. They come into play because they produce exactly the same combinatorics as the outcome of the Wick theorem. That is, the expectation of some monomial with respect to a bosonic or Grassmann, scalar or complex Gaussian measure is given by a determinant or pfaffian or permanent or simply the sum over all pairings. For example, the $2n \times 2n$ determinant in (3.63) can be written as the expectation of $4n$ anticommuting variables or 'fields' with respect to a Grassmann Gaussian measure which is independent of n and it turns out that the remaining series can be rewritten as an exponential which leads to a compact expression, the Grassmann integral representation.

In chapter 9 we estimate sums of diagrams for a fermionic model and show that the sum of all convergent diagrams has a positive radius of convergence. The proof of that, actually already the precise statement of the theorem, makes heavy use of Grassmann integrals. The reason is that, in order to get sensible estimates, we have to get Feynman diagram-like objects, that is, we have to expand the $2n \times 2n$ determinant in (3.63) to a certain amount, but, since we have to use the sign cancellations in that determinant, we are not allowed to expand too much. To handle the algebra of that expansion it turns out that Grassmann integrals are a suitable tool. The key estimate which makes use of sign cancellations is given in Theorem 4.4.9. This theorem, including Theorem 4.4.8, Lemma 4.4.7 and Definition 4.4.6 will not be used before chapter 9.

4.1 Why Grassmann Integration? A Motivating Example

To demonstrate the utility of Grassmann integrals, we consider in this section only a quadratic perturbation instead of a quartic one. In that case the partition function is explicitly known and its perturbation series can be completely resumed. First we will do this without using Grassmann integrals which requires about two pages. Then we show the computation with Grassmann integrals which, once the properties of Grassmann integration are known, fits in three lines.

We consider the Hamiltonian

$$H_\lambda = H_0 + \lambda H' := \tfrac{1}{L^d} \sum_{\mathbf{k}\sigma} e_{\mathbf{k}} a_{\mathbf{k}\sigma}^+ a_{\mathbf{k}\sigma} + \tfrac{\lambda}{L^d} \sum_{\mathbf{k}\sigma} a_{\mathbf{k}\sigma}^+ a_{\mathbf{k}\sigma} \tag{4.1}$$

According to (2.29) its partition function is given by

$$Z(\lambda) = Tr\, e^{-\beta H_\lambda} \big/ Tr\, e^{-\beta H_0} = \prod_{\mathbf{k}\sigma} \tfrac{1+e^{-\beta(e_{\mathbf{k}}+\lambda)}}{1+e^{-\beta e_{\mathbf{k}}}} \tag{4.2}$$

On the other hand, according to chapter 3, the perturbation series for $Z(\lambda)$ is given by

$$
\begin{aligned}
Z(\lambda) &= \sum_{n=0}^{\infty} \tfrac{(-\lambda)^n}{n!} \int_{[0,\beta]^n} ds_1 \cdots ds_n\, \langle TH'(s_1)\cdots H'(s_n)\rangle_0 \\
&= \sum_{n=0}^{\infty} \tfrac{(-\lambda)^n}{n!} \int_0^\beta ds_1 \tfrac{1}{M^d}\sum_{\mathbf{x}_1\sigma_1}\cdots \int_0^\beta ds_n \tfrac{1}{M^d}\sum_{\mathbf{x}_n\sigma_n} \langle T\psi_{\mathbf{x}_1\sigma_1}^+(s_1)\psi_{\mathbf{x}_1\sigma_1}(s_1)\cdots \\
&\qquad\qquad\qquad\qquad\qquad \cdots \psi_{\mathbf{x}_n\sigma_n}^+(s_n)\psi_{\mathbf{x}_n\sigma_n}(s_n)\rangle_0 \\
&= \sum_{n=0}^{\infty} \tfrac{(-\lambda)^n}{n!} \int d\xi_1 \cdots d\xi_n\, \det\,[C(\xi_i,\xi_j)]_{1\le i,j\le n}
\end{aligned} \tag{4.3}
$$

where we abbreviated

$$\int d\xi := \int_0^\beta ds\, \tfrac{1}{M^d}\sum_{\mathbf{x}\sigma} = \lim_{h\to\infty} \tfrac{1}{hM^d}\sum_{s,\mathbf{x},\sigma} \tag{4.4}$$

and in the last equation in (4.4) we wrote the integral as a limit of a Riemannian sum, $s = \tfrac{1}{h}j$, $0 \le j \le \beta h - 1$.

Computation without Grassmann Integrals: We expand the determinant in (4.3):

$$\det\,[C(\xi_i,\xi_j)] = \sum_{\pi\in S_n} \mathrm{sign}\pi\, C(\xi_1,\xi_{\pi 1})\cdots C(\xi_n,\xi_{\pi n})$$

Every permutation can be decomposed into cycles. A 1-cycle corresponds to a factor $C(\xi, \xi)$, a 2-cycle corresponds to $C(\xi, \xi')C(\xi', \xi)$ and an r-cycle corresponds to (eventually relabeling the ξ's)

$$C(\xi_1, \xi_2)C(\xi_2, \xi_3) \cdots C(\xi_{r-1}, \xi_r)C(\xi_r, \xi_1) \tag{4.5}$$

If π consists of b_1 1-cycles, b_2 2-cycles up to b_n n-cycles we say that π is of type

$$t(\pi) = 1^{b_1} 2^{b_2} \cdots n^{b_n} \tag{4.6}$$

where in the last equation $1^{b_1} 2^{b_2} \cdots n^{b_n}$ is not meant as a number but just as an abbreviation for the above statement. Since the total number of permuted elements is n, one has

$$1b_1 + 2b_2 + \cdots nb_n = n \tag{4.7}$$

The number of such permutations is given by

$$\left| \{ \pi \in S_n \,|\, t(\pi) = 1^{b_1} 2^{b_2} \cdots n^{b_n} \} \right| = \frac{n!}{b_1! \cdots b_n! \, 1^{b_1} \cdots n^{b_n}} \tag{4.8}$$

In the denominator on the right hand side of (4.8) of course the number $1^{b_1} \cdots n^{b_n}$ is meant. The sign of such a permutation is given by

$$\text{sign}\pi = (-1)^{\sum_{r=1}^n (b_r - 1)} = (-1)^{n - \sum_r b_r} \quad \text{if } t(\pi) = 1^{b_1} \cdots n^{b_n} \tag{4.9}$$

Now, the basic observation is that, since the ξ-variables are integrated over, every permutation of given type $t(\pi) = 1^{b_1} 2^{b_2} \cdots n^{b_n}$ gives the same contribution

$$\prod_{r=1}^n \left\{ \int d\xi_1 \cdots d\xi_r \, C(\xi_1, \xi_2) \cdots C(\xi_r, \xi_1) \right\}^{b_r} \tag{4.10}$$

Thus, the perturbation series (4.3) equals

$$Z(\lambda) = \sum_{n=0}^{\infty} \frac{(-\lambda)^n}{n!} \sum_{\substack{b_1, \cdots, b_n = 0 \\ \sum_r r b_r = n}}^n \sum_{\substack{\pi \in S_n \\ t(\pi) = 1^{b_1} \cdots n^{b_n}}} (-1)^{n - \sum_r b_r} \times$$

$$\prod_{r=1}^n \left\{ \int d\xi_1 \cdots d\xi_r \, C(\xi_1, \xi_2) \cdots C(\xi_r, \xi_1) \right\}^{b_r}$$

$$= \sum_{n=0}^{\infty} \frac{\lambda^n}{n!} \sum_{\substack{b_1, \cdots, b_n = 0 \\ \sum_r r b_r = n}}^n \frac{n!}{b_1! \cdots b_n! \, 1^{b_1} \cdots n^{b_n}} \prod_{r=1}^n \left\{ -\int d\xi_1 \cdots d\xi_r \, C(\xi_1, \xi_2) \cdots C(\xi_r, \xi_1) \right\}^{b_r}$$

$$= \sum_{n=0}^{\infty} \sum_{\substack{b_1, \cdots, b_n = 0 \\ \sum_r r b_r = n}}^n \prod_{r=1}^n \frac{1}{b_r!} \left\{ -\frac{\lambda^r}{r} \int d\xi_1 \cdots d\xi_r \, C(\xi_1, \xi_2) \cdots C(\xi_r, \xi_1) \right\}^{b_r} \tag{4.11}$$

Let $\underline{b} = (b_1, b_2, \cdots) \in \mathbb{N}^{\mathbb{N}}$. Since

$$\lim_{N\to\infty} \sum_{n=0}^{N} \sum_{\substack{b_1,\cdots,b_n=0 \\ \sum_r rb_r=n}}^{n} F(\underline{b}) = \lim_{N\to\infty} \sum_{b_1,\cdots,b_N=0}^{N} F(\underline{b}) \tag{4.12}$$

we arrive at

$$Z(\lambda) = \prod_{r=1}^{\infty} \sum_{b_r=0}^{\infty} \frac{1}{b_r!} \left\{ -\frac{1}{r} Tr\left[\left(\frac{\lambda}{hM^d}C\right)^r \right] \right\}^{b_r}$$

$$= \prod_{r=1}^{\infty} e^{-\frac{1}{r}Tr[(\frac{\lambda}{hM^d}C)^r]} = e^{-\sum_{r=1}^{\infty} \frac{1}{r}Tr[(\frac{\lambda}{hM^d}C)^r]}$$

$$= e^{Tr\log[Id-\frac{\lambda}{hM^d}C]} = \det\left[Id - \frac{\lambda}{hM^d}C \right]$$

$$= \prod_{k\sigma} \left\{ 1 - \frac{\lambda}{ik_0 - e_{\mathbf{k}}} \right\} \tag{4.13}$$

$$= \prod_{k\sigma} \left\{ \frac{1+e^{-\beta(e_{\mathbf{k}}+\lambda)}}{1+e^{-\beta e_{\mathbf{k}}}} \right\} \tag{4.14}$$

where the last two lines, (4.13) and (4.14), are shown in the following

Lemma 4.1.1 (i) *Let $C = [C(x - x')]_{x,x'}$ denote the matrix with elements*

$$C(x - x') = \frac{1}{\beta L^d} \sum_{k} e^{ik(x-x')} \frac{1}{ik_0 - e_{\mathbf{k}}} \tag{4.15}$$

Let $F = (F_{kx})$ be the unitary matrix of discrete Fourier transform in $d + 1$ dimensional coordinate-temperature space with matrix elements

$$F_{kx} = \frac{1}{\sqrt{(ML)^d h\beta}} e^{-ikx} \tag{4.16}$$

Then

$$\frac{1}{hM^d}(F^*CF)_{k,k'} = \frac{\delta_{k,k'}}{ik_0 - e_{\mathbf{k}}} \tag{4.17}$$

(ii) *The k_0-product has to be evaluated according to the ε-prescription and gives*

$$\prod_{k_0} \left\{ 1 - \frac{\lambda}{ik_0 - e_{\mathbf{k}}} \right\} := \lim_{\varepsilon \nearrow 0} \prod_{k_0} \left\{ 1 - \lambda \frac{e^{-ik_0\varepsilon}}{ik_0 - e_{\mathbf{k}}} \right\} = \frac{1+e^{-\beta(e_{\mathbf{k}}+\lambda)}}{1+e^{-\beta e_{\mathbf{k}}}} \tag{4.18}$$

Proof: (i) We have

$$\frac{1}{hM^d}(F^*CF)_{k,k'} = \frac{1}{(hM^d)^2} \frac{1}{\beta L^d} \sum_{x,x'} e^{-ikx} C(x - x') e^{ik'x'}$$

$$= \frac{1}{(hM^d)^2} \frac{1}{\beta L^d} \sum_{y} e^{-iky} C(y) \sum_{x'} e^{i(k'-k)x'}$$

$$= \frac{1}{hM^d} \sum_{y} e^{-iky} C(y) \, \delta_{k,k'} = \frac{\delta_{k,k'}}{ik_0 - e_{\mathbf{k}}}$$

which proves (i). To obtain (ii), observe that, from (3.59) we have

$$\frac{1}{\beta}\sum_{k_0}^{\epsilon}\frac{1}{ik_0-e_{\mathbf{k}}} := \lim_{\epsilon\nearrow 0}\frac{1}{\beta}\sum_{k_0}\frac{e^{-ik_0\epsilon}}{ik_0-e_{\mathbf{k}}} = \frac{1}{1+e^{\beta e_{\mathbf{k}}}} \tag{4.19}$$

Furthermore

$$\frac{1}{\beta}\sum_{k_0}^{sym}\frac{1}{ik_0-e_{\mathbf{k}}} := \frac{1}{2\beta}\sum_{k_0}\left(\frac{1}{ik_0-e_{\mathbf{k}}}+\frac{1}{-ik_0-e_{\mathbf{k}}}\right)$$

$$= \frac{1}{\beta}\sum_{k_0}\frac{-e_{\mathbf{k}}}{k_0^2+e_{\mathbf{k}}^2} = -\frac{1}{2}\frac{1-e^{-\beta e_{\mathbf{k}}}}{1+e^{-\beta e_{\mathbf{k}}}} \tag{4.20}$$

Thus the product in (4.18) is found to be

$$\prod_{k_0}\left(1-\frac{\lambda}{ik_0-e_{\mathbf{k}}}\right) = \prod_{k_0}^{\epsilon}\left(1-\frac{\lambda}{ik_0-e_{\mathbf{k}}}\right) = e^{-\sum_{r=1}^{\infty}\frac{\lambda^r}{r}\sum_{k_0}^{\epsilon}\frac{1}{(ik_0-e_{\mathbf{k}})^r}}$$

$$= e^{-\sum_{r=1}^{\infty}\frac{\lambda^r}{r}\sum_{k_0}^{sym}\frac{1}{(ik_0-e_{\mathbf{k}})^r}}\,e^{\sum_{k_0}^{sym}\frac{\lambda}{ik_0-e_{\mathbf{k}}}-\sum_{k_0}^{\epsilon}\frac{\lambda}{ik_0-e_{\mathbf{k}}}}$$

$$= \prod_{k_0}^{sym}\left(1-\frac{\lambda}{ik_0-e_{\mathbf{k}}}\right)e^{-\frac{\beta}{2}\lambda}$$

$$= \prod_{k_0>0}\frac{k_0^2+(e_{\mathbf{k}}+\lambda)^2}{k_0^2+e_{\mathbf{k}}^2}\,e^{-\frac{\beta}{2}\lambda} \tag{4.21}$$

Since [37]

$$\prod_{n=0}^{\infty}\left(1+\frac{\lambda}{(2n+1)^2+e^2}\right) = \frac{\cosh\left(\frac{\pi}{2}\sqrt{e^2+\lambda}\right)}{\cosh\frac{\pi}{2}e} \tag{4.22}$$

the product in (4.21) becomes

$$\prod_{k_0\in\frac{\pi}{\beta}(2N+1)}\frac{k_0^2+(e_{\mathbf{k}}+\lambda)^2}{k_0^2+e_{\mathbf{k}}^2}\,e^{-\frac{\beta}{2}\lambda} = \frac{\cosh\frac{\beta}{2}(e_{\mathbf{k}}+\lambda)}{\cosh\frac{\beta}{2}e_{\mathbf{k}}}\,e^{-\frac{\beta}{2}\lambda} = \frac{1+e^{-\beta(e_{\mathbf{k}}+\lambda)}}{1+e^{-\beta e_{\mathbf{k}}}}$$

which proves part (ii). ∎

Computation with Grassmann Integrals: The systematic theory of Grassmann integration is presented in section 4.4. Here we simply use the two basic properties of Grassmann integration, (4.28) and (4.30) below, without proof to reproduce the result (4.14) for the partition function with a quadratic perturbation.

Let

$$\psi_1,\cdots,\psi_N,\bar{\psi}_1,\cdots,\bar{\psi}_N$$

be $2N$ Grassmann variables. That is, we have the anticommutation relations

$$\psi_i\psi_j = -\psi_j\psi_i,\quad \psi_i\bar{\psi}_j = -\bar{\psi}_j\psi_i,\quad \bar{\psi}_i\bar{\psi}_j = -\bar{\psi}_j\bar{\psi}_i \tag{4.23}$$

The reason for writing $\bar{\psi}_1,\cdots,\bar{\psi}_N$ for the second N Grassmann variables instead of $\psi_{N+1},\cdots,\psi_{2N}$ is the following. For the many-electron system we will have to use Grassmann Gaussian integrals where the quadratic part is not of the general form $\sum_{i,j=1}^{2N} \psi_i A_{ij} \psi_j$ in which case the result is the pfaffian of the matrix A, but in our case the quadratic term in the exponential is of the more specific form $\sum_{i,j=1}^{N} \psi_{i+N} B_{ij} \psi_j$ or $\sum_{i,j=1}^{N} \bar{\psi}_i B_{ij} \psi_j$ in which case one obtains the determinant of B. We introduce the Grassmann algebra

$$\mathcal{G}_{2N} = \mathcal{G}_{2N}(\psi_1, \cdots, \psi_N, \bar{\psi}_1, \cdots, \bar{\psi}_N)$$

$$= \left\{ \sum_{I,J \subset \{1,\cdots,N\}} \alpha_{I,J} \prod_{i \in I} \psi_i \prod_{j \in J} \bar{\psi}_j \mid \alpha_{I,J} \in \mathbb{C} \right\} \quad (4.24)$$

The Grassmann integral is a linear functional on $\mathcal{G}_{2N}$.

Definition 4.1.2 *The Grassmann integral $\int \cdot \, \Pi_{i=1}^{N} d\psi_i \, \Pi_{j=1}^{N} d\bar{\psi}_j : \mathcal{G}_{2N} \to \mathbb{C}$ is the linear functional on $\mathcal{G}_{2N}$ defined by*

$$\int \sum_{I,J \subset \{1,\cdots,N\}} \alpha_{I,J} \prod_{i \in I} \psi_i \prod_{j \in J} \bar{\psi}_j \prod_{i=1}^{N} d\psi_i \prod_{j=1}^{N} d\bar{\psi}_j := \alpha_{\{1,\cdots,N\},\{1,\cdots,N\}} \quad (4.25)$$

The notation $\Pi_{i=1}^{N} d\psi_i \, \Pi_{j=1}^{N} d\bar{\psi}_j$ is used to fix a sign. For example, on $\mathcal{G}_4$

$$\int \left(3\psi_1 \bar{\psi}_1 - 4\psi_1 \psi_2 \bar{\psi}_1 \bar{\psi}_2 \right) d\psi_1 d\psi_2 d\bar{\psi}_1 d\bar{\psi}_2 = -4 \quad (4.26)$$

but

$$\int \left(3\psi_1 \bar{\psi}_1 - 4\psi_1 \psi_2 \bar{\psi}_1 \bar{\psi}_2 \right) d\psi_1 d\bar{\psi}_1 d\psi_2 d\bar{\psi}_2 =$$

$$+ 4 \int \psi_1 \bar{\psi}_1 \psi_2 \bar{\psi}_2 \, d\psi_1 d\bar{\psi}_1 d\psi_2 d\bar{\psi}_2 = +4 \quad (4.27)$$

The two most important properties of Grassmann integrals are summarized in the following

Lemma 4.1.3 *Let $A = (A_{ij})_{1 \leq i,j \leq N} \in \mathbb{C}^{N \times N}$ and for some Grassmann valued function $F = F(\psi, \bar{\psi})$ define the exponential by $e^F := \sum_{n=0}^{2N} \frac{1}{n!} F^n$. Then we have*
a)

$$\int e^{-\sum_{i,j=1}^{N} \bar{\psi}_i A_{i,j} \psi_j} \prod_{i=1}^{N} (d\psi_i d\bar{\psi}_i) = \det A \quad (4.28)$$

b) *Let A be invertible, $C = A^{-1}$ be the 'covariance matrix' and*

$$d\mu_C(\psi, \bar{\psi}) := \frac{1}{\det A} e^{-\sum_{i,j=1}^{N} \bar{\psi}_i A_{ij} \psi_j} \prod_{i=1}^{N} (d\psi_i d\bar{\psi}_i) \quad (4.29)$$

Then

$$\int \psi_{i_1} \bar{\psi}_{j_1} \cdots \psi_{i_r} \bar{\psi}_{j_r} \, d\mu_C(\psi, \bar{\psi}) = \det \left[C_{i_k, j_l} \right]_{1 \leq k,l \leq r} \quad (4.30)$$

With these preparations we are now ready to apply the formalism of Grassmann integration to the perturbation series (4.3). We introduce a Grassmann algebra generated by Grassmann variables $\psi(\xi), \bar\psi(\xi)$ which are labeled by

$$\xi = (x_0, \mathbf{x}, \sigma) \in [0, \beta]_{1/h} \times [0, L]^d_{1/M} \times \{\uparrow, \downarrow\} \qquad (4.31)$$

Recall that $[0, \beta]_{1/h}$ denotes the lattice of βh points in $[0, \beta]$ with lattice spacing $1/h$. The covariance matrix is given by

$$C_{\xi,\xi'} = \delta_{\sigma,\sigma'} C(x - x') = \delta_{\sigma,\sigma'} \frac{1}{\beta L^d} \sum_k e^{ik(x-x')} \frac{1}{ik_0 - e_\mathbf{k}} \qquad (4.32)$$

and the Grassmann Gaussian measure reads

$$d\mu_C(\psi, \bar\psi) = \det C \, e^{-\sum_{x\sigma,x'\sigma'} \bar\psi_{x\sigma}[C^{-1}]_{x\sigma,x'\sigma'}\psi_{x'\sigma'}} \prod_{x\sigma}(d\psi_{x\sigma} d\bar\psi_{x\sigma}) \qquad (4.33)$$

We rewrite the $n \times n$ determinant appearing on the right hand side of (4.3) as a Grassmann integral:

$$\det[C(\xi_i, \xi_j)]_{1\le i,j\le n} = \int \psi(\xi_1)\bar\psi(\xi_1) \cdots \psi(\xi_n)\bar\psi(\xi_n) \, d\mu_C(\psi, \bar\psi) \qquad (4.34)$$

Interchanging the Grassmann integral with the sum of the perturbation series, we get

$$
\begin{aligned}
Z(\lambda) &= \int \sum_{n=0}^{\infty} \frac{(-\lambda)^n}{n!} \int d\xi_1 \cdots d\xi_n \, \psi(\xi_1)\bar\psi(\xi_1) \cdots \psi(\xi_n)\bar\psi(\xi_n) \, d\mu_C(\psi, \bar\psi) \\
&= \int \sum_{n=0}^{\infty} \frac{1}{n!} \left(\lambda \int d\xi \, \bar\psi(\xi)\psi(\xi) \right)^n d\mu_C(\psi, \bar\psi) \\
&= \int e^{\frac{\lambda}{hM^d} \sum_{x\sigma} \bar\psi_{x\sigma}\psi_{x\sigma}} d\mu_C(\psi, \bar\psi) \\
&= \det C \int e^{-\sum_{x\sigma,x'\sigma'} \bar\psi_{x\sigma}\left[C^{-1}_{x\sigma,x'\sigma'} - \frac{\lambda}{hM^d}\delta_{x\sigma,x'\sigma'} \right]\psi_{x\sigma}} \prod_{x\sigma}(d\psi_{x\sigma} d\bar\psi_{x\sigma}) \\
&= \det C \, \det \left[C^{-1} - \frac{\lambda}{hM^d} Id \right] \\
&= \det \left[Id - \frac{\lambda}{hM^d} C \right] = \prod_{k\sigma} \left\{ 1 - \frac{\lambda}{ik_0 - e_\mathbf{k}} \right\} \qquad (4.35)
\end{aligned}
$$

which coincides with (4.13). In the fifth line we used (4.28) of Lemma 4.1.3.

4.2 Grassmann Integral Representations

The perturbation series for the quartic model was given by (3.63). Abbreviating again $\xi = (x_0, \mathbf{x}, \sigma) = (x, \sigma)$, this reads

$$
\begin{aligned}
Z(\lambda) &= \sum_{n=0}^{\infty} \frac{(-\lambda)^n}{n!} \int d\xi_1 \cdots d\xi_{2n} \, U(\xi_1 - \xi_2) \cdots U(\xi_{2n-1} - \xi_{2n}) \det\left[C(\xi_i, \xi_j) \right] \\
&= \sum_{n=0}^{\infty} \frac{(-\lambda)^n}{n!} \int d\xi_1 \cdots d\xi_{2n} \prod_{i=1}^{n} U(\xi_{2i-1} - \xi_{2i}) \times \\
&\qquad\qquad \int \psi(\xi_1)\bar\psi(\xi_1) \cdots \psi(\xi_{2n})\bar\psi(\xi_{2n}) \, d\mu_C(\psi, \bar\psi) \\
&= \int \sum_{n=0}^{\infty} \frac{1}{n!} \left(-\lambda \int d\xi d\xi' \psi(\xi)\bar\psi(\xi) U(\xi - \xi')\psi(\xi')\bar\psi(\xi') \right)^n d\mu_C(\psi, \bar\psi) \\
&= \int e^{-\lambda \int d\xi d\xi' \psi(\xi)\bar\psi(\xi) U(\xi-\xi')\psi(\xi')\bar\psi(\xi')} \, d\mu_C(\psi, \bar\psi) \qquad (4.36)
\end{aligned}
$$

The diagrammatic interpretation is as follows. To obtain the coefficient of $(-\lambda)^n/n!$, draw n vertices

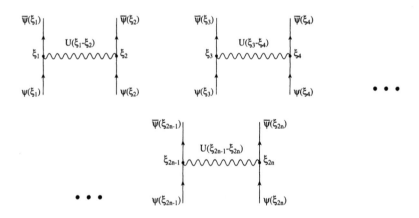

Each vertex corresponds to the expression

$$
\int d\xi d\xi' \bar\psi(\xi)\psi(\xi) U(\xi - \xi')\bar\psi(\xi')\psi(\xi') \qquad (4.37)
$$

Then connect each of the outgoing legs, labelled by the $\bar\psi$'s, with one of the incoming legs, labelled by the ψ's, in all possible ways. This produces

$(2n)!$ terms. The following figure shows a third order contribution for the permutations $\pi = (1, 5, 2, 3, 6, 4)$:

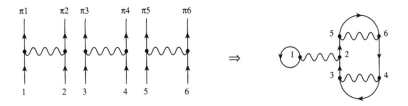

Its value is given by

$$\int d\xi_1 \cdots d\xi_6 U(\xi_1 - \xi_2) U(\xi_3 - \xi_4) U(\xi_5 - \xi_6) \times$$
$$C(\xi_1, \xi_1) C(\xi_2, \xi_5) C(\xi_5, \xi_6) C(\xi_6, \xi_4) C(\xi_4, \xi_3) C(\xi_3, \xi_2)$$

The correlation functions we are interested in are given by (3.72) and (3.73). Let

$$H_0(\{s_{\mathbf{k}\sigma}\}) = \frac{1}{L^d} \sum_{\mathbf{k}\sigma} (e_{\mathbf{k}} + s_{\mathbf{k}\sigma}) a_{\mathbf{k}\sigma}^+ a_{\mathbf{k}\sigma} \tag{4.38}$$

Then

$$Z(\{s_{\mathbf{k}\sigma}\}) = \frac{Tr\, e^{-\beta H_\lambda(\{s_{\mathbf{k}\sigma}\})}}{Tr\, e^{-\beta H_0}} = \frac{Tr\, e^{-\beta H_\lambda(\{s_{\mathbf{k}\sigma}\})}}{Tr\, e^{-\beta H_0(\{s_{\mathbf{k}\sigma}\})}} \frac{Tr\, e^{-\beta H_0(\{s_{\mathbf{k}\sigma}\})}}{Tr\, e^{-\beta H_0}} \tag{4.39}$$

$$= \int e^{-\lambda \int d\xi d\xi' \, \psi(\xi) \bar\psi(\xi) U(\xi - \xi') \psi(\xi') \bar\psi(\xi')} \, d\mu_{C_s}(\psi, \bar\psi) \det C \det[C_s^{-1}]$$

where $C_s(\mathbf{k}\sigma) = \frac{1}{ik_0 - e_{\mathbf{k}} - s_{\mathbf{k}\sigma}}$ and in the last line of (4.39) we used the analog result of (4.35) for the quadratic perturbation given by $\frac{1}{L^d} \sum_{\mathbf{k}\sigma} s_{\mathbf{k}\sigma} a_{\mathbf{k}\sigma}^+ a_{\mathbf{k}\sigma}$. Since

$$d\mu_{C_s}(\psi, \bar\psi) = \det C_s \, e^{-\sum_{x\sigma, x'\sigma'} \bar\psi_{x\sigma}(C_s^{-1})_{x\sigma, x'\sigma'} \psi_{x'\sigma'}} \prod_{x\sigma} (d\psi_{x\sigma} d\bar\psi_{x\sigma}) \tag{4.40}$$

the determinant of C_s cancels out and we end up with

$$Z(\{s_{\mathbf{k}\sigma}\}) = \int e^{-\lambda \int d\xi d\xi' \, \psi(\xi) \bar\psi(\xi) U(\xi - \xi') \psi(\xi') \bar\psi(\xi')} \times \tag{4.41}$$

$$e^{-\sum_{x\sigma, x'\sigma'} \bar\psi_{x\sigma}(C_s^{-1})_{x\sigma, x'\sigma'} \psi_{x'\sigma'}} \det C \prod_{x\sigma} (d\psi_{x\sigma} d\bar\psi_{x\sigma})$$

To proceed we transform to momentum space. Substitution of variables for Grassmann integrals is treated in the following lemma which is also proven in section 4.4

Lemma 4.2.1 *Let $A \in \mathbb{C}^{2N \times 2N}$, and let $\psi = (\psi_1, \cdots, \psi_N)$, $\bar{\psi} = (\bar{\psi}_1, \cdots, \bar{\psi}_N)$, $\eta = (\eta_1, \cdots, \eta_N)$, $\bar{\eta} = (\bar{\eta}_1, \cdots, \bar{\eta}_N)$ be Grassmann variables. Let G be some Grassmann valued function. Then*

$$\int G\left(A(\psi, \bar{\psi})\right) \det A \prod_{i=1}^{N} (d\psi_i d\bar{\psi}_i) = \int G(\eta, \bar{\eta}) \prod_{i=1}^{N} (d\eta_i d\bar{\eta}_i) \quad (4.42)$$

The Fourier transform for Grassmann variables is defined by the following linear transformation:

$$\begin{pmatrix} \psi_{k\sigma} \\ \bar{\psi}_{k\sigma} \end{pmatrix} = \begin{pmatrix} F & 0 \\ 0 & F^* \end{pmatrix} \begin{pmatrix} \psi_{x\sigma} \\ \bar{\psi}_{x\sigma} \end{pmatrix} \quad (4.43)$$

where F is the unitary matrix of discrete Fourier transform given by (4.16). In particular $\det \begin{pmatrix} F & 0 \\ 0 & F^* \end{pmatrix} = 1$. Since the bared ψ's transform with $F^* = \bar{F}$, one may indeed treat a bared ψ as the complex conjugate of ψ. One obtains the following representations.

Theorem 4.2.2 *Let*

$$H_\lambda = \frac{1}{L^d} \sum_{k\sigma} e_k\, a^+_{k\sigma} a_{k\sigma} + \frac{\lambda}{L^{3d}} \sum_{\sigma,\tau} \sum_{k,p,q} V(k-p)\, a^+_{k\sigma} a^+_{q-k,\tau} a_{q-p,\tau} a_{p\sigma} \quad (4.44)$$

and let $\langle a^+_{k\sigma} a_{k\sigma} \rangle$ and $\Lambda(q)$ be given by (3.68) and (3.69). Then there are the following Grassmann integral representations

$$Z(\lambda) = \int e^{-\frac{\lambda}{(\beta L^d)^3} \sum_{\sigma\tau} \sum_{kpq} V(k-p) \bar{\psi}_{k\sigma} \bar{\psi}_{q-k,\tau} \psi_{p\sigma} \psi_{q-p,\tau}} d\mu_C \quad (4.45)$$

$$\frac{1}{L^d} \langle a^+_{k\sigma} a_{k\sigma} \rangle = \frac{1}{\beta} \sum_{k_0 \in \frac{\pi}{\beta}(2\mathbb{Z}+1)} \frac{1}{\beta L^d} \langle \bar{\psi}_{k\sigma} \psi_{k\sigma} \rangle \quad (4.46)$$

$$\Lambda(q) = \frac{1}{\beta} \sum_{q_0 \in \frac{2\pi}{\beta}\mathbb{Z}} \Lambda(q_0, q) \quad (4.47)$$

$$\Lambda(q) = \frac{1}{(\beta L^d)^3} \sum_{\sigma,\tau} \sum_{k,p} V(k-p) \langle \bar{\psi}_{k\sigma} \bar{\psi}_{q-k,\tau} \psi_{p\sigma} \psi_{q-p,\tau} \rangle \quad (4.48)$$

where

$$\langle G(\psi, \bar{\psi}) \rangle = \frac{\int G(\psi, \bar{\psi})\, e^{-\frac{\lambda}{(\beta L^d)^3} \sum_{\sigma\tau} \sum_{kpq} V(k-p) \bar{\psi}_{k\sigma} \bar{\psi}_{q-k\tau} \psi_{p\sigma} \psi_{q-p\tau}} d\mu_C}{\int e^{-\frac{\lambda}{(\beta L^d)^3} \sum_{\sigma\tau} \sum_{kpq} V(k-p) \bar{\psi}_{k\sigma} \bar{\psi}_{q-k\tau} \psi_{p\sigma} \psi_{q-p\tau}} d\mu_C} \quad (4.49)$$

and

$$d\mu_C(\psi, \bar{\psi}) = \prod_{k\sigma} \frac{\beta L^d}{ik_0 - e_k}\, e^{-\frac{1}{\beta L^d} \sum_{k\sigma}(ik_0 - e_k) \bar{\psi}_{k\sigma} \psi_{k\sigma}} \prod_{k\sigma} (d\psi_{k\sigma} d\bar{\psi}_{k\sigma}) \quad (4.50)$$

Proof: We have

$$\psi_{x\sigma} = \frac{1}{\sqrt{(ML)^d \beta h}} \sum_k e^{ikx} \psi_{k\sigma}, \quad \bar{\psi}_{x\sigma} = \frac{1}{\sqrt{(ML)^d \beta h}} \sum_k e^{-ikx} \bar{\psi}_{k\sigma} \quad (4.51)$$

The quartic part becomes

$$\frac{1}{h^2 M^{2d}} \sum_{\sigma,\sigma'} \sum_{x,x'} \bar{\psi}_{x\sigma} \psi_{x\sigma} U(x - x') \bar{\psi}_{x'\sigma'} \psi_{x'\sigma'}$$

$$= \frac{1}{h^2 M^{2d}} \frac{1}{(ML)^{2d}(\beta h)^2} \sum_{\sigma,\sigma'} \sum_{k_1 k_2 k_3 k_4} \sum_{x,x'} e^{-i[(k_1 - k_2)x + (k_3 - k_4)x']} \times$$

$$\bar{\psi}_{k_1\sigma} \psi_{k_2\sigma} U(x - x') \bar{\psi}_{k_3\sigma'} \psi_{k_4\sigma'}$$

$$= \frac{1}{h^2 M^{2d}} \frac{1}{(ML)^{2d}(\beta h)^2} \sum_{\sigma,\sigma'} \sum_{k_1 k_2 k_3 k_4} hM^d U(k_1 - k_2) hM^d \beta L^d \times$$

$$\delta_{k_1 + k_3, k_2 + k_4} \bar{\psi}_{k_1\sigma} \psi_{k_2\sigma} \bar{\psi}_{k_3\sigma'} \psi_{k_4\sigma'}$$

$$= \frac{1}{(hM^d)^2} \frac{1}{(\beta L^d)^2} \sum_{\sigma,\sigma'} \sum_{k_1 k_2 k_3 k_4} \beta L^d \delta_{k_1 + k_3, k_2 + k_4} \bar{\psi}_{k_1\sigma} \psi_{k_2\sigma} \bar{\psi}_{k_3\sigma'} \psi_{k_4\sigma'} \quad (4.52)$$

The quadratic term becomes

$$\sum_{x\sigma, x'\sigma'} \bar{\psi}_{x\sigma} (C^{-1})_{x\sigma, x'\sigma'} \psi_{x'\sigma'} = \sum_{k\sigma, k'\sigma'} \bar{\psi}_{k\sigma} (FC^{-1}F^*)_{k,k'} \psi_{k'\sigma'}$$

$$= \sum_{k\sigma} \bar{\psi}_{k\sigma} \frac{ik_0 - e_{\mathbf{k}}}{hM^d} \psi_{k\sigma} \quad (4.53)$$

and the normalization factor, the determinant of C, is

$$\det C = \prod_{k\sigma} \frac{hM^d}{ik_0 - e_{\mathbf{k}}} \quad (4.54)$$

To arrive at the claimed formulae we make another substitution of variables

$$\psi_{k\sigma} = \sqrt{\frac{hM^d}{\beta L^d}} \eta_{k\sigma}, \quad \bar{\psi}_{k\sigma} = \sqrt{\frac{hM^d}{\beta L^d}} \bar{\eta}_{k\sigma} \quad (4.55)$$

The determinant of this transformation, $\prod_{k\sigma} \frac{hM^d}{\beta L^d}$, changes the normalization factor in (4.54) to the one in (4.50) and (4.50,4.45) follow. To obtain (4.46), we have according to (3.72) and (4.41)

$$\frac{1}{L^d} \langle a_{\mathbf{k}\sigma}^+ a_{\mathbf{k}\sigma} \rangle = \frac{1}{\beta} \frac{\partial}{\partial s_{\mathbf{k}\sigma}} \Big|_{s=0} \log Z(\{s_{\mathbf{k}\sigma}\})$$

$$= \frac{1}{\beta} \frac{\partial}{\partial s_{\mathbf{k}\sigma}} \Big|_{s=0} \log \int e^{-\frac{\lambda}{(\beta L^d)^3} \sum_{\sigma\tau} \sum_{kpq} V(\mathbf{k}-\mathbf{p}) \bar{\psi}_{k\sigma} \bar{\psi}_{q-k,\tau} \psi_{p\sigma} \psi_{q-p,\tau}} \times$$

$$e^{-\frac{1}{\beta L^d} \sum_{k\sigma} (ik_0 - e_{\mathbf{k}} - s_{\mathbf{k}\sigma}) \bar{\psi}_{k\sigma} \psi_{k\sigma}} \prod_{k\sigma} (d\psi_{k\sigma} d\bar{\psi}_{k\sigma})$$

$$= \frac{1}{\beta} \sum_{k_0} \frac{1}{\beta L^d} \langle \bar{\psi}_{k\sigma} \psi_{k\sigma} \rangle \quad (4.56)$$

and similarly according to (3.73)

$$\Lambda(\mathbf{q}) = -\frac{1}{\beta}\frac{\partial}{\partial\lambda_\mathbf{q}}\Big|_{\lambda_\mathbf{q}=\lambda}\log Z(\{\lambda_\mathbf{q}\})$$

$$= -\frac{1}{\beta}\frac{\partial}{\partial\lambda_\mathbf{q}}\Big|_{\lambda_\mathbf{q}=\lambda}\log\int e^{-\frac{1}{(\beta L^d)^3}\sum_{\sigma\tau}\sum_{kpq}\lambda_\mathbf{q}V(\mathbf{k}-\mathbf{p})\bar\psi_{k\sigma}\bar\psi_{q-k,\tau}\psi_{p\sigma}\psi_{q-p,\tau}}\,d\mu_C$$

$$= \frac{1}{\beta}\sum_{q_0}\frac{1}{(\beta L^d)^3}\sum_{\sigma\tau}\sum_{kp}V(\mathbf{k}-\mathbf{p})\langle\bar\psi_{k\sigma}\bar\psi_{q-k,\tau}\psi_{p\sigma}\psi_{q-p,\tau}\rangle \tag{4.57}$$

which proves the theorem. ∎

4.3   Ordinary Gaussian Integrals

In the next two theorems we summarize basic facts about Gaussian integration. The first theorem treats the real case and the second one the complex case.

Theorem 4.3.1 *Let $A \in \mathbb{R}^{n\times n}$ be symmetric and positive definite and let $C = A^{-1}$. For $x, k \in \mathbb{R}^n$ let $\langle k, x\rangle = \sum_{j=1}^n k_j x_j$. Then*

a)

$$\int_{\mathbb{R}^n} e^{-i\langle k,x\rangle}e^{-\frac{1}{2}\langle x,Ax\rangle}d^n x = \frac{(2\pi)^{\frac{n}{2}}}{\sqrt{\det A}}e^{-\frac{1}{2}\langle k,Ck\rangle} \tag{4.58}$$

b) *Let $dP_C(x) := \frac{\sqrt{\det A}}{(2\pi)^{\frac{n}{2}}}e^{-\frac{1}{2}\langle x,Ax\rangle}d^n x$ and let*

$$\langle x_{j_1}\cdots x_{j_m}\rangle_C := \int_{\mathbb{R}^n} x_{j_1}\cdots x_{j_m}\,dP_C(x) \tag{4.59}$$

Then one has, if $C = A^{-1} = (c_{ij})_{1\le i,j\le n}$

$$\langle x_{j_1}\cdots x_{j_m}\rangle_C = \sum_{r=2}^m c_{j_1,j_r}\langle x_{j_2}\cdots\widehat{x_{j_r}}\cdots x_{j_m}\rangle_C \tag{4.60}$$

where $\widehat{x_{j_r}}$ means omission of that factor. In particular,

$$\langle x_{j_1}\cdots x_{j_m}\rangle_C = \begin{cases}\sum_{\text{pairings }\sigma}c_{j_{\sigma 1},j_{\sigma 2}}\cdots c_{j_{\sigma(m-1)},j_{\sigma m}} & \text{for m even}\\ 0 & \text{for m odd}\end{cases} \tag{4.61}$$

where the set of all pairings is given by those permutations $\sigma \in S_m$ for which $\sigma(2i-1) < \sigma(2i)$ and $\sigma(2i-1) < \sigma(2i+1)$ for all i.

Proof: (i) We can diagonalize A by an orthogonal matrix $SAS^T = D = \text{diag}(\lambda_1, \cdots, \lambda_n)$. Change variables $y = Sx$, $d^n y = |\det S| d^n x = d^n x$, $p = Sk$ to get

$$\int e^{-i\langle k,x\rangle} e^{-\frac{1}{2}\langle x, Ax\rangle} d^n x = \int e^{-i\langle Sk, y\rangle} e^{-\frac{1}{2}\langle y, Dy\rangle} d^n y$$

$$= \prod_{i=1}^{n} \sqrt{\tfrac{2\pi}{\lambda_i}} \, e^{-\frac{1}{2\lambda_i} p_i^2}$$

$$= \frac{(2\pi)^{\frac{n}{2}}}{\det D} e^{-\frac{1}{2}\langle p, D^{-1}p\rangle}$$

$$= \frac{(2\pi)^{\frac{n}{2}}}{\det A} e^{-\frac{1}{2}\langle k, A^{-1}k\rangle} \tag{4.62}$$

(ii) We have

$$\langle x_{j_1} \cdots x_{j_m}\rangle_C = \int x_{j_1} \cdots x_{j_m} \, dP_C(x)$$

$$= \frac{1}{(-i)^m} \frac{\partial}{\partial k_{j_1}} \cdots \frac{\partial}{\partial k_{j_m}} \Big|_{k=0} \int e^{-i\langle k,x\rangle} dP_C(x)$$

$$= \frac{1}{(-i)^m} \frac{\partial}{\partial k_{j_1}} \cdots \frac{\partial}{\partial k_{j_m}} \Big|_{k=0} e^{-\frac{1}{2}\langle k, Ck\rangle}$$

$$= \frac{1}{(-i)^m} \frac{\partial}{\partial k_{j_2}} \cdots \frac{\partial}{\partial k_{j_m}} \Big|_{k=0} \left\{ -\sum_{\ell=1}^{n} c_{j_1,\ell} k_\ell \, e^{-\frac{1}{2}\langle k, Ck\rangle} \right\}$$

$$= \sum_{s=1}^{m} \frac{1}{(-i)^{m-2}} \frac{\partial}{\partial k_{j_2}} \cdots \widehat{\frac{\partial}{\partial k_{j_s}}} \cdots \frac{\partial}{\partial k_{j_m}} \Big|_{k=0} c_{j_1,j_s} \, e^{-\frac{1}{2}\langle k, Ck\rangle}$$

$$= \sum_{s=1}^{m} c_{j_1,j_s} \langle x_{j_2} \cdots \widehat{x_{j_s}} \cdots x_{j_m}\rangle_C \tag{4.63}$$

Then (4.61) is obtained from (4.60) by iteration. ∎

Theorem 4.3.2 *Let $A \in \mathbb{C}^{n\times n}$ be self adjoint and positive definite and let $C = A^{-1}$. For $z, k \in \mathbb{C}^n$ let $\langle k, z\rangle = \sum_{j=1}^{n} k_j z_j$ (linear in both arguments). For $z_j = x_j + iy_j$, $\bar{z}_j = x_j - iy_j$ let $dz_j d\bar{z}_j := dx_j dy_j$. Then*

a) *For $k, l \in \mathbb{C}^n$*

$$\int_{\mathbb{R}^{2n}} e^{-i\langle k,z\rangle} e^{-i\langle l,\bar{z}\rangle} e^{-\langle \bar{z}, Az\rangle} d^n z d^n \bar{z} = \frac{\pi^n}{\det A} e^{-\langle k, Cl\rangle} \tag{4.64}$$

b) *Let $dP_C(z, \bar{z}) := \frac{\det A}{\pi^n} e^{-\langle \bar{z}, Az\rangle} d^n z d^n \bar{z}$ and let*

$$\langle z_{i_1} \cdots z_{i_m} \bar{z}_{j_1} \cdots \bar{z}_{j_{m'}}\rangle_C := \int_{\mathbb{R}^{2n}} z_{i_1} \cdots \bar{z}_{j_{m'}} \, dP_C(z, \bar{z}) \tag{4.65}$$

Then one has, if $C = A^{-1} = (c_{ij})_{1 \le i,j \le n}$

$$\langle z_{i_1} \cdots z_{i_m} \bar{z}_{j_1} \cdots \bar{z}_{j_{m'}} \rangle_C = \sum_{r=1}^{m'} c_{i_1,j_r} \langle z_{i_2} \cdots z_{i_m} \bar{z}_{j_1} \cdots \widehat{\bar{z}_{j_r}} \cdots \bar{z}_{j_{m'}} \rangle_C$$

(4.66)

where $\widehat{\bar{z}_{j_r}}$ means omission of that factor. In particular,

$$\langle z_{i_1} \cdots z_{i_m} \bar{z}_{j_1} \cdots \bar{z}_{j_{m'}} \rangle_C = \begin{cases} \sum_{\pi \in S_m} c_{i_1,j_{\pi 1}} \cdots c_{i_m,j_{\pi m}} & \text{for } m = m' \\ 0 & \text{for } m \ne m' \end{cases}$$

(4.67)

Proof: (i) We can diagonalize A with a unitary matrix $U = R + iS$, $UAU^* = D = \text{diag}(\lambda_1, \cdots, \lambda_n)$. Change variables $\xi = u + iv := Uz = (R + iS)(x + iy) = Rx - Sy + i(Sx + Ry)$. Since

$$\det \frac{\partial(u, v)}{\partial(x, y)} = \det \begin{pmatrix} R & -S \\ S & R \end{pmatrix} = |\det(R + iS)|^2 = 1$$

we get with $p = \bar{U}k$, $q = Ul$

$$\int e^{-i\langle k,z \rangle} e^{-i\langle l,\bar{z} \rangle} e^{-\langle \bar{z}, Az \rangle} d^n z d^n \bar{z} = \int e^{-i\langle p,\xi \rangle - i\langle q,\bar{\xi} \rangle} e^{-\langle \bar{\xi}, D\xi \rangle} d^n \xi d^n \bar{\xi}$$

Now for arbitrary complex numbers a and b and positive λ one has

$$\int_{\mathbb{R}^2} e^{-ia(u+iv) - ib(u-iv)} e^{-\lambda(u^2+v^2)} du\, dv$$

$$= \int e^{-i(a+b)u} e^{-\lambda u^2} du \int e^{(a-b)v} e^{-\lambda v^2} dv$$

$$= \frac{\pi}{\lambda} e^{-\frac{1}{4\lambda}(a+b)^2 + \frac{1}{4\lambda}(a-b)^2} = \frac{\pi}{\lambda} e^{-\frac{ab}{\lambda}}$$

(4.68)

which gives

$$\int e^{-i\langle k,z \rangle} e^{-i\langle l,\bar{z} \rangle} e^{-\langle \bar{z}, Az \rangle} d^n z d^n \bar{z}$$

$$= \prod_{i=1}^{n} \frac{\pi}{\lambda_i} e^{-\frac{p_i q_i}{\lambda_i}} = \frac{\pi^n}{\det A} e^{-\langle p, D^{-1}q \rangle} = \frac{\pi^n}{\det A} e^{-\langle k, Cl \rangle}$$

since $\langle p, D^{-1}q \rangle = \langle \bar{U}k, D^{-1}Ul \rangle = \langle k, U^{-1}D^{-1}Ul \rangle = \langle k, A^{-1}l \rangle$.

(ii) We have

$$\langle z_{i_1} \cdots z_{i_m} \bar{z}_{j_1} \cdots \bar{z}_{j_{m'}} \rangle_C$$

$$= \frac{1}{(-i)^{m+m'}} \frac{\partial}{\partial k_{i_1}} \cdots \frac{\partial}{\partial k_{i_m}} \frac{\partial}{\partial l_{j_1}} \cdots \frac{\partial}{\partial l_{j_{m'}}} \Big|_{k=l=0} \int e^{-i\langle k,z \rangle - i\langle l,\bar{z} \rangle} dP_C(z,\bar{z})$$

$$= \frac{1}{(-i)^{m+m'}} \frac{\partial}{\partial k_{i_1}} \cdots \frac{\partial}{\partial k_{i_m}} \frac{\partial}{\partial l_{j_1}} \cdots \frac{\partial}{\partial l_{j_{m'}}} \Big|_{k=l=0} e^{-\langle k,Cl \rangle}$$

$$= \frac{1}{(-i)^{m+m'}} \frac{\partial}{\partial k_{i_2}} \cdots \frac{\partial}{\partial k_{i_m}} \frac{\partial}{\partial l_{j_1}} \cdots \frac{\partial}{\partial l_{j_{m'}}} \Big|_{k=l=0} \left\{ -\sum_{\ell=1}^{n} c_{i_1,\ell} \, l_\ell \, e^{-\langle k,Cl \rangle} \right\}$$

$$= \sum_{s=1}^{m'} \frac{1}{(-i)^{m+m'-2}} \frac{\partial}{\partial k_{i_2}} \cdots \widehat{\frac{\partial}{\partial l_{j_s}}} \cdots \frac{\partial}{\partial l_{j_{m'}}} \Big|_{k=l=0} c_{i_1,j_s} \, e^{-\langle k,Cl \rangle}$$

$$= \sum_{s=1}^{m'} c_{i_1,j_s} \langle z_{i_2} \cdots z_{i_m} \bar{z}_{j_1} \cdots \widehat{\bar{z}_{j_s}} \cdots \bar{z}_{j_{m'}} \rangle_C \qquad (4.69)$$

(4.67) then follows by iteration of (4.66) and the observation that

$$\langle z_{i_1} \cdots z_{i_r} \rangle_C = \frac{1}{(-i)^r} \frac{\partial}{\partial k_{i_1}} \cdots \frac{\partial}{\partial k_{i_r}} \Big|_{k=0} \int e^{-i\langle k,z \rangle} dP(z,\bar{z})$$

$$= \frac{1}{(-i)^r} \frac{\partial}{\partial k_{i_1}} \cdots \frac{\partial}{\partial k_{i_r}} \Big|_{k=0} 1 = 0. \qquad \blacksquare$$

4.4 Theory of Grassmann Integration

In this section we prove some basic facts about Grassmann integration. The first two theorems are the analogs of Theorems 4.3.1 and 4.3.2 for anticommuting variables. The third theorem introduces the 'R-operator' which we will use in chapter 9 to prove bounds on the sum of convergent diagrams. The main advantage of this operation is that it produces the diagrams in such a way that sign cancellations can be implemented quite easily. While the first two theorems are standard material [9], the **R**-operation has been introduced only recently in [22].

For $N < \infty$ let $a_1, \cdots a_N$ be anticommuting variables. That is, we have the relations

$$a_i a_j = -a_j a_i \quad \forall i, j = 1, \cdots, N \qquad (4.70)$$

The Grassmann algebra generated by $a_1, \cdots, a_N$ is given by

$$\mathcal{G}_N = \mathcal{G}_N(a_1, \cdots, a_N) = \left\{ \sum_{I \subset \{1, \cdots, N\}} \alpha_I \prod_{i \in I} a_i \mid \alpha_I \in \mathbb{C} \right\} \qquad (4.71)$$

Definition 4.4.1 *The Grassmann integral* $\int \cdot \Pi_{i=1}^N da_i : \mathcal{G}_N \to \mathbb{C}$ *is the linear functional on* $\mathcal{G}_N$ *defined by*

$$\int \sum_{I \subset \{1,\cdots,N\}} \alpha_I \prod_{i \in I} a_i \prod_{i=1}^N da_i := \alpha_{\{1,\cdots,N\}} \tag{4.72}$$

The notation $\Pi_{i=1}^N da_i$ is used to fix a sign. For example, on $\mathcal{G}_3$

$$\int \Big(3a_1a_2 - 5a_1a_2a_3 \Big) da_1 da_2 da_3 = -5$$

but

$$\int \Big(3a_1a_2 - 5a_1a_2a_3 \Big) da_3 da_2 da_1 = +5 \int a_3 a_2 a_1 \, da_3 da_2 da_1 = +5$$

We start with the following

Lemma 4.4.2 *Let* $a_1, \cdots, a_N$ *and* $b_1, \cdots, b_N$ *be Grassmann variables and suppose that* $b_i = \sum_{j=1}^N T_{ij} a_j$ *where* $T = (T_{ij}) \in \mathbb{C}^{N \times N}$. *Then* $db_1 \cdots db_N = \det T \, da_1 \cdots da_N$. *That is,*

$$\int F\Big(\{ \sum_{j=1}^N T_{ij} a_j \}_{i \in \{1,\cdots,N\}} \Big) \det T \prod_{i=1}^N da_i = \int F(\{b_i\}_{i \in \{1,\cdots,N\}}) \prod_{i=1}^N db_i$$

$$\tag{4.73}$$

for some polynomial function F.

Proof: We have

$$b_1 b_2 \cdots b_N = \sum_{j_1=1}^N T_{1,j_1} a_{j_1} \sum_{j_2=1}^N T_{2,j_2} a_{j_2} \cdots \sum_{j_N=1}^N T_{N,j_N} a_{j_N}$$

$$= \sum_{\pi \in S_n} T_{1,\pi 1} \cdots T_{N,\pi N} \, a_{\pi 1} a_{\pi 2} \cdots a_{\pi N}$$

$$= \sum_{\pi \in S_n} \text{sign} \pi \, T_{1,\pi 1} \cdots T_{N,\pi N} \, a_1 a_2 \cdots a_N$$

$$= \det T \, a_1 a_2 \cdots a_N \tag{4.74}$$

which proves the lemma. ∎

Before we prove the analogs of Theorems 4.3.1 and 4.3.2 for anticommuting variables, we briefly recall some basic properties of pfaffians since Grassmann Gaussian integrals are in general given by pfaffians.

Lemma 4.4.3 *For some complex skew symmetric matrix $A = (a_{ij}) \in \mathbb{C}^{n \times n}$, $A^T = -A$, with $n = 2m$ even, let*

$$\text{Pf} A := \sum_{\text{pairings } \sigma} \text{sign}\sigma \, a_{\sigma 1, \sigma 2} \cdots a_{\sigma(n-1), \sigma n} \qquad (4.75)$$

$$= \frac{1}{2^m m!} \sum_{\pi \in S_n} \text{sign}\pi \, a_{\pi 1, \pi 2} \cdots a_{\pi(n-1), \pi n}$$

be the pfaffian of the matrix A. Here the set of all pairings is given by those permutations $\sigma \in S_n$ for which $\sigma(2i-1) < \sigma(2i)$ and $\sigma(2i-1) < \sigma(2i+1)$ for all i. Then

a) *For any $B \in \mathbb{C}^{n \times n}$ we have $\text{Pf}\left(B^T A B\right) = \det B \, \text{Pf} A$*

b) *$(\text{Pf} A)^2 = \det A$*

c) *If A is invertible, then $\text{Pf}(A^{-1}) = (-1)^m / \text{Pf} A$*

d) *Let $A \in \mathbb{R}^{n \times n}$, $A^T = -A$. Then there is an $S \in SO(n)$ such that*

$$SAS^T = \text{diag}\left(\begin{pmatrix} 0 & \lambda_1 \\ -\lambda_1 & 0 \end{pmatrix}, \cdots, \begin{pmatrix} 0 & \lambda_m \\ -\lambda_m & 0 \end{pmatrix} \right) \qquad (4.76)$$

Proof: We first prove part (d). Since $B := iA$ is self adjoint, $B^* = -iA^T = iA = B$, B and therefore A can be diagonalized and A has pure imaginary eigenvalues. If $i\lambda$, $\lambda \in \mathbb{R}$, is an eigenvalue of A with eigenvector z, then also $-i\lambda$ is an eigenvalue with eigenvalue $\bar{z}$. If $z = x + iy$, then $Az = Ax + iAy = i\lambda(x + iy) = -\lambda y + i\lambda x$ or

$$A \begin{pmatrix} y \\ x \end{pmatrix} = \begin{pmatrix} 0 & \lambda \\ -\lambda & 0 \end{pmatrix} \begin{pmatrix} y \\ x \end{pmatrix} \qquad (4.77)$$

Since $\langle x, x \rangle = \frac{1}{\lambda}\langle x, Ay \rangle = -\frac{1}{\lambda}\langle Ax, y \rangle = \langle y, y \rangle$ we have, since z and $\bar{z}$ are orthogonal (observe that $\langle z, z' \rangle$ is linear in both arguments by definition), $0 = \langle z, z \rangle = \langle x, x \rangle - \langle y, y \rangle + i\langle x, y \rangle + i\langle y, x \rangle = 2i\langle x, y \rangle$ which results in $\langle x, y \rangle = 0$. Thus the matrix with rows $y_1, x_1, \cdots, y_m, x_m$ is orthogonal.

(a) We have $(B^T A B)_{i_1 i_2} = \sum_{j_1, j_2} b_{j_1 i_1} a_{j_1 j_2} b_{j_2 i_2}$ which gives

$$\text{Pf}(B^T A B) = \frac{1}{2^m m!} \sum_{\pi \in S_n} \text{sign}\pi \, (B^T A B)_{\pi 1, \pi 2} \cdots (B^T A B)_{\pi(n-1), \pi n}$$

$$= \frac{1}{2^m m!} \sum_{j_1, \cdots, j_n} \sum_{\pi \in S_n} \text{sign}\pi \, b_{j_1, \pi 1} \cdots b_{j_n, \pi n} a_{j_1, j_2} \cdots a_{j_{n-1}, j_n}$$

$$= \frac{1}{2^m m!} \sum_{j_1, \cdots, j_n} \varepsilon^{j_1 \cdots j_n} \det B \, a_{j_1, j_2} \cdots a_{j_{n-1}, j_n}$$

$$= \det B \, \text{Pf} A \qquad (4.78)$$

(b) Since both sides of the identity are polynomials in a_{ij}, it suffices to prove it for real A. Let $S \in SO(n)$ be the matrix of part (d) which makes A blockdiagonal. Then

$$\text{Pf}A = \det S \, \text{Pf}A = \text{Pf}(SAS^T) = \prod_{i=1}^{m} \text{Pf}\begin{pmatrix} 0 & \lambda_i \\ -\lambda_i & 0 \end{pmatrix} = \prod_{i=1}^{m} \lambda_i \qquad (4.79)$$

Thus

$$(\text{Pf}A)^2 = \prod_{i=1}^{m} \lambda_i^2 = \prod_{i=1}^{m} \det\begin{pmatrix} 0 & \lambda_i \\ -\lambda_i & 0 \end{pmatrix} = \det(SAS^T) = \det A \qquad (4.80)$$

(c) Let D be the blockdiagonal matrix on the right hand side of (4.76). Then $A = S^{-1}DS$ and we have

$$\text{Pf}(A^{-1}) = \text{Pf}(S^{-1}D^{-1}S) = \text{Pf}(D^{-1})$$
$$= \text{Pf}\left[\text{diag}\left(\begin{pmatrix} 0 & -1/\lambda_1 \\ 1/\lambda_1 & 0 \end{pmatrix}, \cdots, \begin{pmatrix} 0 & -1/\lambda_m \\ 1/\lambda_m & 0 \end{pmatrix}\right)\right]$$
$$= \frac{(-1)^m}{\lambda_1 \cdots \lambda_m} = \frac{(-1)^m}{\text{Pf}A} \qquad (4.81)$$

which proves the lemma. ∎

Theorem 4.4.4 Let N be even, $A \in \mathbb{C}^{N \times N}$ be skew symmetric and invertible and let $C = A^{-1}$. Let a_i, b_i be Grassmann variables, $\langle a, b \rangle = \sum_{i=1}^{N} a_i b_i$ and $d^N a = \prod_{i=1}^{N} da_i$. Then

a)

$$\int e^{-i\langle a,b \rangle} e^{-\frac{1}{2}\langle a, Aa \rangle} d^N a = \frac{1}{\text{Pf}C} e^{-\frac{1}{2}\langle b, Cb \rangle} \qquad (4.82)$$

b) Let $d\mu_C(a) = \text{Pf}C \, e^{-\frac{1}{2}\langle a, C^{-1}a \rangle} d^N a$ and let

$$\langle a_{j_1} \cdots a_{j_m} \rangle_C := \int a_{j_1} \cdots a_{j_m} \, d\mu_C(a) \qquad (4.83)$$

Then one has, if $C = (c_{ij})_{1 \le i, j \le N}$,

$$\langle a_{j_1} \cdots a_{j_m} \rangle_C = \sum_{r=2}^{m} (-1)^r c_{j_1, j_r} \langle a_{j_2} \cdots \widehat{a_{j_r}} \cdots a_{j_m} \rangle_C \qquad (4.84)$$

In particular

$$\langle a_{j_1} \cdots a_{j_m} \rangle_C = \begin{cases} \sum_{\text{pairings } \sigma} \text{sign}\sigma \, c_{j_{\sigma 1}, j_{\sigma 2}} \cdots c_{j_{\sigma(m-1)}, j_{\sigma m}} & \text{for } m \text{ even} \\ 0 & \text{for } m \text{ odd} \end{cases}$$
$$= \begin{cases} \text{Pf}\left[(c_{j_r j_s})_{1 \le r, s \le m}\right] & \text{for } m \text{ even} \\ 0 & \text{for } m \text{ odd} \end{cases} \qquad (4.85)$$

where the set of all pairings is given by those permutations $\sigma \in S_m$ for which $\sigma(2i-1) < \sigma(2i)$ and $\sigma(2i-1) < \sigma(2i+1)$ for all i.

Proof: a) Since both sides of (4.82) are rational functions of the matrix entries a_{ij} (in fact, both sides are polynomials in a_{ij}), it suffices to prove (4.82) for real A. For every real skew symmetric matrix A there is an orthogonal matrix $S \in SO(N, \mathbb{R})$ such that

$$SAS^T = \text{diag}\left(\begin{pmatrix} 0 & \lambda_1 \\ -\lambda_1 & 0 \end{pmatrix}, \cdots, \begin{pmatrix} 0 & \lambda_{N/2} \\ -\lambda_{N/2} & 0 \end{pmatrix} \right) \equiv D \tag{4.86}$$

Thus, changing variables to $c = Sa$, $d = Sb$, $d^N c = d^N a$, we get

$$\int e^{\langle a,b \rangle} e^{-\frac{1}{2}\langle a, Aa \rangle} d^N a = \int e^{\langle Sa, Sb \rangle} e^{-\frac{1}{2}\langle Sa, SAS^T Sa \rangle} d^N a$$

$$= \int e^{\langle c,d \rangle} e^{-\frac{1}{2}\langle c, Dc \rangle} d^N c \tag{4.87}$$

Since

$$-\tfrac{1}{2}\langle c, Dc \rangle = -\frac{1}{2}\sum_{i=1}^{N/2}(c_{2i-1}, c_{2i})\begin{pmatrix} 0 & \lambda_i \\ -\lambda_i & 0 \end{pmatrix}\begin{pmatrix} c_{2i-1} \\ c_{2i} \end{pmatrix} = -\sum_{i=1}^{N/2}\lambda_i c_{2i-1}c_{2i} \tag{4.88}$$

and because of

$$\int e^{d_1 c_1 + d_2 c_2} e^{-\lambda c_1 c_2} dc_1 dc_2$$

$$= \int \left\{ 1 + d_1 c_1 + d_2 c_2 + \tfrac{1}{2}(d_1 c_1 + d_2 c_2)^2 \right\}(1 - \lambda c_1 c_2)\, dc_1 dc_2$$

$$= -\lambda + \tfrac{1}{2}\int (d_1 c_1 d_2 c_2 + d_2 c_2 d_1 c_1)\, dc_1 dc_2$$

$$= -\lambda - d_1 d_2 = -\lambda(1 + \tfrac{1}{\lambda}d_1 d_2) = -\lambda e^{\frac{d_1 d_2}{\lambda}}$$

$$= -\lambda e^{\frac{1}{2}(d_1, d_2)\begin{pmatrix} 0 & 1/\lambda \\ -1/\lambda & 0 \end{pmatrix}\begin{pmatrix} d_1 \\ d_2 \end{pmatrix}}$$

$$= -\lambda e^{-\frac{1}{2}(d_1, d_2)\begin{pmatrix} 0 & \lambda \\ -\lambda & 0 \end{pmatrix}^{-1}\begin{pmatrix} d_1 \\ d_2 \end{pmatrix}} \tag{4.89}$$

we get

$$\int e^{\langle a,b \rangle} e^{-\frac{1}{2}\langle a, Aa \rangle} d^N a = \int e^{\langle c,d \rangle} e^{-\frac{1}{2}\langle c, Dc \rangle} d^N c$$

$$= \prod_{i=1}^{N/2}\left\{ -\lambda_i\, e^{-\frac{1}{2}(d_{2i-1}, d_{2i})\begin{pmatrix} 0 & \lambda_i \\ -\lambda_i & 0 \end{pmatrix}^{-1}\begin{pmatrix} d_{2i-1} \\ d_{2i} \end{pmatrix}} \right\}$$

$$= (-1)^{N/2}\,\text{Pf}D\, e^{-\frac{1}{2}\langle d, D^{-1}d \rangle}$$

$$= (-1)^{N/2}\,\text{Pf}[S^T DS]\, e^{-\frac{1}{2}\langle b, S^{-1}D^{-1}Sb \rangle}$$

$$= (-1)^{N/2}\,\text{Pf}A\, e^{-\frac{1}{2}\langle b, A^{-1}b \rangle}$$

$$= \tfrac{1}{\text{Pf}(A^{-1})}\, e^{-\frac{1}{2}\langle b, A^{-1}b \rangle} \tag{4.90}$$

which proves part (a). To obtain part (b), observe that

$$\frac{\partial}{\partial b_{j_m}} e^{\langle b,a \rangle} = a_{j_m} e^{\langle b,a \rangle} = e^{\langle b,a \rangle} a_{j_m},$$

$$\frac{\partial}{\partial b_{j_{m-1}}} \frac{\partial}{\partial b_{j_m}} e^{\langle b,a \rangle} = \frac{\partial}{\partial b_{j_{m-1}}} e^{\langle b,a \rangle} a_{j_m} = e^{\langle b,a \rangle} a_{j_{m-1}} a_{j_m}$$

$$\frac{\partial}{\partial b_{j_1}} \cdots \frac{\partial}{\partial b_{j_m}} e^{\langle b,a \rangle} = e^{\langle b,a \rangle} a_{j_1} \cdots a_{j_m} = a_{j_1} \cdots a_{j_m} e^{\langle b,a \rangle}$$

which gives

$$\langle a_{j_1} \cdots a_{j_m} \rangle_C$$

$$= \frac{\partial}{\partial b_{j_1}} \cdots \frac{\partial}{\partial b_{j_m}} \Big|_{b=0} \int e^{\langle b,a \rangle} d\mu_C(a)$$

$$= \frac{\partial}{\partial b_{j_1}} \cdots \frac{\partial}{\partial b_{j_m}} \Big|_{b=0} e^{-\frac{1}{2}\langle b,Cb \rangle}$$

$$= (-1)^{m-1} \frac{\partial}{\partial b_{j_2}} \cdots \frac{\partial}{\partial b_{j_m}} \left\{ \frac{\partial}{\partial b_{j_1}} \left(-\frac{1}{2} \sum_{r,s=1}^{N} c_{rs} b_r b_s \right) \right\} e^{-\frac{1}{2}\langle b,Cb \rangle} \Big|_{b=0}$$

$$= (-1)^{m-1} \frac{\partial}{\partial b_{j_2}} \cdots \frac{\partial}{\partial b_{j_m}} \left(-\frac{1}{2} \sum_{s=1}^{N} c_{j_1 s} b_s + \frac{1}{2} \sum_{r=1}^{N} c_{r j_1} b_r \right) e^{-\frac{1}{2}\langle b,Cb \rangle} \Big|_{b=0}$$

$$= (-1)^{m-1} \frac{\partial}{\partial b_{j_2}} \cdots \frac{\partial}{\partial b_{j_m}} \left(-\sum_{s=1}^{N} c_{j_1 s} b_s \right) e^{-\frac{1}{2}\langle b,Cb \rangle} \Big|_{b=0}$$

$$= \sum_{\ell=2}^{m} (-1)^{m-1}(-1)^{m-\ell} \frac{\partial}{\partial b_{j_2}} \cdots \widehat{\frac{\partial}{\partial b_{j_\ell}}} \cdots \frac{\partial}{\partial b_{j_m}} (-c_{j_1 j_\ell}) e^{-\frac{1}{2}\langle b,Cb \rangle} \Big|_{b=0}$$

$$= \sum_{\ell=2}^{m} (-1)^{\ell} c_{j_1 j_\ell} \frac{\partial}{\partial b_{j_2}} \cdots \widehat{\frac{\partial}{\partial b_{j_\ell}}} \cdots \frac{\partial}{\partial b_{j_m}} e^{-\frac{1}{2}\langle b,Cb \rangle} \Big|_{b=0}$$

$$= \sum_{\ell=2}^{m} (-1)^{\ell} c_{j_1 j_\ell} \langle a_{j_2} \cdots \widehat{a_{j_\ell}} \cdots a_{j_m} \rangle_C \qquad (4.91)$$

which proves (4.84). Then (4.85) is obtained by iterating (4.84). ∎

In the next theorem, we specialize the above theorem to the case that the matrix A has the block structure $A = \begin{pmatrix} 0 & B \\ -B^T & 0 \end{pmatrix}$ in which case the exponential becomes $e^{-\langle a,B\bar{a} \rangle}$ if we redefine $a := (a_1, \cdots, a_{N/2})$ and $\bar{a} = (a_{N/2+1}, \cdots, a_N)$.

Theorem 4.4.5 Let $B \in \mathbb{C}^{N \times N}$ be invertible, $C := B^{-1}$, and let $a_1, \cdots, a_N$, $\bar{a}_1, \cdots, \bar{a}_N$ and $b_1, \cdots, b_N, \bar{b}_1, \cdots, \bar{b}_N$ be anticommuting variables. Let $\langle \bar{b}, a \rangle = \sum_{i=1}^{N} \bar{b}_i a_i$ (linear in both arguments). Then

a)

$$\int e^{\langle \bar{b},a \rangle + \langle b,\bar{a} \rangle} e^{-\langle a,B\bar{a} \rangle} d^N a d^N \bar{a} = \det B \, e^{-\langle b,C\bar{b} \rangle} \qquad (4.92)$$

b) Let $d\mu_C(a, \bar{a}) := \frac{1}{\det B} e^{-\langle a, B\bar{a}\rangle} d^N a \, d^N \bar{a}$ and let

$$\langle \bar{a}_{i_1} \cdots \bar{a}_{i_m} a_{j_1} \cdots a_{j_{m'}} \rangle_C := \int \bar{a}_{i_1} \cdots \bar{a}_{i_m} a_{j_1} \cdots a_{j_{m'}} \, d\mu_C(a, \bar{a}) \quad (4.93)$$

Then

$$\langle \bar{a}_{i_1} \cdots \bar{a}_{i_m} a_{j_1} \cdots a_{j_{m'}} \rangle_C =$$

$$\sum_{r=1}^{m'} (-1)^{m-1+r} c_{i_1 j_r} \langle \bar{a}_{i_2} \cdots \bar{a}_{i_m} a_{j_1} \cdots \widehat{a_{j_r}} \cdots a_{j_{m'}} \rangle_C \quad (4.94)$$

In particular, $\langle \bar{a}_{i_1} \cdots \bar{a}_{i_m} a_{j_1} \cdots a_{j_{m'}} \rangle_C = 0$ if $m \neq m'$ and

$$\langle \bar{a}_{i_1} a_{j_1} \cdots \bar{a}_{i_m} a_{j_m} \rangle_C = \det \left[(c_{i_r j_s})_{1 \le r, s \le m} \right] \quad (4.95)$$

Proof: Let $c = (a, \bar{a})$, $d = (\bar{b}, b)$ and $A = \begin{pmatrix} 0 & B \\ -B^T & 0 \end{pmatrix}$. Then $\langle a, B\bar{a}\rangle = \frac{1}{2}\langle c, Ac \rangle$, $\langle \bar{b}, a \rangle + \langle b, \bar{a} \rangle = \langle d, c \rangle$ and we get according to Theorem 4.4.4

$$\int e^{\langle \bar{b}, a \rangle + \langle b, \bar{a} \rangle} e^{-\langle a, B\bar{a}\rangle} d^N a \, d^N \bar{a} = \int e^{\langle d, c \rangle} e^{-\frac{1}{2}\langle c, Ac \rangle} d^{2N} c$$

$$= \mathrm{Pf} A \, e^{-\frac{1}{2}\langle d, A^{-1}d \rangle}$$

$$= \det B \, e^{-\langle b, C\bar{b}\rangle} \quad (4.96)$$

since $\mathrm{Pf} \begin{pmatrix} 0 & B \\ -B^T & 0 \end{pmatrix} = \det B$ and

$$\langle d, A^{-1}d \rangle = \left\langle (\bar{b}, b), \begin{pmatrix} 0 & -B^{T-1} \\ B^{-1} & 0 \end{pmatrix} \begin{pmatrix} \bar{b} \\ b \end{pmatrix} \right\rangle = 2\langle b, B^{-1}\bar{b} \rangle$$

which proves (a). To obtain (b), let $\tilde{C} = A^{-1} = \begin{pmatrix} 0 & -B^{T-1} \\ B^{-1} & 0 \end{pmatrix} = \begin{pmatrix} 0 & -C^T \\ C & 0 \end{pmatrix}$. Then, for $1 \le i, j \le N$ (observe that c_i is a Grassmann variable and c_{ij} is the matrix element of C)

$$\tilde{C}_{i,j} = 0, \quad \tilde{C}_{i,N+j} = -c_{j,i}, \quad \tilde{C}_{N+i,j} = c_{i,j}, \quad \tilde{C}_{N+i,N+j} = 0 \quad (4.97)$$

and we get

$$\langle \bar{a}_{i_1} \cdots \bar{a}_{i_m} a_{j_1} \cdots a_{j_{m'}} \rangle_C = \langle c_{N+i_1} \cdots c_{N+i_m} c_{j_1} \cdots c_{j_{m'}} \rangle_{\tilde{C}}$$

$$= \sum_{r=1}^{m'} (-1)^{m-1+r} \tilde{C}_{N+i_1, j_r} \langle c_{N+i_2} \cdots c_{N+i_m} c_{j_1} \cdots \widehat{c_{j_r}} \cdots c_{j_{m'}} \rangle_{\tilde{C}}$$

$$= \sum_{r=1}^{m'} (-1)^{m-1+r} c_{i_1, j_r} \langle \bar{a}_{i_2} \cdots \bar{a}_{i_m} a_{j_1} \cdots \widehat{a_{j_r}} \cdots a_{j_{m'}} \rangle_C \quad (4.98)$$

Finally (4.95) follows from

$$\langle \bar{a}_{i_1} a_{j_1} \cdots \bar{a}_{i_m} a_{j_m} \rangle_C = \sum_{r=1}^{m} c_{i_1 j_r} \langle a_{j_1} \bar{a}_{i_2} a_{j_2} \cdots \bar{a}_{i_m} a_{j_m} \rangle_C$$

$$= \sum_{r=1}^{m} (-1)^{r-1} c_{i_1 j_r} \langle \bar{a}_{i_2} a_{j_1} \cdots \bar{a}_{i_r} a_{j_{r-1}} \bar{a}_{i_{r+1}} a_{j_{r+1}} \cdots \bar{a}_{i_m} a_{j_m} \rangle_C$$

which is the expansion of the determinant with respect to the first row. ∎

We now introduce the **R** operation of [22]. To do so, we briefly recall the idea of Wick ordering. A Gaussian integral of some monomial $a_{i_1} \cdots a_{i_{2r}}$ is given by the sum of all pairings or contractions:

$$\int a_{i_1} \cdots a_{i_{2r}} d\mu_C(a) = \sum_{\text{pairings } \sigma} \text{sign}\sigma \, c_{i_{\sigma 1}, i_{\sigma 2}} \cdots c_{i_{\sigma(2r-1)}, i_{\sigma(2r)}}$$

$$= \text{Pf}\left[(c_{i_k, i_\ell})_{1 \le k, \ell \le 2r}\right] \qquad (4.99)$$

Graphically one may represent every a_{i_k} by a half line or 'leg' and then $\int \cdot d\mu_C$ means to contract the $2r$ half lines to r lines in all possible $(2r-1)!!$ ways. To each line which is obtained by pairing the halflines a_{i_k} and a_{i_ℓ} we assign the matrix element c_{i_k, i_ℓ}. Now given two monomials $\Pi_{i \in I} a_i$ and $\Pi_{j \in J} a_j$, $I, J \subset \{1, \cdots, N\}$, it is useful to have an operation which produces the sum of contractions between a_i and $a_{i'}$ 'fields' and between a_i and a_j fields but no contractions between a_j and $a_{j'}$ fields, $i, i' \in I$, $j, j' \in J$. This is exactly what Wick ordering $:\Pi_{j \in J} a_j:$ is doing,

$$\int \prod_{i \in I} a_i : \prod_{j \in J} a_j : d\mu_C(a) = \text{Pf}\left[\begin{pmatrix} (c_{ii'})_{i,i' \in I} & (c_{ij'})_{i \in I, j' \in J} \\ (c_{ji'})_{j \in J, i' \in I} & 0 \end{pmatrix}\right] \qquad (4.100)$$

Usually (4.100) is accomplished by defining $:\Pi_{j \in J} a_j:$ as a suitable linear combination of monomials of lower degree and matrix elements $c_{jj'}$, for example $:a_j a_{j'}: = a_j a_{j'} - c_{jj'}$. Here we use the approach of [22] which simply doubles the number of Grassmann variables.

Definition 4.4.6 *Let $C \in \mathbb{C}^{N \times N}$ be invertible and skew symmetric, $C^\sharp :=$ $\begin{pmatrix} C & C \\ C & 0 \end{pmatrix}$, let $a_1, \cdots, a_N, a_1^\sharp, \cdots, a_N^\sharp$ be anticommuting variables, $a_i a_j = -a_j a_i$, $a_i a_j^\sharp = -a_j^\sharp a_i$, $a_i^\sharp a_j^\sharp = -a_j^\sharp a_i^\sharp$, and let $d\mu_{C^\sharp}(a, a^\sharp)$ be the Grassmann Gaussian measure on $\mathcal{G}_{2N}(a, a^\sharp)$. Then, for some polynomials f and g, we define Wick ordering according to*

$$\int_{\mathcal{G}_N} f(a) \, :g(a): d\mu_C(a) := \int_{\mathcal{G}_{2N}} f(a) \, g(a^\sharp) \, d\mu_{C^\sharp}(a, a^\sharp) \qquad (4.101)$$

As a consequence, we have the following

Lemma 4.4.7 *Let f and g be some polynomials. Then*

$$\int_{\mathcal{G}_N} f(a)\, g(a)\, d\mu_C(a) = \int_{\mathcal{G}_N} \int_{\mathcal{G}_N} f(a) \; {:} g(a+b){:}_a \, d\mu_C(b)\, d\mu_C(a) \quad (4.102)$$

where $: \cdot :_a$ *means Wick ordering with respect to the a-variables, see (4.103) below.*

Proof: By definition,

$$\int_{\mathcal{G}_N} \int_{\mathcal{G}_N} f(a) \; {:} g(a+b){:}_a \, d\mu_C(b)\, d\mu_C(a) =$$

$$\int_{\mathcal{G}_{2N}} \int_{\mathcal{G}_N} f(a)\, g(a^\sharp + b)\, d\mu_C(b)\, d\mu_{C^\sharp}(a, a^\sharp) \quad (4.103)$$

We claim that for every polynomial $h = h(a, b)$

$$\int_{\mathcal{G}_{2N}} \int_{\mathcal{G}_N} h(a, a^\sharp + b)\, d\mu_C(b)\, d\mu_{C^\sharp}(a, a^\sharp) = \int_{\mathcal{G}_N} h(a, a)\, d\mu_C(a) \quad (4.104)$$

To prove this, it suffices to prove (4.104) for the generating function $h(a, b) = e^{\langle a,c \rangle + \langle b,d \rangle}$. Then the left hand side of (4.104) becomes

$$\int_{\mathcal{G}_{2N}} \int_{\mathcal{G}_N} e^{\langle a,c \rangle + \langle a^\sharp + b,d \rangle} \, d\mu_C(b)\, d\mu_{C^\sharp}(a, a^\sharp)$$

$$= \int_{\mathcal{G}_N} e^{\langle b,d \rangle}\, d\mu_C(b) \int_{\mathcal{G}_{2N}} e^{\langle a,c \rangle + \langle a^\sharp,d \rangle}\, d\mu_{C^\sharp}(a, a^\sharp)$$

$$= e^{-\frac{1}{2}\langle d,Cd \rangle}\, e^{-\frac{1}{2}\langle (c,d),\left(\begin{smallmatrix} C & C \\ C & 0 \end{smallmatrix}\right)\left(\begin{smallmatrix} c \\ d \end{smallmatrix}\right) \rangle}$$

$$= e^{-\frac{1}{2}\langle (c,d),\left(\begin{smallmatrix} C & C \\ C & C \end{smallmatrix}\right)\left(\begin{smallmatrix} c \\ d \end{smallmatrix}\right) \rangle} = e^{-\frac{1}{2}\langle c+d, C(c+d) \rangle}$$

$$= \int_{\mathcal{G}_N} e^{\langle a,c+d \rangle}\, d\mu_C(a) \quad (4.105)$$

which proves (4.104). Then (4.103) follows by the choice $h(a, b) = f(a)g(b)$. ∎

In Theorem (4.2.2) we found that the Grassmann integral representations for the correlation functions are of the form $\int f(a)\, e^{W(a)} d\mu_C(a) / \int e^{W(a)} d\mu_C(a)$. In chapter 9 on renormalization group methods we will generate the sum of all diagrams up to n'th order by an n-fold application of the following

Theorem 4.4.8 *Let $W = W(a)$ be some polynomial of Grassmann variables, let $d\mu_C$ be a Gaussian measure on the Grassmann algebra $\mathcal{G}_N(a)$ and let $Z = \int e^{W(a)} d\mu_C(a)$. For some polynomial f let $(\mathbf{R}f)(a) \in \mathcal{G}_N(a)$ be given by*

$$(\mathbf{R}f)(a) = \int f(b) \; :e^{W(b+a)-W(a)} - 1:_b \; d\mu_C(b) \tag{4.106}$$

$$= \sum_{n=1}^{N} \frac{1}{n!} \int f(b) \; :[W(b+a) - W(a)]^n:_b \; d\mu_C(b)$$

Then we have

$$\frac{1}{Z} \int f(a) \, e^{W(a)} d\mu_C(a) \; =$$

$$\int f(a) \, d\mu_C(a) \; + \; \frac{1}{Z} \int (\mathbf{R}f)(a) \, e^{W(a)} d\mu_C(a) \tag{4.107}$$

Proof: We have with Lemma 4.4.7

$$\frac{1}{Z} \int f(a) \, e^{W(a)} d\mu_C(a) = \frac{1}{Z} \int f(a) \; :e^{W(a+b)}:_a \; d\mu_C(b) \, d\mu_C(a)$$

$$= \frac{1}{Z} \int f(a) \; :e^{W(a+b)-W(b)}:_a \; e^{W(b)} d\mu_C(b) \, d\mu_C(a)$$

$$= \frac{1}{Z} \int f(a) \; :e^{W(a+b)-W(b)} - 1:_a \; d\mu_C(a) \, e^{W(b)} d\mu_C(b)$$

$$+ \; \frac{1}{Z} \int f(a) \, e^{W(b)} d\mu_C(b) \, d\mu_C(a)$$

$$= \frac{1}{Z} \int (\mathbf{R}f)(b) \, e^{W(b)} d\mu_C(b) \; + \; \int f(a) \, d\mu_C(a)$$

which proves the theorem. ∎

Finally we write down Gram's inequality which will be the key estimate in the proof of Theorem 9.2.2 where we show that the sum of all convergent diagrams has in fact a small positive radius of convergence instead of being only an asymptotic series. We use the version as it is given in the appendix of [22].

Theorem 4.4.9 *Let $(b_1, \cdots, b_{2N}) := (a_1, \cdots, a_N, \bar{a}_1, \cdots, \bar{a}_N)$ be anticommuting variables, let $C \in \mathbb{C}^{N \times N}$ be invertible, let $S = \begin{pmatrix} 0 & C \\ -C^T & 0 \end{pmatrix}$ and let*

$$d\mu_S(b) = \mathrm{Pf}\, S \; e^{-\frac{1}{2}\langle b, S^{-1} b \rangle} d^{2N} b = \det C \; e^{-\langle a, C^{-1} \bar{a} \rangle} d^N a \, d^N \bar{a} \tag{4.108}$$

Suppose in addition that there is a complex Hilbert space $\mathcal{H}$ and elements $v_i, w_i \in \mathcal{H}$, $i = 1, \cdots, N$ such that

$$C = \big(\langle v_i, w_j \rangle_{\mathcal{H}} \big)_{1 \le i,j \le N} \tag{4.109}$$

and

$$\|v_i\|_{\mathcal{H}}, \|w_j\|_{\mathcal{H}} \le \Lambda \tag{4.110}$$

for some positive constant $\Lambda \in \mathbb{R}$. Then, for some subsets $I, J \subset \{1, \cdots, 2N\}$

$$\left| \int \prod_{i \in I} b_i \, d\mu_S(b) \right| \le \Lambda^{|I|} \tag{4.111}$$

$$\left| \int \prod_{i \in I} b_i : \prod_{j \in J} b_j : d\mu_S(b) \right| \le \begin{cases} (\sqrt{2}\,\Lambda)^{|I|+|J|} & \text{if } |J| \le |I| \\ 0 & \text{if } |J| > |I| \end{cases} \tag{4.112}$$

Proof: To prove the first inequality, suppose that $I = \{i_1, \cdots, i_r\}$. Then

$$\int b_{i_1} \cdots b_{i_r} \, d\mu_S(b) = \mathrm{Pf}\left[(S_{i_k i_\ell})_{1 \le k, \ell \le r} \right] \tag{4.113}$$

where

$$S_{i_k i_\ell} = \begin{cases} 0 & \text{if } 1 \le i_k, i_\ell \le N \\ C_{i_k, i_\ell - N} & \text{if } 1 \le i_k \le N \text{ and } N < i_\ell \le 2N \\ -C_{i_\ell, i_k - N} & \text{if } N < i_k \le 2N \text{ and } 1 \le i_\ell \le N \\ 0 & \text{if } N < i_k \le 2N \text{ and } N < i_\ell \le 2N \end{cases} \tag{4.114}$$

More concisely,

$$\int b_{i_1} \cdots b_{i_r} \, d\mu_S(b) = \mathrm{Pf}\begin{pmatrix} 0 & U \\ -U^T & 0 \end{pmatrix} \tag{4.115}$$

where $U = (U_{k\ell})$ is the $\rho := \max\{k | i_k \le N\}$ by $r - \rho$ matrix with elements

$$U_{k\ell} = C_{i_k, i_{\ell+\rho} - N} = \langle v_{i_k}, w_{i_{\ell+\rho} - N} \rangle_{\mathcal{H}} \tag{4.116}$$

Since

$$\mathrm{Pf}\begin{pmatrix} 0 & U \\ -U^T & 0 \end{pmatrix} = \begin{cases} 0 & \text{if } \rho \neq r - \rho \\ (-1)^{\frac{\rho(\rho-1)}{2}} \det U & \text{if } \rho = r - \rho \end{cases} \tag{4.117}$$

we get, for $r = 2\rho$, using Gram's inequality for determinants

$$\left| \int b_{i_1} \cdots b_{i_r} \, d\mu_S(b) \right| = \left| \det \left(\langle v_{i_k}, w_{i_{\ell+\rho} - N} \rangle_{\mathcal{H}} \right) \right|$$

$$\le \prod_{k=1}^{\rho} \|v_{i_k}\|_{\mathcal{H}} \|w_{i_{\ell+\rho} - N}\|_{\mathcal{H}} \le \Lambda^{2\rho} \tag{4.118}$$

which proves (4.111). To prove (4.112), recall that according to definition 4.4.6

$$\int_{\mathcal{G}_{2N}} \prod_{i \in I} b_i : \prod_{j \in J} b_j : d\mu_S(b) = \int_{\mathcal{G}_{4N}} \prod_{i \in I} b_i \prod_{j \in J} b_j^\sharp \, d\mu_{S^\sharp}(b, b^\sharp) \tag{4.119}$$

where

$$S^\sharp = \begin{pmatrix} S & S \\ S & 0 \end{pmatrix} = \begin{pmatrix} 0 & C & 0 & C \\ -C^T & 0 & -C^T & 0 \\ 0 & C & 0 & 0 \\ -C^T & 0 & 0 & 0 \end{pmatrix} \qquad (4.120)$$

This matrix is conjugated by the permutation matrix

$$\begin{pmatrix} Id & 0 & 0 & 0 \\ 0 & 0 & Id & 0 \\ 0 & Id & 0 & 0 \\ 0 & 0 & 0 & Id \end{pmatrix}$$

to

$$\begin{pmatrix} 0 & C^\sharp \\ -C^{\sharp T} & 0 \end{pmatrix} = \begin{pmatrix} 0 & 0 & C & C \\ 0 & 0 & C & 0 \\ -C^T & -C^T & 0 & 0 \\ -C^T & 0 & 0 & 0 \end{pmatrix} \qquad (4.121)$$

Also, for $1 \le i \le 2N$ define the vectors $v_i^\sharp, w_i^\sharp \in \mathcal{H} \oplus \mathcal{H}$ by

$$v_i^\sharp = \begin{cases} (v_i, 0) & \text{for } 1 \le i \le N \\ (v_i, v_i) & \text{for } N < i \le 2N \end{cases}$$

$$w_i^\sharp = \begin{cases} (0, w_i) & \text{for } 1 \le i \le N \\ (w_i, 0) & \text{for } N < i \le 2N \end{cases}$$

Then

$$C^\sharp = \left(\langle v_i^\sharp, w_j^\sharp \rangle_{\mathcal{H} \oplus \mathcal{H}} \right)_{1 \le i, j \le 2N} \qquad (4.122)$$

and

$$\| v_i^\sharp \|_{\mathcal{H} \oplus \mathcal{H}}, \; \| w_j^\sharp \|_{\mathcal{H} \oplus \mathcal{H}} \le \sqrt{2}\,\Lambda \qquad (4.123)$$

for all $1 \le i, j \le 2N$. Thus (4.112) has been reduced to (4.111) for the matrix

$$\begin{pmatrix} 0 & C^\sharp \\ -C^{\sharp T} & 0 \end{pmatrix}$$

the Hilbert space $\mathcal{H} \oplus \mathcal{H}$ and the vectors $v_i^\sharp, w_j^\sharp$. ∎

Chapter 5

Bosonic Functional Integral Representation

In this chapter we apply a Hubbard-Stratonovich transformation to the Grassmann integral representation to obtain a bosonic functional integral representation. The main reason for this is not only that we would like to have integrals with usual commuting variables amenable to standard calculus techniques, but is merely the following one. By writing down the perturbation series for the partition function, evaluating the n'th order coefficients with Wick's theorem and rewriting the resulting determinant as a Grassmann integral, we arrived at the following representation for the two-point function $G(k) = \langle a_k^+ a_k \rangle$

$$G(k) = \frac{\int \bar\psi_k \psi_k \, e^{-\lambda \sum \bar\psi\bar\psi\psi\psi} e^{-\sum(ip_0 - e_\mathbf{p})\bar\psi\psi} \, d\psi d\bar\psi}{\int e^{-\lambda \sum \bar\psi\bar\psi\psi\psi} e^{-\sum(ip_0 - e_\mathbf{p})\bar\psi\psi} \, d\psi d\bar\psi} \tag{5.1}$$

If we would have started with a bosonic model, the representation for $G(k)$ would look exactly the same as (5.1) with the exception that the integral in (5.1) would be a usual one, with commuting variables. Regardless whether the variables in (5.1) are commuting or anticommuting, in both cases we can apply a Hubbard-Stratonovich transformation and in both cases one obtains a result which looks more or less as follows:

$$G(k) = \frac{\int \left[(ip_0 - e_\mathbf{p})\delta_{p,p'} + \sqrt{\lambda}\,\phi_{p-p'} \right]_{k,k}^{-1} e^{-V_{\text{eff}}(\phi)} \, d\phi}{\int e^{-V_{\text{eff}}(\phi)} \, d\phi} \tag{5.2}$$

with an effective potential

$$V_{\text{eff}}(\phi) = |\phi|^2 + \epsilon \log \frac{\det\left[(ip_0 - e_\mathbf{p})\delta_{p,p'} + \sqrt{\lambda}\,\phi_{p-p'} \right]}{\det\left[(ip_0 - e_\mathbf{p})\delta_{p,p'} \right]} \tag{5.3}$$

Thus, we have to compute an inverse matrix element and to average it with respect to a normalized measure which is determined by the effective potential (5.3). The statistics, bosonic or fermionic, only shows up in the effective potential through a sign, namely $\epsilon = -1$ for fermions and $\epsilon = +1$ for (complex) bosons ($\epsilon = 1/2$ for scalar bosons). Besides this unifying feature the representation (5.2) is a natural starting point for a saddle point analysis. Observe that the effective potential basically behaves as $|\phi|^2 + \epsilon \log[Id + \phi]$. Thus, it should be bounded from below and have one or more global minima.

In fact, in section 5.2 we rigorously prove this for the many-electron system with attractive delta interaction. Therefore, as a first approximation one may try the following

$$G(k) = \frac{\int F_k(\phi)\, e^{-V_{\text{eff}}(\phi)}\, d\phi}{\int e^{-V_{\text{eff}}(\phi)}\, d\phi} = \frac{\int F_k(\phi)\, e^{-[V_{\text{eff}}(\phi) - V_{\text{eff}}(\phi_{\min})]}\, d\phi}{\int e^{-[V_{\text{eff}}(\phi) - V_{\text{eff}}(\phi_{\min})]}\, d\phi} \approx F_k(\phi_{\min}) \quad (5.4)$$

or, if the global minimum $\phi_{\min}$ is not unique, $G(k) \approx \int F_k(\phi_{\min}) d\phi_{\min}$ where the integral is an average over all global minima.

Approximation (5.4) would be exact if there would be an extra parameter in front of the effective potential in the exponents in (5.4) which would tend to infinity. In fact, the effective potential is proportional to the volume, but this is due to the number of integration variables, $\sum_k |\phi_k|^2 \sim \beta L^d \int dk |\phi_k|^2$, and if this is the case, usually fluctuations around the minimum are important. However, as we will see in chapter 6, the BCS approximation has the effect that we can separate off a volume factor in front of the effective potential and the number of integration variables does not depend on the volume. Thus, in the BCS approximation the saddle point approximation becomes exact. In section 6.2 we show that this already holds for the quartic BCS model, it is not necessary to make the quadratic mean field approximation.

In the next section, we derive the representation (5.2) for the many-electron system and in section 5.2 we determine the global minimum of the full effective potential with attractive delta interaction and show that it corresponds to the BCS configuration. We also compute the second order Taylor expansion around that minimum. The representation (5.2) will also be the starting point for the analysis in chapter 10.

5.1   The Hubbard-Stratonovich Transformation

In this section we derive bosonic functional integral representations for the quantities we are interested in, for the partition function, the momentum distribution and for our four-point function $\Lambda(q)$. We specialize to the case of a delta interaction in coordinate space, $V(\mathbf{x}-\mathbf{y}) = \lambda\delta(\mathbf{x}-\mathbf{y})$ or $V(\mathbf{k}-\mathbf{p}) = \lambda$ in momentum space. A more general interaction is discussed in chapter 6 where we also consider BCS theory with higher ℓ-wave terms. There we expand $V(\mathbf{k} - \mathbf{p})$ into spherical harmonics on the Fermi surface and include the p- and d-wave terms.

We shortly recall the Grassmann integral representations from which the bosonic functional integral representations will be derived below. According to Theorem 4.2.2 we had

$$Z(\lambda) = Tr\, e^{-\beta H_\lambda} / Tr\, e^{-\beta H_0} = \int e^{-\frac{\lambda}{(\beta L^d)^3} \sum_{kpq} \bar\psi_{k\uparrow} \bar\psi_{q-k,\downarrow} \psi_{p\uparrow} \psi_{q-p,\downarrow}} d\mu_C \quad (5.5)$$

$$\tfrac{1}{L^d} \langle a^+_{\mathbf{k}\sigma} a_{\mathbf{k}\sigma} \rangle = \tfrac{1}{\beta} \sum_{k_0 \in \frac{\pi}{\beta}(2\mathbb{Z}+1)} \tfrac{1}{\beta L^d} \langle \bar\psi_{k\sigma} \psi_{k\sigma} \rangle \quad (5.6)$$

$$\tfrac{\lambda}{L^{3d}} \sum_{\mathbf{k},\mathbf{p}} \langle a^+_{\mathbf{k}\uparrow} a^+_{\mathbf{q-k},\downarrow} a_{\mathbf{q-p},\downarrow} a_{\mathbf{p}\uparrow} \rangle = \tfrac{1}{\beta} \sum_{q_0 \in \frac{2\pi}{\beta}\mathbb{Z}} \Lambda(q_0, \mathbf{q}) \quad (5.7)$$

$$\Lambda(q) = \tfrac{\lambda}{(\beta L^d)^3} \sum_{k,p} \langle \bar\psi_{k\uparrow} \bar\psi_{q-k,\downarrow} \psi_{p\uparrow} \psi_{q-p,\downarrow} \rangle \quad (5.8)$$

where

$$\langle G(\psi, \bar\psi) \rangle = \frac{\int G(\psi, \bar\psi)\, e^{-\frac{\lambda}{(\beta L^d)^3} \sum_{kpq} \bar\psi_{k\uparrow} \bar\psi_{q-k\downarrow} \psi_{p\uparrow} \psi_{q-p\downarrow}} d\mu_C}{\int e^{-\frac{\lambda}{(\beta L^d)^3} \sum_{kpq} \bar\psi_{k\uparrow} \bar\psi_{q-k\downarrow} \psi_{p\uparrow} \psi_{q-p\downarrow}} d\mu_C} \quad (5.9)$$

and

$$d\mu_C(\psi, \bar\psi) = \prod_{k\sigma} \tfrac{\beta L^d}{ik_0 - e_k}\, e^{-\frac{1}{\beta L^d} \sum_{k\sigma} (ik_0 - e_k) \bar\psi_{k\sigma} \psi_{k\sigma}} \prod_{k\sigma} d\psi_{k\sigma} d\bar\psi_{k\sigma} \quad (5.10)$$

The main idea of the Hubbard-Stratonovich transformation is to make the quartic action in the Grassmann integral (5.9) quadratic by the use of the identity (5.11) below and then to integrate out the Grassmann fields.

Lemma 5.1.1 *Let $a, b \in \mathbb{C}$ or commuting elements of some Grassmann algebra. Let $\phi = u + iv$, $\bar\phi = u - iv$ and $d\phi\, d\bar\phi := du\, dv$. Then*

$$e^{ab} = \int_{\mathbb{R}^2} e^{a\phi + b\bar\phi}\, e^{-|\phi|^2}\, \tfrac{d\phi\, d\bar\phi}{\pi} \quad (5.11)$$

Proof: We have

$$\int_{\mathbb{R}^2} e^{a\phi + b\bar\phi}\, e^{-|\phi|^2}\, \tfrac{d\phi\, d\bar\phi}{\pi} = \tfrac{1}{\sqrt{\pi}} \int_{\mathbb{R}} e^{(a+b)u} e^{-u^2} du\, \tfrac{1}{\sqrt{\pi}} \int_{\mathbb{R}} e^{i(a-b)v} e^{-v^2} dv$$

$$= e^{\frac{(a+b)^2}{4}} e^{\frac{-(a-b)^2}{4}} = e^{ab} \quad \blacksquare$$

The lemma above will be applied in the following form:

$$e^{-\sum_q a_q b_q} = \int_{\mathbb{R}^{2N}} e^{i \sum_q (a_q \phi_q + b_q \bar\phi_q)}\, e^{-\sum_q |\phi_q|^2} \prod_q \tfrac{d\phi_q\, d\bar\phi_q}{\pi} \quad (5.12)$$

where N is the number of momenta $q = (q_0, \mathbf{q})$ and

$$a_q = \left(\frac{\lambda}{(\beta L^d)^3}\right)^{\frac{1}{2}} \sum_k \psi_{k\uparrow} \psi_{q-k\downarrow}, \qquad b_q = \left(\frac{\lambda}{(\beta L^d)^3}\right)^{\frac{1}{2}} \sum_k \bar{\psi}_{k\uparrow} \bar{\psi}_{q-k\downarrow} \quad (5.13)$$

The result is

Theorem 5.1.2 *Let* $a_k = ik_0 - e_{\mathbf{k}}$, $\kappa = \beta L^d$, $g = \sqrt{\lambda}$ *and let* $dP(\phi)$ *be the normalized measure*

$$dP(\{\phi_q\}) = \frac{1}{Z} \det \begin{bmatrix} a_k \delta_{k,p} & \frac{ig}{\sqrt{\kappa}} \bar{\phi}_{p-k} \\ \frac{ig}{\sqrt{\kappa}} \phi_{k-p} & a_{-k} \delta_{k,p} \end{bmatrix} e^{-\sum_q |\phi_q|^2} \prod_q d\phi_q d\bar{\phi}_q \quad (5.14)$$

Then there are the following integral representations

$$Z(\lambda) = \int \frac{\det \begin{bmatrix} a_k \delta_{k,p} & \frac{ig}{\sqrt{\kappa}} \bar{\phi}_{p-k} \\ \frac{ig}{\sqrt{\kappa}} \phi_{k-p} & a_{-k} \delta_{k,p} \end{bmatrix}}{\det \begin{bmatrix} a_k \delta_{k,p} & 0 \\ 0 & a_{-k} \delta_{k,p} \end{bmatrix}} e^{-\sum_q |\phi_q|^2} \prod_q \frac{d\phi_q d\bar{\phi}_q}{\pi} \quad (5.15)$$

$$\frac{1}{\kappa} \langle \bar{\psi}_{t\sigma} \psi_{t\sigma} \rangle = \int \begin{bmatrix} a_k \delta_{k,p} & \frac{ig}{\sqrt{\kappa}} \bar{\phi}_{p-k} \\ \frac{ig}{\sqrt{\kappa}} \phi_{k-p} & a_{-k} \delta_{k,p} \end{bmatrix}^{-1}_{t\sigma,t\sigma} dP(\phi) \quad (5.16)$$

$$\Lambda(q) = \int |\phi_q|^2 \, dP(\phi) - 1 \quad (5.17)$$

Proof: We have

$$\frac{1}{\kappa} \langle \bar{\psi}_{t\sigma} \psi_{t\sigma} \rangle = \frac{\partial}{\partial s_{t\sigma}}\Big|_{s=0, \lambda_q = \lambda} \log Z(\{s_{k\sigma}\}, \{\lambda_q\}) \quad (5.18)$$

$$\Lambda_q = -\lambda \frac{\partial}{\partial \lambda_q}\Big|_{s=0, \lambda_q = \lambda} \log Z(\{s_{k\sigma}\}, \{\lambda_q\}) \quad (5.19)$$

if we define

$$Z(\{s_{k\sigma}\}, \{\lambda_q\}) = \int e^{-\frac{1}{\kappa^3} \sum_{kpq} \lambda_q \bar{\psi}_{k\uparrow} \bar{\psi}_{q-k,\downarrow} \psi_{p\uparrow} \psi_{q-p,\downarrow}} \times \quad (5.20)$$

$$\prod_{k\sigma} \frac{\kappa}{ik_0 - e_{\mathbf{k}}} e^{-\frac{1}{\kappa} \sum_{k\sigma} (a_k - s_{k\sigma}) \bar{\psi}_{k\sigma} \psi_{k\sigma}} \prod_{k\sigma} d\psi_{k\sigma} d\bar{\psi}_{k\sigma}$$

Using (5.12) and (5.13) we get

$$e^{-\frac{1}{\kappa^3} \sum_{kpq} \lambda_q \bar{\psi}_{k\uparrow} \bar{\psi}_{q-k,\downarrow} \psi_{p\uparrow} \psi_{q-p,\downarrow}} = \quad (5.21)$$

$$\int e^{i \sum_q \left[\phi_q \left(\frac{\lambda_q}{\kappa^3}\right)^{\frac{1}{2}} \sum_k \psi_{k\uparrow} \psi_{q-k\downarrow} + \bar{\phi}_q \left(\frac{\lambda_q}{\kappa^3}\right)^{\frac{1}{2}} \sum_k \bar{\psi}_{k\uparrow} \bar{\psi}_{q-k\downarrow}\right]} e^{-\sum_q |\phi_q|^2} \prod_q \frac{d\phi_q d\bar{\phi}_q}{\pi}$$

which gives

$$Z(\{s_{k\sigma}\}, \{\lambda_q\}) \tag{5.22}$$

$$= \int\int \exp\left\{ -\frac{1}{\kappa} \sum_{k,p} \left[\psi_{k\uparrow} i \frac{g_{k-p}}{\sqrt{\kappa}} \phi_{k-p} \psi_{-p\downarrow} + \bar\psi_{-p\uparrow} i \frac{g_{p-k}}{\sqrt{\kappa}} \bar\phi_{p-k} \bar\psi_{k\downarrow} + \right. \right.$$

$$\left. \left. + \bar\psi_{k\uparrow}(a_k - s_{k\uparrow})\delta_{k,p}\psi_{p\uparrow} + \bar\psi_{-k\downarrow}(a_{-k} - s_{-k\downarrow})\delta_{k,p}\psi_{-p\downarrow} \right] \right\} \times$$

$$\prod_{k\sigma} \frac{\kappa}{a_k} \prod_{k\sigma} d\psi_{k\sigma} d\bar\psi_{k\sigma} \, e^{-\sum_q |\phi_q|^2} \prod_q \frac{d\phi_q d\bar\phi_q}{\pi}$$

The exponent in (5.22) can be written as

$$-\frac{1}{\kappa} \sum_{k,p} (\psi_{k\uparrow}, \bar\psi_{-k\downarrow}) \begin{pmatrix} -(a_k - s_{k\uparrow})\delta_{k,p} & i\frac{g_{k-p}}{\sqrt{\kappa}}\phi_{k-p} \\ -i\frac{g_{p-k}}{\sqrt{\kappa}}\bar\phi_{p-k} & (a_{-k} - s_{-k\downarrow})\delta_{k,p} \end{pmatrix} \begin{pmatrix} \bar\psi_{p\uparrow} \\ \psi_{-p\downarrow} \end{pmatrix}$$

Because of

$$d\psi_{k\uparrow} d\bar\psi_{k\uparrow} d\psi_{-k\downarrow} d\bar\psi_{-k\downarrow} = -d\psi_{k\uparrow} d\bar\psi_{k\uparrow} d\bar\psi_{-k\downarrow} d\psi_{-k\downarrow}$$

the Grassmann integral gives

$$\det\left[\frac{1}{\kappa} \begin{pmatrix} (a_k - s_{k\uparrow})\delta_{k,p} & i\frac{g_{k-p}}{\sqrt{\kappa}}\phi_{k-p} \\ i\frac{g_{p-k}}{\sqrt{\kappa}}\bar\phi_{p-k} & (a_{-k} - s_{-k\downarrow})\delta_{k,p} \end{pmatrix} \right] \prod_{k\sigma} \frac{\kappa}{a_k} = \tag{5.23}$$

$$\det\begin{bmatrix} (a_k - s_{k\uparrow})\delta_{k,p} & i\frac{g_{k-p}}{\sqrt{\kappa}}\phi_{k-p} \\ i\frac{g_{p-k}}{\sqrt{\kappa}}\bar\phi_{p-k} & (a_{-k} - s_{-k\downarrow})\delta_{k,p} \end{bmatrix} \Big/ \det\begin{bmatrix} a_k\delta_{k,p} & 0 \\ 0 & a_{-k}\delta_{k,p} \end{bmatrix}$$

which results in

$$Z(\{s_{k\sigma}\}, \{\lambda_q\}) = \tag{5.24}$$

$$\int \frac{\det\begin{bmatrix} (a_k - s_{k\uparrow})\delta_{k,p} & i\frac{g_{k-p}}{\sqrt{\kappa}}\phi_{k-p} \\ i\frac{g_{p-k}}{\sqrt{\kappa}}\bar\phi_{p-k} & (a_{-k} - s_{-k\downarrow})\delta_{k,p} \end{bmatrix}}{\det\begin{bmatrix} a_k\delta_{k,p} & 0 \\ 0 & a_{-k}\delta_{k,p} \end{bmatrix}} e^{-\sum_q |\phi_q|^2} \prod_q \frac{d\phi_q d\bar\phi_q}{\pi}$$

Now, using

$$\frac{d}{dt} \det\begin{bmatrix} | & & | \\ \vec{x}_1(t) & \cdots & \vec{x}_n(t) \\ | & & | \end{bmatrix} = \sum_{i=1}^n \det\begin{bmatrix} | & & | & & | \\ \vec{x}_1(t) & \cdots & \frac{d}{dt}\vec{x}_i(t) & \cdots & \vec{x}_n(t) \\ | & & | & & | \end{bmatrix} \tag{5.25}$$

and observing that, if we fix a column labelled by p and letting k label the entries of the column vector,

$$-\frac{d}{ds_{t\uparrow}} \begin{pmatrix} (a_k - s_{k\uparrow})\delta_{k,p} \\ i\frac{g}{\sqrt{\kappa}}\bar\phi_{p-k} \end{pmatrix} = \begin{pmatrix} \delta_{k,t}\delta_{k,p} \\ 0 \end{pmatrix} \tag{5.26}$$

we get by Cramer's rule

$$\frac{\frac{d}{ds_{t\sigma}}\big|_{s=0} \det \begin{bmatrix} (a_k - s_{k\uparrow})\delta_{k,p} & i\frac{g}{\sqrt{\kappa}}\phi_{k-p} \\ i\frac{g}{\sqrt{\kappa}}\bar{\phi}_{p-k} & (a_{-k} - s_{-k\downarrow})\delta_{k,p} \end{bmatrix}}{\det \begin{bmatrix} (a_k - s_{k\uparrow})\delta_{k,p} & i\frac{g}{\sqrt{\kappa}}\phi_{k-p} \\ i\frac{g}{\sqrt{\kappa}}\bar{\phi}_{p-k} & (a_{-k} - s_{-k\downarrow})\delta_{k,p} \end{bmatrix}} = \begin{bmatrix} a_k\delta_{k,p} & i\frac{g}{\sqrt{\kappa}}\phi_{k-p} \\ i\frac{g}{\sqrt{\kappa}}\bar{\phi}_{p-k} & a_{-k}\delta_{k,p} \end{bmatrix}^{-1}_{t\sigma,t\sigma}$$

which results in (5.16). To obtain (5.17), we substitute variables

$$Z(\{\lambda_q\}) = \int \frac{\det \begin{bmatrix} a_k\delta_{k,p} & i\frac{1}{\sqrt{\kappa}}\phi_{k-p} \\ i\frac{1}{\sqrt{\kappa}}\bar{\phi}_{p-k} & a_{-k}\delta_{k,p} \end{bmatrix}}{\det \begin{bmatrix} a_k\delta_{k,p} & 0 \\ 0 & a_{-k}\delta_{k,p} \end{bmatrix}} e^{-\sum_q \frac{|\phi_q|^2}{\lambda_q}} \prod_q \frac{d\phi_q d\bar{\phi}_q}{\pi\lambda_q} \tag{5.27}$$

to obtain

$$-\Lambda_q = \lambda \frac{\partial}{\partial \lambda_q}\big|_{\lambda_q=\lambda} \log Z(\{\lambda_q\})$$

$$= \lambda \frac{\partial}{\partial \lambda_q}\big|_{\lambda_q=\lambda} \log \int \frac{\det \begin{bmatrix} a_k\delta_{k,p} & i\frac{1}{\sqrt{\kappa}}\phi_{k-p} \\ i\frac{1}{\sqrt{\kappa}}\bar{\phi}_{p-k} & a_{-k}\delta_{k,p} \end{bmatrix}}{\det \begin{bmatrix} a_k\delta_{k,p} & 0 \\ 0 & a_{-k}\delta_{k,p} \end{bmatrix}} e^{-\sum_q \frac{|\phi_q|^2}{\lambda_q}} \prod_q \frac{d\phi_q d\bar{\phi}_q}{\pi\lambda_q}$$

$$= \lambda \frac{\int \left(\frac{|\phi_q|^2}{\lambda_q^2} - \frac{1}{\lambda_q}\right) \det \begin{bmatrix} a_k\delta_{k,p} & i\frac{1}{\sqrt{\kappa}}\phi_{k-p} \\ i\frac{1}{\sqrt{\kappa}}\bar{\phi}_{p-k} & a_{-k}\delta_{k,p} \end{bmatrix} e^{-\sum_q \frac{|\phi_q|^2}{\lambda_q}} \prod_q \frac{d\phi_q d\bar{\phi}_q}{\pi\lambda_q}}{\int \det \begin{bmatrix} a_k\delta_{k,p} & i\frac{1}{\sqrt{\kappa}}\phi_{k-p} \\ i\frac{1}{\sqrt{\kappa}}\bar{\phi}_{p-k} & a_{-k}\delta_{k,p} \end{bmatrix} e^{-\sum_q \frac{|\phi_q|^2}{\lambda_q}} \prod_q \frac{d\phi_q d\bar{\phi}_q}{\pi\lambda_q}}$$

$$= \frac{\int (|\phi_q|^2 - 1) \det \begin{bmatrix} a_k\delta_{k,p} & i\frac{g}{\sqrt{\kappa}}\phi_{k-p} \\ i\frac{g}{\sqrt{\kappa}}\bar{\phi}_{p-k} & a_{-k}\delta_{k,p} \end{bmatrix} e^{-\sum_q |\phi_q|^2} \prod_q \frac{d\phi_q d\bar{\phi}_q}{\pi}}{\int \det \begin{bmatrix} a_k\delta_{k,p} & i\frac{g}{\sqrt{\kappa}}\phi_{k-p} \\ i\frac{g}{\sqrt{\kappa}}\bar{\phi}_{p-k} & a_{-k}\delta_{k,p} \end{bmatrix} e^{-\sum_q |\phi_q|^2} \prod_q \frac{d\phi_q d\bar{\phi}_q}{\pi}}$$

$$= \int (|\phi_q|^2 - 1) \, dP(\phi)$$

which proves the theorem. ∎

5.2   The Effective Potential

In this section we find the global minimum of the full effective potential for the many-electron system with delta interaction and compute the second order Taylor expansion around it. The results of this section are basically the content of [45].

In the preceding section we showed that the $\langle \bar{\psi}\psi \rangle$ two-point function for the many-electron system has the following representation

$$\frac{1}{\kappa}\langle \bar{\psi}_{t\sigma}\psi_{t\sigma}\rangle = \frac{\int G(t\sigma, \phi)\, e^{-V_{\text{eff}}(\phi)} \prod_q d\phi_q d\bar{\phi}_q}{\int e^{-V_{\text{eff}}(\phi)} \prod_q d\phi_q d\bar{\phi}_q} \tag{5.28}$$

where

$$G(t\sigma, \phi) = \begin{bmatrix} a_k\delta_{k,p} & \frac{ig}{\sqrt{\kappa}}\,\bar{\phi}_{p-k} \\ \frac{ig}{\sqrt{\kappa}}\,\phi_{k-p} & a_{-k}\delta_{k,p} \end{bmatrix}^{-1}_{t\sigma, t\sigma} \tag{5.29}$$

and the effective potential is given by

$$V_{\text{eff}}(\{\phi_q\}) = \sum_q |\phi_q|^2 - \log\det \begin{bmatrix} \delta_{k,p} & \frac{ig}{\sqrt{\kappa}}\,\frac{\bar{\phi}_{p-k}}{a_k} \\ \frac{ig}{\sqrt{\kappa}}\,\frac{\phi_{k-p}}{a_{-k}} & \delta_{k,p} \end{bmatrix} \tag{5.30}$$

To obtain (5.30) from (5.14), we divided the determinant and the normalization factor Z by $\det \begin{bmatrix} a_k\delta_{k,p} & 0 \\ 0 & a_{-k}\delta_{k,p} \end{bmatrix}$. In general, V_{eff} is complex. Its real part is given by

$$\text{Re}\, V_{\text{eff}}(\{\phi_q\}) = \sum_q |\phi_q|^2 - \log\left|\det \begin{bmatrix} \delta_{k,p} & \frac{ig}{\sqrt{\kappa}}\,\frac{\bar{\phi}_{p-k}}{a_k} \\ \frac{ig}{\sqrt{\kappa}}\,\frac{\phi_{k-p}}{a_{-k}} & \delta_{k,p} \end{bmatrix}\right| \tag{5.31}$$

Apparently the evaluation of (5.28) is a complicated problem. Let $V_{\text{eff}}^{\text{min}}$ be the global minimum of $V_{\text{eff}}(\{\phi_q\})$. Since

$$\frac{\int G(t\sigma, \phi)\, e^{-V_{\text{eff}}(\phi)} \prod_q d\phi_q d\bar{\phi}_q}{\int e^{-V_{\text{eff}}(\phi)} \prod_q d\phi_q d\bar{\phi}_q} = \frac{\int G(t\sigma, \phi)\, e^{-(V_{\text{eff}}(\phi)-V_{\text{eff}}^{\text{min}})} \prod_q d\phi_q d\bar{\phi}_q}{\int e^{-(V_{\text{eff}}(\phi)-V_{\text{eff}}^{\text{min}})} \prod_q d\phi_q d\bar{\phi}_q}$$

the relevant contributions to the functional integral should come from configurations $\{\phi_q\}$ which minimize the real part of the effective potential. Thus as a first approximation one may start with the saddle point approximation which simply evaluates (5.28) by substituting the integrand by its value at the global minimum, or, if this is not unique, averages the integrand over all configurations which minimize the effective potential. In the following theorem we rigorously prove that the global minimum of the real part of the effective potential of the many-electron system with attractive delta interaction is given by

$$\phi_q = \sqrt{\beta L^d}\, r_0 e^{i\theta}\, \delta_{q,0} \tag{5.32}$$

where θ is an arbitrary phase and $|\Delta|^2 := \lambda r_0^2$ is a solution of the BCS gap equation

$$\frac{\lambda}{\beta L^d} \sum_k \frac{1}{k_0^2 + e_k^2 + |\Delta|^2} = 1 \tag{5.33}$$

In the next chapter we discuss the BCS model and show that the saddle point approximation with the minimum (5.32) corresponds to the BCS approximation.

Theorem 5.2.1 *Let* ReV *be the real part of the effective potential for the many-electron system with attractive delta interaction given by (5.31). Let* e_k *satisfy* $e_k = e_{-k}$ *and:* $\forall k\ e_k = e_{k+q} \Rightarrow q = 0$. *Then all global minima of* ReV *are given by*

$$\phi_q = \delta_{q,0} \sqrt{\beta L^d}\, r_0\, e^{i\theta}, \qquad \theta \in [0, 2\pi] \quad arbitrary \tag{5.34}$$

where r_0 *is a solution of the BCS equation (5.33) or, equivalently, the global minimum of the function*

$$V_{\text{BCS}}(\rho) = V_{\text{eff}}(\{\delta_{q,0}\sqrt{\beta L^d}\rho\, e^{i\theta}\}) \tag{5.35}$$

$$= \beta L^d \rho^2 - \sum_k \log\left[1 + \frac{\lambda \rho^2}{k_0^2 + e_k^2}\right] = \beta L^d \rho^2 - \sum_k \log\left[\frac{\cosh(\frac{\beta}{2}\sqrt{e_k^2 + \lambda \rho^2})}{\cosh\frac{\beta}{2}e_k}\right]$$

More specifically, there is the bound

$$\text{ReV}(\{\phi_q\}) \geq \tag{5.36}$$

$$V_{\text{BCS}}(\|\phi\|) - \min_k \left\{ \log\left[\prod_{q\neq 0}\left(1 - \frac{|\frac{\lambda}{\beta L^d}\sum_p \phi_p \bar{\phi}_{p+q}|^2}{(|a_k|^2 + \lambda\|\phi\|^2)(|a_{k-q}|^2 + \lambda\|\phi\|^2)}\right)^{\frac{1}{2}}\right]\right.$$

$$\left. + \log\left[\prod_{q\neq 0}\left(1 - \frac{\frac{\lambda}{\beta L^d}|\phi_q|^2|a_k - a_{k-q}|^2}{(|a_k|^2 + \lambda\|\phi\|^2)(|a_{k-q}|^2 + \lambda\|\phi\|^2)}\right)^{\frac{1}{2}}\right]\right\}$$

where $\|\phi\|^2 := \frac{1}{\beta L^d}\sum_q |\phi_q|^2$ *and* $|a_k|^2 := k_0^2 + e_k^2$. *In particular,*

$$\text{ReV}(\{\phi_q\}) \geq V_{\text{BCS}}(\|\phi\|) \tag{5.37}$$

since the products in (5.36) are less or equal 1.

Proof: Suppose first that (5.36) holds. For each q, the round brackets in (5.36) are between 0 and 1 which means that $-\log(\Pi_q \cdots)$ is positive. Thus

$$\mathrm{Re}V(\{\phi_q\}) \geq V_{\mathrm{BCS}}(\|\phi\|) \geq V_{\mathrm{BCS}}(r_0) \tag{5.38}$$

which proves that $\phi_q = \delta_{q,0}\sqrt{\beta L^d}\, r_0\, e^{i\theta}$ are indeed global minima of $\mathrm{Re}V$. On the other hand, if a configuration $\{\phi_q\}$ is a global minimum, then the logarithms in (5.36) must be zero for all k which in particular means that for all $q \neq 0$

$$\sum_p \phi_p \bar{\phi}_{p+q} = 0 \tag{5.39}$$

and for all k and $q \neq 0$

$$|\phi_q|^2 |a_k - a_{k-q}|^2 = |\phi_q|^2 [q_0^2 + (e_{\mathbf{k}} - e_{\mathbf{k}-\mathbf{q}})^2] = 0 \tag{5.40}$$

which implies $\phi_q = 0$ for all $q \neq 0$. It remains to prove (5.36).

To this end, we write (recall that $\kappa = \beta L^d$ and $C_k = \frac{1}{a_k} = \frac{1}{ik_0 - e_{\mathbf{k}}}$)

$$\det \begin{bmatrix} \mathrm{Id} & \frac{ig}{\sqrt{\kappa}} \bar{C}\phi^* \\ \frac{ig}{\sqrt{\kappa}} C\phi & \mathrm{Id} \end{bmatrix} = \det \begin{bmatrix} | & | \\ \vec{b}_k & \vec{b}'_{k'} \\ | & | \end{bmatrix} \tag{5.41}$$

where (k, k' fixed, p labels the vector components)

$$\vec{b}_k = \begin{pmatrix} \delta_{k,p} \\ \\ \frac{ig}{\sqrt{\kappa}} \frac{\phi_{k-p}}{a_k} \end{pmatrix} \qquad \vec{b}'_{k'} = \begin{pmatrix} \frac{ig}{\sqrt{\kappa}} \frac{\bar{\phi}_{p-k'}}{\bar{a}_{k'}} \\ \\ \delta_{k',p} \end{pmatrix} \tag{5.42}$$

If $|\vec{b}_k|$ denotes the euclidean norm of $\vec{b}_k$, then we have

$$|\vec{b}_k|^2 = 1 + \frac{\lambda}{\beta L^d} \sum_p \frac{|\phi_{k-p}|^2}{|a_k|^2} = 1 + \lambda \frac{\|\phi\|^2}{|a_k|^2} = |\vec{b}'_k|^2 \tag{5.43}$$

Therefore one obtains, if $\vec{e}_k = \frac{\vec{b}_k}{|\vec{b}_k|}$, $\vec{e}'_k = \frac{\vec{b}'_k}{|\vec{b}'_k|}$

$$\det \begin{bmatrix} | & | \\ \vec{b}_k & \vec{b}'_{k'} \\ | & | \end{bmatrix} = \prod_k \left\{ 1 + \lambda \frac{\|\phi\|^2}{|a_k|^2} \right\} \det \begin{bmatrix} | & | \\ \vec{e}_k & \vec{e}'_{k'} \\ | & | \end{bmatrix} \tag{5.44}$$

From this the inequality (5.38) already follows since the determinant on the right hand side of (5.44) is less or equal 1. To obtain (5.36), we choose a fixed but arbitrary momentum t and orthogonalize all vectors $\vec{e}_k$, $\vec{e}'_{k'}$ in the determinant with respect to e_t. That is, we write

$$\det \begin{bmatrix} | & | & | \\ \vec{e}_t & \vec{e}_k & \vec{e}'_{k'} \\ | & | & | \end{bmatrix} = \det \begin{bmatrix} | & | & | \\ \vec{e}_t & \vec{e}_k - (\vec{e}_k, \vec{e}_t)\vec{e}_t & \vec{e}'_{k'} - (\vec{e}'_{k'}, \vec{e}_t)\vec{e}_t \\ | & | & | \end{bmatrix} \tag{5.45}$$

Finally we apply Hadamard's inequality,

$$|\det F| \leq \prod_{j=1}^{n} |\vec{f}_j| = \left\{ \prod_{j=1}^{n} \sum_{i=1}^{n} |f_{ij}|^2 \right\}^{\frac{1}{2}} \tag{5.46}$$

if $F = (f_{ij})_{1 \leq i,j \leq n}$ is a complex matrix, to the determinant on the right hand side of (5.45). Since

$$|\vec{e}_k - (\vec{e}_k, \vec{e}_t)\vec{e}_t|^2 = 1 - |(\vec{e}_k, \vec{e}_t)|^2$$

one obtains

$$\left| \det \left[\begin{array}{ccc} | & & | \\ \vec{e}_k - (\vec{e}_k, \vec{e}_t)\vec{e}_t & \vec{e}'_{k'} - (\vec{e}'_{k'}, \vec{e}_t)\vec{e}_t \\ | & & | \end{array} \right] \right|$$

$$\leq |\vec{e}_t| \prod_{\substack{k \in M_\nu \\ k \neq t}} (1 - |(\vec{e}_k, \vec{e}_t)|^2)^{\frac{1}{2}} \prod_{k \in M_\nu} (1 - |(\vec{e}'_k, \vec{e}_t)|^2)^{\frac{1}{2}} \tag{5.47}$$

or with (5.41) and (5.44)

$$\mathrm{Re}V(\{\phi_q\}) = \sum_q |\phi_q|^2 - \log \left| \det \left[\begin{array}{cc} Id & \frac{ig}{\sqrt{\kappa}} \phi^* \bar{C} \\ \frac{ig}{\sqrt{\kappa}} \phi C & Id \end{array} \right] \right| \tag{5.48}$$

$$= \kappa \|\phi\|^2 - \sum_k \log \left\{ 1 + \lambda \frac{\|\phi\|^2}{|a_k|^2} \right\} - \log \left| \det \left[\begin{array}{cc} | & | \\ \vec{e}_k & \vec{e}'_{k'} \\ | & | \end{array} \right] \right|$$

$$= V_{\mathrm{BCS}}(\|\phi\|) - \log \left| \det \left[\begin{array}{cc} | & | \\ \vec{e}_k & \vec{e}'_{k'} \\ | & | \end{array} \right] \right|$$

$$\geq V_{\mathrm{BCS}}(\|\phi\|) - \log \left\{ \prod_{\substack{k \in M_\nu \\ k \neq t}} (1 - |(\vec{e}_k, \vec{e}_t)|^2)^{\frac{1}{2}} \prod_{k \in M_\nu} (1 - |(\vec{e}'_k, \vec{e}_t)|^2)^{\frac{1}{2}} \right\}$$

Finally one has

$$(\vec{b}_k, \vec{b}_t) = \sum_p \frac{ig}{\sqrt{\kappa}} \frac{\phi_{k-p}}{a_k} \overline{\frac{ig}{\sqrt{\kappa}} \frac{\phi_{t-p}}{a_t}}$$

which gives

$$|(\vec{e}_k, \vec{e}_t)|^2 = \frac{\left| \frac{\lambda}{\kappa} \sum_p \phi_{k-p} \bar{\phi}_{t-p} \right|^2}{(|a_k|^2 + \lambda \|\phi\|^2)(|a_t|^2 + \lambda \|\phi\|^2)} = \frac{\left| \frac{\lambda}{\kappa} \sum_p \phi_p \bar{\phi}_{t-k+p} \right|^2}{(|a_k|^2 + \lambda \|\phi\|^2)(|a_t|^2 + \lambda \|\phi\|^2)} \tag{5.49}$$

and

$$(\vec{b}'_k, \vec{b}_t) = \frac{ig}{\sqrt{\kappa}} \frac{\bar{\phi}_{t-k}}{\bar{a}_k} + \overline{\frac{ig}{\sqrt{\kappa}} \frac{\phi_{t-k}}{a_t}} = \frac{ig}{\sqrt{\kappa}} \bar{\phi}_{t-k} \left(\frac{1}{\bar{a}_k} - \frac{1}{a_t} \right) \tag{5.50}$$

which gives

$$|(\vec{e}'_k, \vec{e}_t)|^2 = \frac{\frac{\lambda}{\kappa} |\phi_{t-k}|^2 |a_t - a_k|^2}{(|a_k|^2 + \lambda \|\phi\|^2)(|a_t|^2 + \lambda \|\phi\|^2)} \tag{5.51}$$

Substituting (5.49) and (5.51) in (5.48) gives, substituting $k \to q = t - k$

$$\mathrm{Re}V(\{\phi_q\}) \tag{5.52}$$

$$\geq V_{\mathrm{BCS}}(\|\phi\|) - \log\left\{ \prod_{\substack{k \in M_\nu \\ k \neq t}} (1 - |(\vec{e}_k, \vec{e}_t)|^2)^{\frac{1}{2}} \prod_{k \in M_\nu} (1 - |(\vec{e}_k', \vec{e}_t)|^2)^{\frac{1}{2}} \right\}$$

$$\overset{k=t-q}{=} V_{\mathrm{BCS}}(\|\phi\|) - \log\left\{ \prod_{q \neq 0} (1 - |(\vec{e}_{t-q}, \vec{e}_t)|^2)^{\frac{1}{2}} \prod_{q \neq 0} (1 - |(\vec{e}_{t-q}', \vec{e}_t)|^2)^{\frac{1}{2}} \right\}$$

Since t was arbitrary, we can take the maximum of the right hand side of (5.52) with respect to t which proves the theorem. ∎

Thus in the saddle point approximation we obtain

$$\frac{1}{\kappa}\langle \bar{\psi}_{k\uparrow} \psi_{k\uparrow} \rangle = \frac{1}{2\pi} \int_0^{2\pi} d\theta \left[\begin{array}{cc} a_k \delta_{k,p} & igr_0 e^{-i\theta} \delta_{p-k,0} \\ igr_0 e^{i\theta} \delta_{k-p,0} & a_{-k} \delta_{k,p} \end{array} \right]_{k\uparrow, k\uparrow}^{-1}$$

$$= \frac{a_{-k}}{|a_k|^2 + \lambda r_0^2} = \frac{-ik_0 - e_{\mathbf{k}}}{k_0^2 + e_{\mathbf{k}}^2 + |\Delta|^2} \tag{5.53}$$

which gives, at zero temperature,

$$\frac{1}{L^d}\langle a_{\mathbf{k}\uparrow}^+ a_{\mathbf{k}\uparrow} \rangle = \lim_{\epsilon \nearrow 0} \int \frac{dk_0}{2\pi} \frac{(-ik_0 - e_{\mathbf{k}})e^{ik_0\epsilon}}{k_0^2 + e_{\mathbf{k}}^2 + |\Delta|^2}$$

$$= \frac{1}{2}\left(1 - \frac{e_{\mathbf{k}}}{\sqrt{e_{\mathbf{k}}^2 + |\Delta|^2}} \right) \tag{5.54}$$

This coincides with the result obtained from the BCS model which we discuss in the next chapter.

To compute corrections to the saddle point approximation one Taylor expands the effective potential around the global minimum up to second order. In this approximation the integration measure becomes Gaussian and quantities like $\Lambda(q) = \langle |\phi_q|^2 \rangle$ can be computed. Usually it is not clear to what extent this gives the right answer. In fact, in section 10.4 we will argue that the result for $\Lambda(q)$ obtained in this way is most likely a wrong one in one (and two) dimensions, while it seems to be the right answer in three space dimensions. In the following theorem the second order Taylor expansion of the effective potential for the many-electron system with attractive delta interaction around the global minimum (5.34) is given.

Theorem 5.2.2 *Let V be the effective potential (5.30), let $\kappa = \beta L^d$ and let*

$$\xi_q = \phi_q - \delta_{q,0}\sqrt{\kappa}\, r_0\, e^{i\theta_0} = \begin{cases} (\rho_0 - \sqrt{\kappa}\, r_0)e^{i\theta_0} & \text{for } q = 0 \\ \rho_q e^{i\theta_q} & \text{for } q \neq 0. \end{cases} \tag{5.55}$$

Then

$$V(\{\phi_q\}) = V_{\min} + 2\beta_0 \left(\rho_0 - \sqrt{\kappa}\, r_0\right)^2 + \sum_{q \neq 0} (\alpha_q + i\gamma_q)\rho_q^2 \qquad (5.56)$$

$$+ \tfrac{1}{2} \sum_{q \neq 0} \beta_q \, |e^{-i\theta_0}\phi_q + e^{i\theta_0}\bar{\phi}_{-q}|^2 + O(\xi^3)$$

where, if $E_k^2 = k_0^2 + e_{\mathbf{k}}^2 + \lambda r_0^2$,

$$\alpha_q = \tfrac{1}{2}\tfrac{\lambda}{\kappa} \sum_k \frac{q_0^2 + (e_{\mathbf{k}} - e_{\mathbf{k}-\mathbf{q}})^2}{E_k^2 E_{k-q}^2} > 0, \quad \beta_q = \tfrac{\lambda}{\kappa} \sum_k \frac{\lambda r_0^2}{E_k^2 E_{k-q}^2} > 0, \qquad (5.57)$$

$$\gamma_q = -\tfrac{\lambda}{\kappa} \sum_k \frac{k_0 e_{\mathbf{k}-\mathbf{q}} - (k_0 - q_0)e_{\mathbf{k}}}{E_k^2 E_{k-q}^2} \in \mathbb{R} \qquad (5.58)$$

and

$$V_{\min} = \kappa \left(r_0^2 - \tfrac{1}{\kappa} \sum_{\mathbf{k}} \log \left[\frac{\cosh(\tfrac{\beta}{2}\sqrt{e_{\mathbf{k}}^2 + \lambda r_0^2})}{\cosh \tfrac{\beta}{2} e_{\mathbf{k}}} \right] \right). \qquad (5.59)$$

Proof: We abbreviate $\kappa = \beta L^d$ and write

$$V(\{\phi_q\}) = \sum_q |\phi_q|^2 - \log \frac{\det \begin{bmatrix} A & \frac{ig}{\sqrt{\kappa}}\phi^* \\ \frac{ig}{\sqrt{\kappa}}\phi & \bar{A} \end{bmatrix}}{\det \begin{bmatrix} A & 0 \\ 0 & \bar{A} \end{bmatrix}} \qquad (5.60)$$

where $A = C^{-1} = (\delta_{k,p} a_k)_{k,p \in M_\kappa}$ and $a_k := 1/C_k = ik_0 - e_{\mathbf{k}}$. Then

$$V(\{\phi_q\}) - V(\{\sqrt{\kappa}\,\delta_{q,0}\, r_0 e^{i\theta_0}\}) =$$

$$\sum_q \rho_q^2 - \kappa r_0^2 - \log \frac{\det \begin{bmatrix} A & \frac{ig}{\sqrt{\kappa}}\phi^* \\ \frac{ig}{\sqrt{\kappa}}\phi & \bar{A} \end{bmatrix}}{\det \begin{bmatrix} A & igr_0\, e^{-i\theta_0} \\ igr_0\, e^{i\theta_0} & \bar{A} \end{bmatrix}} \qquad (5.61)$$

where $igr_0\, e^{i\theta_0} \equiv igr_0\, e^{i\theta_0}\, Id$ in the determinant above. Since

$$\begin{bmatrix} A & igr_0\, e^{-i\theta_0} \\ igr_0\, e^{i\theta_0} & \bar{A} \end{bmatrix}^{-1} = \begin{bmatrix} \frac{\bar{a}_k \delta_{k,p}}{|a_k|^2 + \lambda r_0^2} & -\frac{igr_0\, e^{-i\theta_0} \delta_{k,p}}{|a_k|^2 + \lambda r_0^2} \\ -\frac{igr_0\, e^{i\theta_0} \delta_{k,p}}{|a_k|^2 + \lambda r_0^2} & \frac{a_k \delta_{k,p}}{|a_k|^2 + \lambda r_0^2} \end{bmatrix}$$

$$\equiv \frac{1}{|a|^2 + \lambda r_0^2} \begin{bmatrix} \bar{A} & -igr_0\, e^{-i\theta_0} \\ -igr_0\, e^{i\theta_0} & A \end{bmatrix} \qquad (5.62)$$

and because of

$$
\begin{bmatrix} A & \frac{ig}{\sqrt{\kappa}}\phi^* \\ \frac{ig}{\sqrt{\kappa}}\phi & \bar{A} \end{bmatrix} = \begin{bmatrix} A & igr_0\, e^{-i\theta_0} \\ igr_0\, e^{i\theta_0} & \bar{A} \end{bmatrix} +
$$

$$
\begin{bmatrix} 0 & \frac{ig}{\sqrt{\kappa}}\phi^* - igr_0\, e^{-i\theta_0} \\ \frac{ig}{\sqrt{\kappa}}\phi - igr_0\, e^{i\theta_0} & 0 \end{bmatrix}
$$

$$
= \begin{bmatrix} A & ig\bar{\gamma} \\ ig\gamma & \bar{A} \end{bmatrix} + \begin{bmatrix} 0 & ig\,\xi^* \\ ig\,\xi & 0 \end{bmatrix} \tag{5.63}
$$

where $\gamma = r_0\, e^{i\theta_0}$ and $\xi = (\xi_{k-p})_{k,p}$ is given by (5.55), the quotient of determinants in (5.60) is given by

$$
\det\left[Id + \frac{1}{|a|^2+\lambda r_0^2}\begin{pmatrix} \bar{A} & -ig\bar{\gamma} \\ -ig\gamma & A \end{pmatrix}\begin{pmatrix} 0 & ig\,\xi^* \\ ig\,\xi & 0 \end{pmatrix}\right]
$$

$$
= \det\left[Id + \begin{pmatrix} \lambda\frac{\bar{\gamma}}{|a|^2+\lambda r_0^2}\,\xi & ig\frac{\bar{A}}{|a|^2+\lambda r_0^2}\,\xi^* \\ ig\frac{A}{|a|^2+\lambda r_0^2}\,\xi & \lambda\frac{\gamma}{|a|^2+\lambda r_0^2}\,\xi^* \end{pmatrix}\right] \tag{5.64}
$$

Since

$$
\log\det[Id+B] = Tr\log[Id+B] = \sum_{n=1}^{\infty} \frac{(-1)^{n+1}}{n} Tr B^n
$$

$$
= Tr\, B - \tfrac{1}{2}Tr\, B^2 + \tfrac{1}{3}Tr\, B^3 - + \cdots \tag{5.65}
$$

one obtains to second order in ξ:

$$
\log\det\left[Id + \begin{pmatrix} \lambda\frac{\bar{\gamma}}{|a|^2+\lambda r_0^2}\,\xi & ig\frac{\bar{A}}{|a|^2+\lambda r_0^2}\,\xi^* \\ ig\frac{A}{|a|^2+\lambda r_0^2}\,\xi & \lambda\frac{\gamma}{|a|^2+\lambda r_0^2}\,\xi^* \end{pmatrix}\right] \tag{5.66}
$$

$$
= Tr\begin{pmatrix} \lambda\frac{\bar{\gamma}}{|a|^2+\lambda r_0^2}\,\xi & ig\frac{\bar{A}}{|a|^2+\lambda r_0^2}\,\xi^* \\ ig\frac{A}{|a|^2+\lambda r_0^2}\,\xi & \lambda\frac{\gamma}{|a|^2+\lambda r_0^2}\,\xi^* \end{pmatrix}
$$

$$
- \tfrac{1}{2}Tr\left\{ \begin{pmatrix} \lambda\frac{\bar{\gamma}}{|a|^2+\lambda r_0^2}\,\xi & ig\frac{\bar{A}}{|a|^2+\lambda r_0^2}\,\xi^* \\ ig\frac{A}{|a|^2+\lambda r_0^2}\,\xi & \lambda\frac{\gamma}{|a|^2+\lambda r_0^2}\,\xi^* \end{pmatrix}^2 \right\} + O(\xi^3)
$$

$$
= Tr\frac{\lambda\bar{\gamma}}{|a|^2+\lambda r_0^2}\,\xi + Tr\frac{\lambda\gamma}{|a|^2+\lambda r_0^2}\,\xi^*
$$

$$
- \tfrac{1}{2}\left\{ Tr\frac{\lambda\bar{\gamma}}{|a|^2+\lambda r_0^2}\,\xi\frac{\lambda\bar{\gamma}}{|a|^2+\lambda r_0^2}\,\xi + Tr\frac{ig\bar{A}}{|a|^2+\lambda r_0^2}\,\xi^*\frac{igA}{|a|^2+\lambda r_0^2}\,\xi \right.
$$

$$
\left. + Tr\frac{\lambda\gamma}{|a|^2+\lambda r_0^2}\,\xi^*\frac{\lambda\gamma}{|a|^2+\lambda r_0^2}\,\xi^* + Tr\frac{igA}{|a|^2+\lambda r_0^2}\,\xi\frac{ig\bar{A}}{|a|^2+\lambda r_0^2}\,\xi^* \right\} + O(\xi^3)
$$

One has

$$
\left(\frac{\lambda\bar{\gamma}}{|a|^2+\lambda r_0^2}\,\xi\right)_{k,p} = \frac{\lambda\bar{\gamma}}{|a_k|^2+\lambda r_0^2}\,\xi_{k,p}\,, \quad \left(\frac{\lambda\gamma}{|a|^2+\lambda r_0^2}\,\xi^*\right)_{k,p} = \frac{\lambda\gamma}{|a_k|^2+\lambda r_0^2}\,\bar{\xi}_{p,k}
$$

$$\left(\frac{\lambda\bar{\gamma}}{|a|^2+\lambda r_0^2} \, \xi \, \frac{\lambda\bar{\gamma}}{|a|^2+\lambda r_0^2} \, \xi \right)_{k,k} = \sum_p \frac{\lambda\bar{\gamma}}{|a_k|^2+\lambda r_0^2} \, \xi_{k,p} \, \frac{\lambda\bar{\gamma}}{|a_p|^2+\lambda r_0^2} \, \xi_{p,k} \qquad (5.67)$$

$$= \left(\frac{\lambda\bar{\gamma}\,\xi_{k,k}}{|a_k|^2+\lambda r_0^2} \right)^2 + \sum_{\substack{p\\p\neq k}} \frac{\lambda\bar{\gamma}}{|a_k|^2+\lambda r_0^2} \, \frac{\lambda\bar{\gamma}}{|a_p|^2+\lambda r_0^2} \, \xi_{k,p}\,\xi_{p,k}$$

$$= \frac{1}{\kappa} \left(\frac{\lambda r_0(\rho_0-\sqrt{\kappa}r_0)}{|a_k|^2+\lambda r_0^2} \right)^2 + \frac{1}{\kappa} \sum_{q\neq 0} \frac{\lambda\bar{\gamma}}{|a_k|^2+\lambda r_0^2} \, \frac{\lambda\bar{\gamma}}{|a_{k-q}|^2+\lambda r_0^2} \, \phi_q\,\phi_{-q}$$

$$\left(\frac{\lambda\gamma}{|a|^2+\lambda r_0^2} \, \xi^* \, \frac{\lambda\gamma}{|a|^2+\lambda r_0^2} \, \xi^* \right)_{k,k} = \sum_p \frac{\lambda\gamma}{|a_k|^2+\lambda r_0^2} \, \bar{\xi}_{p,k} \, \frac{\lambda\gamma}{|a_p|^2+\lambda r_0^2} \, \bar{\xi}_{k,p} \qquad (5.68)$$

$$= \left(\frac{\lambda\gamma\,\bar{\xi}_{k,k}}{|a_k|^2+\lambda r_0^2} \right)^2 + \sum_{\substack{p\\p\neq k}} \frac{\lambda\gamma}{|a_k|^2+\lambda r_0^2} \, \frac{\lambda\gamma}{|a_p|^2+\lambda r_0^2} \, \bar{\xi}_{k,p}\,\bar{\xi}_{p,k}$$

$$= \frac{1}{\kappa} \left(\frac{\lambda r_0(\rho_0-\sqrt{\kappa}r_0)}{|a_k|^2+\lambda r_0^2} \right)^2 + \frac{1}{\kappa} \sum_{q\neq 0} \frac{\lambda\gamma}{|a_k|^2+\lambda r_0^2} \, \frac{\lambda\gamma}{|a_{k-q}|^2+\lambda r_0^2} \, \bar{\phi}_q\,\bar{\phi}_{-q}$$

and

$$\left(\frac{igA}{|a|^2+\lambda r_0^2} \, \xi \right)_{k,p} = \frac{iga_k}{|a_k|^2+\lambda r_0^2} \, \xi_{k,p} \,, \qquad \left(\frac{ig\bar{A}}{|a|^2+\lambda r_0^2} \, \xi^* \right)_{k,p} = \frac{ig\bar{a}_k}{|a_k|^2+\lambda r_0^2} \, \bar{\xi}_{p,k}$$

$$\left(\frac{igA}{|a|^2+\lambda r_0^2} \, \xi \, \frac{ig\bar{A}}{|a|^2+\lambda r_0^2} \, \xi^* \right)_{k,k} = \sum_p \frac{iga_k}{|a_k|^2+\lambda r_0^2} \, \xi_{k,p} \, \frac{ig\bar{a}_p}{|a_p|^2+\lambda r_0^2} \, \bar{\xi}_{k,p} \qquad (5.69)$$

$$= -\lambda \frac{|a_k|^2}{(|a_k|^2+\lambda r_0^2)^2} \, |\xi_{k,k}|^2 - \lambda \sum_{\substack{p\\p\neq k}} \frac{a_k}{|a_k|^2+\lambda r_0^2} \, \xi_{k,p} \, \frac{\bar{a}_p}{|a_p|^2+\lambda r_0^2} \, \bar{\xi}_{k,p}$$

$$= -\frac{\lambda}{\kappa} \frac{|a_k|^2}{(|a_k|^2+\lambda r_0^2)^2} \, (\rho_0-\sqrt{\kappa}r_0)^2 - \frac{\lambda}{\kappa} \sum_{q\neq 0} \frac{a_k}{|a_k|^2+\lambda r_0^2} \, \frac{\bar{a}_{k-q}}{|a_{k-q}|^2+\lambda r_0^2} \, \rho_q^2$$

$$\left(\frac{ig\bar{A}}{|a|^2+\lambda r_0^2} \, \xi^* \, \frac{igA}{|a|^2+\lambda r_0^2} \, \xi \right)_{k,k} = \sum_p \frac{ig\bar{a}_k}{|a_k|^2+\lambda r_0^2} \, \bar{\xi}_{p,k} \, \frac{iga_p}{|a_p|^2+\lambda r_0^2} \, \xi_{p,k} \qquad (5.70)$$

$$= -\lambda \frac{|a_k|^2}{(|a_k|^2+\lambda r_0^2)^2} \, |\xi_{k,k}|^2 - \lambda \sum_{\substack{p\\p\neq k}} \frac{\bar{a}_k}{|a_k|^2+\lambda r_0^2} \, \bar{\xi}_{p,k} \, \frac{a_p}{|a_p|^2+\lambda r_0^2} \, \xi_{p,k}$$

$$= -\frac{\lambda}{\kappa} \frac{|a_k|^2}{(|a_k|^2+\lambda r_0^2)^2} \, (\rho_0-\sqrt{\kappa}r_0)^2 - \frac{\lambda}{\kappa} \sum_{q\neq 0} \frac{\bar{a}_k}{|a_k|^2+\lambda r_0^2} \, \frac{a_{k+q}}{|a_{k+q}|^2+\lambda r_0^2} \, \rho_q^2$$

Therefore (5.66) becomes

$$
\log\det\left[Id + \begin{pmatrix} \lambda\frac{\bar{\gamma}}{|a|^2+\lambda r_0^2}\,\xi & ig\frac{\bar{A}}{|a|^2+\lambda r_0^2}\,\xi^* \\ ig\frac{A}{|a|^2+\lambda r_0^2}\,\xi & \lambda\frac{\gamma}{|a|^2+\lambda r_0^2}\,\xi^* \end{pmatrix}\right]
$$

$$
= Tr\frac{\lambda\bar{\gamma}}{|a|^2+\lambda r_0^2}\,\xi + Tr\frac{\lambda\gamma}{|a|^2+\lambda r_0^2}\,\xi^*
$$

$$
-\frac{1}{2}\left\{ Tr\frac{\lambda\bar{\gamma}}{|a|^2+\lambda r_0^2}\,\xi\,\frac{\lambda\bar{\gamma}}{|a|^2+\lambda r_0^2}\,\xi + Tr\frac{ig\bar{A}}{|a|^2+\lambda r_0^2}\,\xi^*\,\frac{igA}{|a|^2+\lambda r_0^2}\,\xi\right.
$$

$$
\left. + Tr\frac{\lambda\gamma}{|a|^2+\lambda r_0^2}\,\xi^*\,\frac{\lambda\gamma}{|a|^2+\lambda r_0^2}\,\xi^* + Tr\frac{igA}{|a|^2+\lambda r_0^2}\,\xi\,\frac{ig\bar{A}}{|a|^2+\lambda r_0^2}\,\xi^* \right\} + O(\xi^3)
$$

$$
= 2\frac{\lambda}{\kappa}\sum_k \frac{\sqrt{\kappa}\,r_0(\rho_0-\sqrt{\kappa}\,r_0)}{|a_k|^2+\lambda r_0^2}
$$

$$
-\frac{1}{2}\left\{ \frac{1}{\kappa}\sum_k\left(\frac{\lambda r_0(\rho_0-\sqrt{\kappa}\,r_0)}{|a_k|^2+\lambda r_0^2}\right)^2 + \sum_{q\neq 0}\frac{1}{\kappa}\sum_k\frac{\lambda\bar{\gamma}}{|a_k|^2+\lambda r_0^2}\frac{\lambda\bar{\gamma}}{|a_{k-q}|^2+\lambda r_0^2}\phi_q\,\phi_{-q}\right.
$$

$$
+\frac{1}{\kappa}\sum_k\left(\frac{\lambda r_0(\rho_0-\sqrt{\kappa}\,r_0)}{|a_k|^2+\lambda r_0^2}\right)^2 + \sum_{q\neq 0}\frac{1}{\kappa}\sum_k\frac{\lambda\gamma}{|a_k|^2+\lambda r_0^2}\frac{\lambda\gamma}{|a_{k-q}|^2+\lambda r_0^2}\bar{\phi}_q\,\bar{\phi}_{-q}
$$

$$
-\frac{\lambda}{\kappa}\sum_k\frac{|a_k|^2}{(|a_k|^2+\lambda r_0^2)^2}(\rho_0-\sqrt{\kappa}\,r_0)^2 - \sum_{q\neq 0}\frac{\lambda}{\kappa}\sum_k\frac{\bar{a}_k}{|a_k|^2+\lambda r_0^2}\frac{a_{k+q}}{|a_{k+q}|^2+\lambda r_0^2}\rho_q^2
$$

$$
\left.-\frac{\lambda}{\kappa}\sum_k\frac{|a_k|^2}{(|a_k|^2+\lambda r_0^2)^2}(\rho_0-\sqrt{\kappa}\,r_0)^2 - \sum_{q\neq 0}\frac{\lambda}{\kappa}\sum_k\frac{a_k}{|a_k|^2+\lambda r_0^2}\frac{\bar{a}_{k-q}}{|a_{k-q}|^2+\lambda r_0^2}\rho_q^2\right\}
$$

$$
+ O(\xi^3) \tag{5.71}
$$

Using the BCS equation (5.33), $\frac{\lambda}{\kappa}\sum_k\frac{1}{|a_k|^2+\lambda r_0^2}=1$ and abbreviating

$$
E_k^2 := |a_k|^2 + \lambda r_0^2 = k_0^2 + e_{\mathbf{k}}^2 + \lambda r_0^2
$$

this becomes

$$
2\sqrt{\kappa}\,r_0(\rho_0-\sqrt{\kappa}\,r_0)\cdot 1 - (\rho_0-\sqrt{\kappa}\,r_0)^2\frac{\lambda}{\kappa}\sum_k\frac{\lambda r_0^2-|a_k|^2}{E_k^4}
$$

$$
+\sum_{q\neq 0}\rho_q^2\left\{\frac{\lambda}{\kappa}\sum_k\frac{\bar{a}_k a_{k-q}}{E_k^2 E_{k-q}^2}\right\} - \sum_{q\neq 0}Re\left(e^{-2i\theta_0}\phi_q\phi_{-q}\right)\left\{\frac{\lambda}{\kappa}\sum_k\frac{\lambda r_0^2}{E_k^2 E_{k-q}^2}\right\}
$$

$$
= 2\sqrt{\kappa}\,r_0(\rho_0-\sqrt{\kappa}\,r_0)\cdot 1 + (\rho_0-\sqrt{\kappa}\,r_0)^2 - (\rho_0-\sqrt{\kappa}\,r_0)^2 2\frac{\lambda}{\kappa}\sum_k\frac{\lambda r_0^2}{E_k^4}
$$

$$
+\sum_{q\neq 0}\rho_q^2\left\{\frac{\lambda}{\kappa}\sum_k\frac{\bar{a}_k a_{k-q}}{E_k^2 E_{k-q}^2}\right\} - \sum_{q\neq 0}Re\left(e^{-2i\theta_0}\phi_q\phi_{-q}\right)\left\{\frac{\lambda}{\kappa}\sum_k\frac{\lambda r_0^2}{E_k^2 E_{k-q}^2}\right\}
$$

$$
= \rho_0^2 - \kappa r_0^2 - (\rho_0-\sqrt{\kappa}\,r_0)^2 2\frac{\lambda}{\kappa}\sum_k\frac{\lambda r_0^2}{E_k^4} \tag{5.72}
$$

$$
+\sum_{q\neq 0}\rho_q^2\left\{\frac{\lambda}{\kappa}\sum_k\frac{\bar{a}_k a_{k-q}}{E_k^2 E_{k-q}^2}\right\} - \sum_{q\neq 0}Re\left(e^{-2i\theta_0}\phi_q\phi_{-q}\right)\left\{\frac{\lambda}{\kappa}\sum_k\frac{\lambda r_0^2}{E_k^2 E_{k-q}^2}\right\}
$$

Therefore one obtains, recalling that $\xi_{k,p} = \frac{1}{\sqrt{\kappa}}\phi_{k-p} - \gamma\delta_{k,p}$,

$$V(\{\phi_q\}) - V(\{\sqrt{\kappa}\,\delta_{q,0}\,r_0 e^{i\theta_0}\})$$

$$= \sum_q \rho_q^2 - \kappa\,r_0^2 - \log\det\left[Id + \begin{pmatrix} \lambda\frac{\bar{\gamma}}{|a|^2+\lambda r_0^2}\xi & ig\frac{\bar{A}}{|a|^2+\lambda r_0^2}\xi^* \\ ig\frac{A}{|a|^2+\lambda r_0^2}\xi & \lambda\frac{\gamma}{|a|^2+\lambda r_0^2}\xi^* \end{pmatrix}\right]$$

$$= \sum_{q\neq 0}\rho_q^2 + (\rho_0 - \sqrt{\kappa}r_0)^2 2\frac{\lambda}{\kappa}\sum_k\frac{\lambda r_0^2}{E_k^4} - \sum_{q\neq 0}\rho_q^2\left\{\frac{\lambda}{\kappa}\sum_k\frac{\bar{a}_k a_{k-q}}{E_k^2 E_{k-q}^2}\right\}$$

$$+ \sum_{q\neq 0}\mathrm{Re}\left(e^{-2i\theta_0}\phi_q\phi_{-q}\right)\left\{\frac{\lambda}{\kappa}\sum_k\frac{\lambda r_0^2}{E_k^2 E_{k-q}^2}\right\} \tag{5.73}$$

Consider the coefficient of $\sum_{q\neq 0}\rho_q^2$. It is given by

$$1 - \frac{\lambda}{\kappa}\sum_k\frac{\bar{a}_k a_{k-q}}{E_k^2 E_{k-q}^2} = \frac{1}{2}(1+1) - \frac{1}{2}\frac{\lambda}{\kappa}\sum_k\frac{2\bar{a}_k a_{k-q}}{E_k^2 E_{k-q}^2}$$

$$= \frac{1}{2}\left(\frac{\lambda}{\kappa}\sum_k\frac{|a_k|^2+\lambda r_0^2}{E_k^2 E_{k-q}^2} + \frac{\lambda}{\kappa}\sum_k\frac{|a_{k-q}|^2+\lambda r_0^2}{E_k^2 E_{k-q}^2}\right) - \frac{1}{2}\frac{\lambda}{\kappa}\sum_k\frac{2\bar{a}_k a_{k-q}}{E_k^2 E_{k-q}^2}$$

$$= \frac{1}{2}\frac{\lambda}{\kappa}\sum_k\frac{a_k\bar{a}_k - a_k\bar{a}_{k-q} - \bar{a}_k a_{k-q} + a_{k-q}\bar{a}_{k-q}}{E_k^2 E_{k-q}^2}$$

$$+ \frac{\lambda}{\kappa}\sum_k\frac{\lambda r_0^2}{E_k^2 E_{k-q}^2} - \frac{1}{2}\frac{\lambda}{\kappa}\sum_k\frac{\bar{a}_k a_{k-q} - a_k\bar{a}_{k-q}}{E_k^2 E_{k-q}^2}$$

$$= \frac{1}{2}\frac{\lambda}{\kappa}\sum_k\frac{(a_k - a_{k-q})(\bar{a}_k - \bar{a}_{k-q})}{E_k^2 E_{k-q}^2} + \frac{\lambda}{\kappa}\sum_k\frac{\lambda r_0^2}{E_k^2 E_{k-q}^2} - i\frac{\lambda}{\kappa}\sum_k\frac{\mathrm{Im}(\bar{a}_k a_{k-q})}{E_k^2 E_{k-q}^2}$$

$$= \frac{1}{2}\frac{\lambda}{\kappa}\sum_k\frac{q_0^2+(e_k-e_{k-q})^2}{E_k^2 E_{k-q}^2} + \frac{\lambda}{\kappa}\sum_k\frac{\lambda r_0^2}{E_k^2 E_{k-q}^2} - i\frac{\lambda}{\kappa}\sum_k\frac{k_0 e_{k-q} - (k_0-q_0)e_k}{E_k^2 E_{k-q}^2}$$

$$= \alpha_q + i\gamma_q + \beta_q \tag{5.74}$$

Inserting (5.74) in (5.73), one gets

$$V(\{\phi_q\}) - V(\{\sqrt{\kappa}\,\delta_{q,0}\,r_0 e^{i\theta_0}\})$$

$$= (\rho_0 - \sqrt{\kappa}r_0)^2 2\beta_0 + \sum_{q\neq 0}\rho_q^2\left\{1 - \frac{\lambda}{\kappa}\sum_k\frac{2\bar{a}_k a_{k-q}}{E_k^2 E_{k-q}^2}\right\}$$

$$+ \sum_{q\neq 0}\mathrm{Re}\left(e^{-2i\theta_0}\phi_q\phi_{-q}\right)\left\{\frac{\lambda}{\kappa}\sum_k\frac{\lambda r_0^2}{E_k^2 E_{k-q}^2}\right\}$$

$$= (\rho_0 - \sqrt{\kappa}r_0)^2 2\beta_0 + \sum_{q\neq 0}\rho_q^2(\alpha_q + i\gamma_q)$$

$$+ \sum_{q\neq 0}\rho_q^2\beta_q + \sum_{q\neq 0}\frac{e^{-2i\theta_0}\phi_q\phi_{-q} + e^{2i\theta_0}\bar{\phi}_q\bar{\phi}_{-q}}{2}\beta_q \tag{5.75}$$

Since $\beta_q = \beta_{-q}$, the last two q-sums in (5.75) may be combined to give

$$\sum_{q\neq 0} \frac{\rho_q^2 + \rho_{-q}^2}{2} \beta_q + \sum_{q\neq 0} \frac{e^{-2i\theta_0}\phi_q\phi_{-q} + e^{2i\theta_0}\bar{\phi}_q\bar{\phi}_{-q}}{2} \beta_q$$

$$= \frac{1}{2}\sum_{q\neq 0} (\phi_q\bar{\phi}_q + \phi_{-q}\bar{\phi}_{-q} + e^{-2i\theta_0}\phi_q\phi_{-q} + e^{2i\theta_0}\bar{\phi}_q\bar{\phi}_{-q})\beta_q$$

$$= \frac{1}{2}\sum_{q\neq 0} (e^{-i\theta_0}\phi_q + e^{i\theta_0}\bar{\phi}_{-q})(e^{i\theta_0}\bar{\phi}_q + e^{-i\theta_0}\phi_{-q})\beta_q$$

$$= \frac{1}{2}\sum_{q\neq 0} |e^{-i\theta_0}\phi_q + e^{i\theta_0}\bar{\phi}_{-q}|^2 \beta_q \qquad (5.76)$$

which proves the theorem ∎

Finally we consider the effective potential in the presence of a small $U(1)$ symmetry breaking external field. That is, we add to our original Hamiltonian the term

$$\frac{1}{L^d} \sum_{\mathbf{k}} \left(\bar{r}a^+_{\mathbf{k}\uparrow}a^+_{-\mathbf{k}\downarrow} + ra_{-\mathbf{k}\downarrow}a_{\mathbf{k}\uparrow} \right) \qquad (5.77)$$

where $r = |r|\,e^{i\alpha} \in \mathbb{C}$. In section 6.2 below we show that in that case the effective potential can be transformed to

$$U_r(\{\phi_q\}) = u_0^2 + \left(v_0 + \sqrt{\kappa}\,\frac{|r|}{g}\right)^2 + \sum_{q\neq 0} |\phi_q|^2 - \log \frac{\det \begin{bmatrix} a_k\delta_{k,p} & \frac{ig}{\sqrt{\kappa}}\,\bar{\tilde{\phi}}_{p-k} \\ \frac{ig}{\sqrt{\kappa}}\,\tilde{\phi}_{k-p} & a_{-k}\delta_{k,p} \end{bmatrix}}{\det \begin{bmatrix} a_k\delta_{k,p} & 0 \\ 0 & a_{-k}\delta_{k,p} \end{bmatrix}}$$

$$(5.78)$$

where

$$\tilde{\phi}_q := \begin{cases} \phi_q & \text{for } q \neq 0 \\ e^{i\alpha}\phi_0 & \text{for } q = 0 \end{cases} \qquad (5.79)$$

We obtain the following

Corollary 5.2.3 *Let U_r be the effective potential (5.78) where $\tilde{\phi}$ is given by (5.79). Then:*
(i) The global minimum of $\mathrm{Re}\,U_r(\{\phi_q\})$ is unique and is given by

$$\phi_q^{\min} = \delta_{q,0}\sqrt{\kappa}\,iy_0 \qquad (5.80)$$

where $y_0 = y_0(|r|)$ is the unique global minimum of the function $V_{\mathrm{BCS},r} : \mathbb{R} \to \mathbb{R}$,

$$V_{\mathrm{BCS},r}(y) := U_r\left(u_0 = 0,\ v_0 = \sqrt{\kappa}\,y;\ \phi_q = 0 \text{ for } q \neq 0\right)$$

$$= \kappa\left\{\left(y + \frac{|r|}{g}\right)^2 - \frac{1}{\kappa}\sum_{k}\log\left[1 + \frac{\lambda y^2}{k_0^2 + e_k^2}\right]\right\} \qquad (5.81)$$

(ii) *The second order Taylor expansion of U_r around $\phi^{\min}$ is given by*

$$U_r(\{\phi_q\}) = U_{r,\min} + 2\beta_0(v_0 - \sqrt{\kappa}y_0)^2 \tag{5.82}$$

$$+ \sum_{q \neq 0}(\alpha_q + i\gamma_q)|\phi_q|^2 + \tfrac{1}{2}\sum_{q \neq 0}\beta_q|e^{-i\alpha}\phi_q - e^{i\alpha}\bar{\phi}_{-q}|^2$$

$$+ \frac{|r|}{g|y_0|}\left(u_0^2 + (v_0 - \sqrt{\kappa}y_0)^2 + \sum_{q \neq 0}|\phi_q|^2\right) + O((\phi - \phi^{\min})^3)$$

where $U_{r,\min} := U_r(\{\phi_q^{\min}\})$ *and the coefficients* α_q, β_q *and* γ_q *are given by* *(5.57, 5.58) of Theorem 5.2.2 but* E_k *in this case is given by* $E_k^2 = |a_k|^2 + \lambda y_0^2 = k_0^2 + e_k^2 + \lambda y_0^2$.

Remark: Of course one has $\lim_{|r| \to 0} \lambda y_0(|r|)^2 = \lambda r_0^2 = \Delta^2$ where $\pm r_0$ is the global minimum of $V_{\mathrm{BCS},r=0}$.

Proof: **(i)** As in the proof of Theorem 5.2.1 one shows that

$$\log\left|\det\begin{bmatrix} Id & \frac{ig}{\sqrt{\kappa}}C\tilde{\phi}^* \\ \frac{ig}{\sqrt{\kappa}}\tilde{C}\tilde{\phi} & Id \end{bmatrix}\right| \leq \sum_k \log\left[1 + \frac{\frac{\lambda}{\kappa}\sum_q|\phi_q|^2}{|a_k|^2}\right]$$

$$=: \sum_k \log\left[1 + \frac{\lambda(x^2+y^2)}{|a_k|^2}\right] \tag{5.83}$$

where we abbreviated

$$x^2 := \tfrac{1}{\kappa}\left(u_0^2 + \sum_{q \neq 0}|\phi_q|^2\right), \quad y^2 := \tfrac{1}{\kappa}v_0^2 \tag{5.84}$$

Thus

$$\mathrm{Re}U_r(\{\phi_q\}) \geq \left(v_0 + \sqrt{\kappa}\frac{|r|}{g}\right)^2 + u_0^2 + \sum_{q \neq 0}|\phi_q|^2 - \sum_k \log\left[1 + \frac{\lambda(x^2+y^2)}{|a_k|^2}\right]$$

$$=: \kappa W_r(x, y) \tag{5.85}$$

where

$$W_r(x, y) = x^2 + \left(y + \frac{|r|}{g}\right)^2 - \tfrac{1}{\kappa}\sum_k \log\left[1 + \frac{\lambda(x^2+y^2)}{|a_k|^2}\right] \tag{5.86}$$

The global minimum of W_r is unique and given by $x = 0$ and $y = y_0$ where y_0 is the unique global minimum of (5.81). Since U_r at $u_0 = 0$, $v_0 = \sqrt{\kappa}y$; $\phi_q = 0$ for $q \neq 0$ equals $V_{\mathrm{BCS},r}(y)$, part (i) follows.

(ii) Part (ii) is proven in the same way as Theorem 5.2.2. One has

$$U_r(\{\phi_q\}) - U_{r,\min}$$

$$= u_0^2 + \sum_{q\neq 0} |\phi_q|^2 + \left(v_0 + \sqrt{\kappa}\,\tfrac{|r|}{g}\right)^2 - \left(\sqrt{\kappa}\,y_0 + \sqrt{\kappa}\,\tfrac{|r|}{g}\right)^2$$

$$- \log\left\{ \det\begin{bmatrix} a_k\delta_{k,p} & \frac{ig}{\sqrt{\kappa}}\bar{\phi}_{p-k} \\ \frac{ig}{\sqrt{\kappa}}\tilde{\phi}_{k-p} & a_{-k}\delta_{k,p} \end{bmatrix} \Big/ \det\begin{bmatrix} a_k\delta_{k,p} & \frac{ig}{\sqrt{\kappa}}e^{-i\alpha}(-i)y_0\delta_{k,p} \\ \frac{ig}{\sqrt{\kappa}}e^{i\alpha}iy_0\delta_{k,p} & a_{-k}\delta_{k,p} \end{bmatrix} \right\}$$

$$= u_0^2 + \sum_{q\neq 0} |\phi_q|^2 + \left(v_0 - \sqrt{\kappa}\,y_0\right)^2 + 2(v_0 - \sqrt{\kappa}\,y_0)\left(\sqrt{\kappa}\,y_0 + \sqrt{\kappa}\,\tfrac{|r|}{g}\right)$$

$$- \log\det\left[Id + \left(\begin{matrix} a_k\delta_{k,p} & \frac{ig}{\sqrt{\kappa}}e^{-i\alpha}(-i)y_0\delta_{k,p} \\ \frac{ig}{\sqrt{\kappa}}e^{i\alpha}iy_0\delta_{k,p} & a_{-k}\delta_{k,p} \end{matrix}\right)^{-1} \times \right.$$

$$\left. \left(\begin{matrix} 0 & \frac{ig}{\sqrt{\kappa}}\bar{\xi}_{p-k} \\ \frac{ig}{\sqrt{\kappa}}\xi_{k-p} & 0 \end{matrix}\right)\right]$$

where in this case

$$\xi_{k-p} := \tilde{\phi}_{k-p} - \sqrt{\kappa}\,e^{i\alpha}\,iy_0\,\delta_{k,p} \tag{5.87}$$

The expression $\log\det[Id + \cdots]$ is expanded as in the proof of Theorem 5.2.2. One obtains, if $E_k^2 := |a_k|^2 + \lambda y_0^2$,

$$\log\det[Id + \cdots] = 2\sqrt{\kappa}\,y_0(v_0 - \sqrt{\kappa}\,y_0)\tfrac{\lambda}{\kappa}\sum_k \tfrac{1}{E_k^2}$$

$$+ \tfrac{\lambda}{\kappa}\sum_k \frac{\lambda y_0^2\left(u_0^2 - (v_0 - \sqrt{\kappa}\,y_0)^2\right)}{E_k^4} + \tfrac{\lambda}{\kappa}\sum_k \frac{|a_k|^2\left(u_0^2 + (v_0 - \sqrt{\kappa}\,y_0)^2\right)}{E_k^4}$$

$$+ \tfrac{1}{2}\tfrac{\lambda}{\kappa}\sum_{q\neq 0}\sum_k \left(\frac{a_k\bar{a}_{k-q}}{E_k^2 E_{k-q}^2} + \frac{\bar{a}_k a_{k+q}}{E_k^2 E_{k+q}^2}\right)|\phi_q|^2$$

$$- \tfrac{1}{2}\tfrac{\lambda}{\kappa}\sum_{q\neq 0}\sum_k \left(\frac{\lambda y_0^2 e^{2i\alpha}}{E_k^2 E_{k+q}^2}\bar{\phi}_q\bar{\phi}_{-q} + \frac{\lambda y_0^2 e^{-2i\alpha}}{E_k^2 E_{k-q}^2}\phi_q\phi_{-q}\right) \tag{5.88}$$

Since y_0 is a minimum of $V_{BCS,r}$, one has the BCS equation

$$2\left(y_0 + \tfrac{|r|}{g}\right) - \tfrac{\lambda}{\kappa}\sum_k \frac{2y_0}{E_k^2} = 0 \quad\Leftrightarrow\quad \tfrac{\lambda}{\kappa}\sum_k \frac{1}{E_k^2} = 1 - \frac{|r|}{g|y_0|} \tag{5.89}$$

Using this, one gets (observe that y_0 is negative)

$$U_r(\{\phi_q\}) = U_{r,\min} + \tfrac{|r|}{g|y_0|}\left(u_0^2 + (v_0 - \sqrt{\kappa}\,y_0)^2\right)$$

$$+ 2\tfrac{\lambda}{\kappa}\sum_k \frac{\lambda y_0^2(v_0 - \sqrt{\kappa}\,y_0)^2}{E_k^4} + \sum_{q\neq 0}\left\{1 - \tfrac{\lambda}{\kappa}\sum_k \frac{\bar{a}_k a_{k-q}}{E_k^2 E_{k-q}^2}\right\}|\phi_q|^2$$

$$+ \tfrac{1}{2}\tfrac{\lambda}{\kappa}\sum_{q\neq 0}\sum_k \left(\frac{\lambda y_0^2 e^{2i\alpha}}{E_k^2 E_{k+q}^2}\bar{\phi}_q\bar{\phi}_{-q} + \frac{\lambda y_0^2 e^{-2i\alpha}}{E_k^2 E_{k-q}^2}\phi_q\phi_{-q}\right) \tag{5.90}$$

Using the BCS equation (5.89) again, one obtains (compare (5.74))

$$1 - \frac{\lambda}{\kappa} \sum_k \frac{\bar{a}_k a_{k-q}}{E_k^2 E_{k-q}^2} = \alpha_q + i\gamma_q + \beta_q + \frac{|r|}{g|y_0|} \tag{5.91}$$

Substituting this in (5.90) and rearranging as in the proof of Theorem 5.2.2 proves part (ii). ■

Chapter 6

BCS Theory and Spontaneous Symmetry Breaking

The Hamiltonian for the many-electron system in finite volume $[0, L]^d$ is given by $H = H_0 + H_{int}$ where

$$H_0 = \frac{1}{L^d} \sum_{\mathbf{k}\sigma} e_{\mathbf{k}} a^+_{\mathbf{k}\sigma} a_{\mathbf{k}\sigma} \tag{6.1}$$

$e_{\mathbf{k}} = \frac{\mathbf{k}^2}{2m} - \mu$, and

$$H_{int} = \frac{1}{L^{3d}} \sum_{\sigma,\tau} \sum_{\mathbf{k},\mathbf{p},\mathbf{q}} V(\mathbf{k} - \mathbf{p}) a^+_{\mathbf{k}\sigma} a^+_{\mathbf{q}-\mathbf{k},\tau} a_{\mathbf{q}-\mathbf{p},\tau} a_{\mathbf{p}\sigma} \tag{6.2}$$

As usual, the interacting part may be represented by the following diagram:

$$H_{int} = \qquad \text{(diagram: lines labelled } p,\sigma \text{ and } q-p,\tau \text{ above; } k,\sigma \text{ and } q-k,\tau \text{ below; wavy line } V(k-p))$$

Since there is conservation of momentum, there are three independent momenta. One may consider three natural limiting cases with two independent momenta:

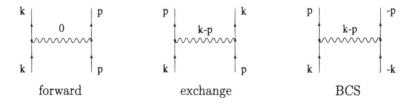

forward   exchange   BCS

In chapter 8 where we prove rigorous bounds on Feynman diagrams we show that the most singular contributions to the perturbation series come from the

$\mathbf{q} = 0$ term in (6.2). This corresponds to the third diagram in the above figure. To retain only this term, the $\mathbf{q} = 0$ term of H_{int} given by (6.2), is the basic approximation of BCS theory. Furthermore $V(\mathbf{k} - \mathbf{p})$ is substituted by its value on the Fermi surface, $V(\mathbf{k} - \mathbf{p}) \approx V(\hat{\mathbf{k}} - \hat{\mathbf{p}})$ where $\hat{\mathbf{k}} = k_F \frac{\mathbf{k}}{\|\mathbf{k}\|}$, $k_F = \sqrt{2m\mu}$, and the momenta $\mathbf{k}, \mathbf{p}$ are restricted to values close to the Fermi surface, $|e_{\mathbf{k}}|, |e_{\mathbf{p}}| \leq \omega_D$, where the cutoff ω_D is referred to as the Debye frequency. Thus the BCS approximation reads

$$H_{\text{int}} \approx H_{\text{BCS}} \tag{6.3}$$

where

$$H_{\text{BCS}} = \frac{1}{L^{3d}} \sum_{\sigma,\tau} \sum_{\substack{\mathbf{k},\mathbf{p} \\ |e_{\mathbf{k}}|,|e_{\mathbf{p}}| \leq \omega_D}} V(\hat{\mathbf{k}} - \hat{\mathbf{p}}) \, a^+_{\mathbf{k}\sigma} a^+_{-\mathbf{k},\tau} a_{-\mathbf{p},\tau} a_{\mathbf{p}\sigma} \tag{6.4}$$

We first make a comment on the volume factors. In (6.4) we retained only the $\mathbf{q} = 0$ term of (6.2), but we did not cancel a volume factor. It is not obvious that (6.4) is still proportional to the volume. However, in the case of the forward approximation which is defined by putting $\mathbf{k} = \mathbf{p}$ in (6.2) without canceling a volume factor,

$$H_{\text{int}} \approx H_{\text{forw}} := \frac{1}{L^{3d}} \sum_{\sigma,\tau} \sum_{\mathbf{k},\mathbf{p}} V(0) a^+_{\mathbf{k}\sigma} a^+_{\mathbf{p},\tau} a_{\mathbf{p},\tau} a_{\mathbf{k}\sigma} \tag{6.5}$$

there is an easy argument which shows that (6.5) is still proportional to the volume (for constant density). Namely, on a fixed N-particle space $\mathcal{F}_N$ the interacting part H_{int} is a multiplication operator given by

$$H_{\text{int}|\mathcal{F}_N} = \frac{1}{2} \sum_{\substack{i,j=1 \\ i \neq j}}^{N} V(\mathbf{x}_i - \mathbf{x}_j) \tag{6.6}$$

Let $\delta_{\mathbf{y}}(\mathbf{x}) = \delta(\mathbf{x} - \mathbf{y})$. Then $\varphi(\mathbf{x}_1, \cdots, \mathbf{x}_N) := \delta_{\mathbf{y}_1} \wedge \cdots \wedge \delta_{\mathbf{y}_N}(\mathbf{x}_1, \cdots, \mathbf{x}_N)$ is an eigenfunction of H_{int} with eigenvalue

$$E = \frac{1}{2} \sum_{\substack{i,j=1 \\ i \neq j}}^{N} V(\mathbf{y}_i - \mathbf{y}_j) = \frac{1}{2L^d} \sum_{\substack{i,j=1 \\ i \neq j}}^{N} \sum_{\mathbf{q}} e^{-i(\mathbf{y}_i - \mathbf{y}_j)\mathbf{q}} V(\mathbf{q}) \tag{6.7}$$

which is, for $V(\mathbf{x}) \in L^1$, proportional to N or to the volume for constant density. One finds that φ is also an eigenvector of the forward term, $H_{\text{forw}}\varphi = E_{\text{forw}}\varphi$ where E_{forw} is obtained from (6.7) by putting $\mathbf{q} = 0$ without canceling a volume factor,

$$E_{\text{forw}} = \frac{1}{2L^d} \sum_{\substack{i,j=1 \\ i \neq j}}^{N} V(\mathbf{q} = 0) \tag{6.8}$$

which is also proportional to the volume.

Although the model defined by the Hamiltonian H_{BCS} in (6.4) is still quartic in the annihilation and creation operators, it can be solved explicitly, without making a quadratic mean field approximation as it is usually done. The important point is that the approximation 'putting $\mathbf{q} = 0$ without canceling a volume factor' has the effect that in the bosonic functional integral representation the volume factors enter the formulae in such a way that in the thermodynamic limit the integration variables are forced to take values at the global minimum of the effective potential. That is, fluctuations around the minimum configuration are suppressed for $L^d \to \infty$.

The interaction $V(\hat{\mathbf{k}} - \hat{\mathbf{p}})$ in (6.4) can be expanded into spherical harmonics:

$$V(\hat{\mathbf{k}} - \hat{\mathbf{p}}) = \begin{cases} \frac{1}{2}\sum_{\ell=-\infty}^{\infty} \lambda_{|\ell|}\, e^{i\ell\varphi_{\mathbf{k}}} e^{-i\ell\varphi_{\mathbf{p}}} + \frac{\lambda_0}{2} & \text{for } d = 2 \\ \sum_{\ell=0}^{\infty}\sum_{m=-\ell}^{\ell} \lambda_\ell \bar{Y}_{\ell m}(\hat{\mathbf{k}}/k_F) Y_{\ell m}(\hat{\mathbf{p}}/k_F) & \text{for } d = 3 \end{cases} \tag{6.9}$$

where $\mathbf{k} = |\mathbf{k}|(\cos\varphi_{\mathbf{k}}, \sin\varphi_{\mathbf{k}})$ for $d = 2$. We start with the easiest case where only the $\ell = 0$ term in (6.9) is retained. That is, we approximate

$$H_{\text{BCS}} \approx H_{\text{BCS}}^{\ell=0} \tag{6.10}$$

where

$$H_{\text{BCS}}^{\ell=0} = \frac{\lambda}{L^{3d}} \sum_{\substack{\mathbf{k},\mathbf{p} \\ |e_{\mathbf{k}}|,|e_{\mathbf{p}}|\leq\omega_D}} a_{\mathbf{k}\uparrow}^+ a_{-\mathbf{k},\downarrow}^+ a_{-\mathbf{p},\downarrow} a_{\mathbf{p}\uparrow} \tag{6.11}$$

where we put $\lambda = 2\lambda_0$, the factor of 2 coming from spin sums. In the next section we present the standard quadratic mean field formalism which approximates (6.11) by a quadratic mean field Hamiltonian which then can be diagonalized with a Bogoliubov transformation. In section 6.2 we show that already the quartic model defined by (6.11) is explicitly solvable and we discuss symmetry breaking. We find that the correlation functions of the model defined by (6.11) coincide with those of the quadratic mean field model. However, if higher angular momentum terms in (6.9) are taken into account, the quadratic mean field formalism does not necessarily give the right answer. This is discussed in section 6.3.

6.1 The Quadratic Mean Field Model

In this section we present the standard quadratic BCS mean field formalism for a constant (s-wave) interaction as it can be found in many books on many-body theory or superconductivity [57], [31]. We use the formulation of [23].

Let $H_0 + H_{\mathrm{BCS}}^{\ell=0}$ be the quartic BCS Hamiltonian given by (6.11) and (6.1). We add and subtract expectation values of products of field operators $\langle a_{\mathbf{k}\uparrow}^+ a_{-\mathbf{k}\downarrow}^+ \rangle$, $\langle a_{-\mathbf{p}\downarrow} a_{\mathbf{p}\uparrow} \rangle$ which will be defined later selfconsistently. Finally we neglect terms which are quadratic in the differences $aa - \langle aa \rangle$, $a^+ a^+ - \langle a^+ a^+ \rangle$ which are supposed to be small:

$$
\begin{aligned}
H_{\mathrm{BCS}}^{\ell=0} &= \frac{\lambda}{L^{3d}} \sum_{\substack{\mathbf{k},\mathbf{p} \\ |e_{\mathbf{k}}|,|e_{\mathbf{p}}| \leq \omega_D}} a_{\mathbf{k}\uparrow}^+ a_{-\mathbf{k}\downarrow}^+ a_{-\mathbf{p}\downarrow} a_{\mathbf{p}\uparrow} \\
&= \frac{\lambda}{L^{3d}} \sum_{\substack{\mathbf{k},\mathbf{p} \\ |e_{\mathbf{k}}|,|e_{\mathbf{p}}| \leq \omega_D}} \left([a_{\mathbf{k}\uparrow}^+ a_{-\mathbf{k}\downarrow}^+ - \langle a_{\mathbf{k}\uparrow}^+ a_{-\mathbf{k},\downarrow}^+ \rangle] + \langle a_{\mathbf{k}\uparrow}^+ a_{-\mathbf{k}\downarrow}^+ \rangle \right) \times \\
&\qquad\qquad \left([a_{-\mathbf{p}\downarrow} a_{\mathbf{p}\uparrow} - \langle a_{-\mathbf{p}\downarrow} a_{\mathbf{p}\uparrow} \rangle] + \langle a_{-\mathbf{p}\downarrow} a_{\mathbf{p}\uparrow} \rangle \right) \\
&\approx \frac{\lambda}{L^{3d}} \sum_{\substack{\mathbf{k},\mathbf{p} \\ |e_{\mathbf{k}}|,|e_{\mathbf{p}}| \leq \omega_D}} \left\{ [a_{\mathbf{k}\uparrow}^+ a_{-\mathbf{k}\downarrow}^+ - \langle a_{\mathbf{k}\uparrow}^+ a_{-\mathbf{k},\downarrow}^+ \rangle] \langle a_{-\mathbf{p}\downarrow} a_{\mathbf{p}\uparrow} \rangle \right. \\
&\qquad\qquad \left. + \langle a_{\mathbf{k}\uparrow}^+ a_{-\mathbf{k},\downarrow}^+ \rangle [a_{-\mathbf{p}\downarrow} a_{\mathbf{p}\uparrow} - \langle a_{-\mathbf{p}\downarrow} a_{\mathbf{p}\uparrow} \rangle] + \langle a_{\mathbf{k}\uparrow}^+ a_{-\mathbf{k}\downarrow}^+ \rangle \langle a_{-\mathbf{p}\downarrow} a_{\mathbf{p}\uparrow} \rangle \right\} \\
&= \frac{\lambda}{L^{3d}} \sum_{\substack{\mathbf{k},\mathbf{p} \\ |e_{\mathbf{k}}|,|e_{\mathbf{p}}| \leq \omega_D}} \left\{ a_{\mathbf{k}\uparrow}^+ a_{-\mathbf{k}\downarrow}^+ \langle a_{-\mathbf{p}\downarrow} a_{\mathbf{p}\uparrow} \rangle + \langle a_{\mathbf{k}\uparrow}^+ a_{-\mathbf{k}\downarrow}^+ \rangle a_{-\mathbf{p}\downarrow} a_{\mathbf{p}\uparrow} \right. \\
&\qquad\qquad \left. - \langle a_{\mathbf{k}\uparrow}^+ a_{-\mathbf{k}\downarrow}^+ \rangle \langle a_{-\mathbf{p}\downarrow} a_{\mathbf{p}\uparrow} \rangle \right\} \\
&= \frac{\lambda}{L^d} \sum_{\substack{\mathbf{k} \\ |e_{\mathbf{k}}| \leq \omega_D}} \left\{ a_{\mathbf{k}\uparrow}^+ a_{-\mathbf{k}\downarrow}^+ \Delta + \bar{\Delta} \, a_{-\mathbf{p}\downarrow} a_{\mathbf{p}\uparrow} \right\} - \lambda L^d |\Delta|^2 =: H_{\mathrm{MF}} \qquad (6.12)
\end{aligned}
$$

where we put

$$
\Delta = \frac{1}{L^d} \sum_{\mathbf{p}} \frac{1}{L^d} \langle a_{-\mathbf{p}\downarrow} a_{\mathbf{p}\uparrow} \rangle \qquad (6.13)
$$

and the expectation value has to be taken with respect to H_{MF}. Thus, the quadratic mean field approximation consists in

$$
H_{\mathrm{int}} \approx H_{\mathrm{MF}} = \frac{\lambda}{L^d} \sum_{\substack{\mathbf{k} \\ |e_{\mathbf{k}}| \leq \omega_D}} \left\{ a_{\mathbf{k}\uparrow}^+ a_{-\mathbf{k},\downarrow}^+ \Delta + \bar{\Delta} \, a_{-\mathbf{k}\downarrow} a_{\mathbf{k}\uparrow} \right\} - \lambda L^d |\Delta|^2 \qquad (6.14)
$$

where Δ has to be determined selfconsistently by the BCS equation

$$
\Delta = \frac{\lambda}{L^{2d}} \sum_{\substack{\mathbf{p} \\ |e_{\mathbf{p}}| \leq \omega_D}} \frac{Tr\,[a_{-\mathbf{p}\downarrow} a_{\mathbf{p}\uparrow} \, e^{-\beta(H_0 + H_{\mathrm{MF}})}]}{Tr\, e^{-\beta(H_0 + H_{\mathrm{MF}})}} \qquad (6.15)
$$

This mean field model is diagonalized in the following

Theorem 6.1.1 *Let* $H = H_0 + H_{MF} = \sum_{\mathbf{k}} h_{\mathbf{k}}$ *where*

$$h_{\mathbf{k}} = \frac{1}{L^d} e_{\mathbf{k}} \left(a^+_{\mathbf{k}\uparrow} a_{\mathbf{k}\uparrow} + a^+_{-\mathbf{k}\downarrow} a_{-\mathbf{k}\downarrow} \right) \tag{6.16}$$
$$- \frac{\lambda}{L^d} \chi(|e_{\mathbf{k}}| \le \omega_D) \left(a^+_{\mathbf{k}\uparrow} a^+_{-\mathbf{k}\downarrow} \Delta + \bar{\Delta}\, a_{-\mathbf{k}\downarrow} a_{\mathbf{k}\uparrow} \right) - \lambda L^d |\Delta|^2$$

Then one has

a) $[h_{\mathbf{k}}, h_{\mathbf{k}'}] = 0 \ \forall \mathbf{k}, \mathbf{k}'$, *that is, the* $h_{\mathbf{k}}$ *can be simultaneously diagonalized.*

b) *Let*

$$t_{\mathbf{k}} = \begin{cases} 0 & \text{for } e_{\mathbf{k}} > \omega_D \\ \frac{1}{2} \arctan_{(-\pi,0)}\left(-\frac{\Delta}{e_{\mathbf{k}}} \right) & \text{for } |e_{\mathbf{k}}| \le \omega_D \\ -\frac{\pi}{2} & \text{for } e_{\mathbf{k}} < -\omega_D \end{cases} \tag{6.17}$$

and

$$S_{\mathbf{k}} := \frac{1}{L^d} \left(a^+_{\mathbf{k}\uparrow} a^+_{-\mathbf{k}\downarrow} - a_{-\mathbf{k}\downarrow} a_{\mathbf{k}\uparrow} \right), \quad U_{\mathbf{k}} := e^{t_{\mathbf{k}} S_{\mathbf{k}}} \tag{6.18}$$

Then $[U_{\mathbf{k}}, U_{\mathbf{k}'}] = 0 \ \forall \mathbf{k}, \mathbf{k}'$ *and*

$$U_{\mathbf{k}} h_{\mathbf{k}} U^+_{\mathbf{k}} = \frac{1}{L^d} E_{\mathbf{k}} \left(a^+_{\mathbf{k}\uparrow} a_{\mathbf{k}\uparrow} + a^+_{-\mathbf{k}\downarrow} a_{-\mathbf{k}\downarrow} \right) - (E_{\mathbf{k}} - e_{\mathbf{k}}) \tag{6.19}$$

where

$$E_{\mathbf{k}} := \begin{cases} \sqrt{e_{\mathbf{k}}^2 + |\Delta|^2} & \text{for } |e_{\mathbf{k}}| \le \omega_D \\ |e_{\mathbf{k}}| & \text{for } |e_{\mathbf{k}}| > \omega_D \end{cases} \tag{6.20}$$

c) *Let* $U := \prod_{\mathbf{k}} U_{\mathbf{k}}$. *Then*

$$UHU^+ = \frac{1}{L^d} \sum_{\mathbf{k}\sigma} E_{\mathbf{k}}\, a^+_{\mathbf{k}\sigma} a_{\mathbf{k}\sigma} - \sum_{\mathbf{k}} (E_{\mathbf{k}} - e_{\mathbf{k}}) \tag{6.21}$$

and $E_0 := \inf \sigma(H) = -\sum_{\mathbf{k}} (E_{\mathbf{k}} - e_{\mathbf{k}})$ *and* $E_1 \ge E_0 + |\Delta| \ \forall E_1 \in \sigma(H)$, $E_1 \ne E_0$.

d) *The ground state* $\Omega_0 = U^+ \mathbf{1}$ *of* H *is given by*

$$\Omega_0 = \prod_{\substack{\mathbf{k} \\ |e_{\mathbf{k}}| \le \omega_D}} \left(\cos t_{\mathbf{k}} - \frac{1}{L^d} \sin t_{\mathbf{k}}\, a^+_{\mathbf{k}\uparrow} a^+_{-\mathbf{k}\downarrow} \right) \prod_{\substack{\mathbf{k} \\ e_{\mathbf{k}} < -\omega_D}} \left(\frac{1}{L^d} a^+_{\mathbf{k}\uparrow} a^+_{-\mathbf{k}\downarrow} \right) \mathbf{1} \tag{6.22}$$

where $\mathbf{1} = (1, 0, 0, \cdots) \in \oplus_n \mathcal{F}_n$ *is the vacuum state.*

e) *The BCS-equation (6.15) reads*

$$\frac{\lambda}{L^d} \sum_{\substack{\mathbf{k} \\ |e_{\mathbf{k}}| \le \omega_D}} \frac{\tanh(\beta E_{\mathbf{k}}/2)}{2E_{\mathbf{k}}} = 1 \tag{6.23}$$

and for attractive $\lambda > 0$ there is the nonanalytic zero temperature solution

$$\Delta(T = 0) \approx 2\omega_D\, e^{-\frac{1}{n_d\lambda}} \qquad (6.24)$$

where $n_d = \omega_{d-1}mk_F^{d-2}/(2\pi)^d$, ω_d being the surface of the d-dimensional unit sphere.

f) *The momentum distribution is given by*

$$\frac{1}{L^d}\langle a_{\mathbf{k}\sigma}^+ a_{\mathbf{k}\sigma}\rangle = \frac{1}{2}\left(1 - e_{\mathbf{k}}\frac{\tanh(\beta E_{\mathbf{k}}/2)}{E_{\mathbf{k}}}\right)$$

$$\overset{\beta\to\infty}{\longrightarrow} \frac{1}{2}\left(1 - \frac{e_{\mathbf{k}}}{E_{\mathbf{k}}}\right) \qquad (6.25)$$

and the anomalous $\langle aa\rangle$, $\langle a^+a^+\rangle$ expectations become nonzero and are given by

$$\frac{1}{L^d}\langle a_{-\mathbf{k}\downarrow} a_{\mathbf{k}\uparrow}\rangle = \frac{\Delta}{2E_{\mathbf{k}}}\tanh(\beta E_{\mathbf{k}}/2) \qquad (6.26)$$

and $\langle a_{\mathbf{k}\uparrow}^+ a_{-\mathbf{k}\downarrow}^+\rangle = \overline{\langle a_{-\mathbf{k}\downarrow} a_{\mathbf{k}\uparrow}\rangle}$.

Proof: a) This follows from

$$[AB, CD] = A\{B,C\}D - AC\{B,D\} + \{A,C\}DB - C\{A,D\}B \qquad (6.27)$$

and

$$\mathbf{k} \neq \mathbf{k}' \;\Rightarrow\; \{a_{\mathbf{k}\uparrow}^{(+)} \text{ or } a_{-\mathbf{k}\downarrow}^{(+)},\, a_{\mathbf{k}'\uparrow}^{(+)} \text{ or } a_{-\mathbf{k}'\downarrow}^{(+)}\} = 0 \qquad (6.28)$$

b) Since $S_{\mathbf{k}}^+ = -S_{\mathbf{k}}$ we have $U_{\mathbf{k}}^+ = e^{-t_{\mathbf{k}}S_{\mathbf{k}}} = U_{\mathbf{k}}^{-1}$ and $U_{\mathbf{k}}$ is unitary. Because of (6.27,6.28) all $S_{\mathbf{k}}$ commute. Furthermore

$$U_{\mathbf{k}}h_{\mathbf{k}}U_{\mathbf{k}}^+ = \frac{1}{L^d}e_{\mathbf{k}}\big(a_{\mathbf{k}\uparrow}^+(t_{\mathbf{k}})a_{\mathbf{k}\uparrow}(t_{\mathbf{k}}) + a_{-\mathbf{k}\downarrow}^+(t_{\mathbf{k}})a_{-\mathbf{k}\downarrow}(t_{\mathbf{k}})\big) \qquad (6.29)$$
$$- \frac{\lambda}{L^d}\chi(|e_{\mathbf{k}}| \leq \omega_D)\big(a_{\mathbf{k}\uparrow}^+(t_{\mathbf{k}})a_{-\mathbf{k}\downarrow}^+(t_{\mathbf{k}})\Delta + \bar{\Delta}\,a_{-\mathbf{k}\downarrow}(t_{\mathbf{k}})a_{\mathbf{k}\uparrow}(t_{\mathbf{k}})\big)$$

where for some operator A

$$A(t) := e^{tS_{\mathbf{k}}}Ae^{-tS_{\mathbf{k}}} \qquad (6.30)$$

Since

$$\frac{d}{dt}A(t) = e^{tS_{\mathbf{k}}}[S_{\mathbf{k}}, A]e^{-tS_{\mathbf{k}}} \qquad (6.31)$$

we have

$$\frac{d}{dt}\begin{pmatrix} a_{\mathbf{k}\uparrow}(t) \\ a_{-\mathbf{k}\downarrow}^+(t) \\ a_{-\mathbf{k}\downarrow}(t) \\ a_{\mathbf{k}\uparrow}^+(t) \end{pmatrix} = e^{tS_{\mathbf{k}}}\frac{1}{L^d}\left[a_{\mathbf{k}\uparrow}^+ a_{-\mathbf{k}\downarrow}^+ - a_{-\mathbf{k}\downarrow}a_{\mathbf{k}\uparrow},\, \begin{pmatrix} a_{\mathbf{k}\uparrow} \\ a_{-\mathbf{k}\downarrow}^+ \\ a_{-\mathbf{k}\downarrow} \\ a_{\mathbf{k}\uparrow}^+ \end{pmatrix}\right]e^{-tS_{\mathbf{k}}} \qquad (6.32)$$

Using $[AB, C] = A\{B, C\} - \{A, C\}B$ the commutator is found to be

$$\frac{1}{L^d}\left[a^+_{\mathbf{k}\uparrow}a^+_{-\mathbf{k}\downarrow} - a_{-\mathbf{k}\downarrow}a_{\mathbf{k}\uparrow}, \begin{pmatrix} a_{\mathbf{k}\uparrow} \\ a^+_{-\mathbf{k}\downarrow} \\ a_{-\mathbf{k}\downarrow} \\ a^+_{\mathbf{k}\uparrow} \end{pmatrix}\right] = \begin{pmatrix} -a^+_{-\mathbf{k}\downarrow} \\ a_{\mathbf{k}\uparrow} \\ a^+_{\mathbf{k}\uparrow} \\ -a_{-\mathbf{k}\downarrow} \end{pmatrix}$$

$$= \begin{pmatrix} 0 & -1 \\ 1 & 0 \\ & & 0 & 1 \\ & & -1 & 0 \end{pmatrix}\begin{pmatrix} a_{\mathbf{k}\uparrow} \\ a^+_{-\mathbf{k}\downarrow} \\ a_{-\mathbf{k}\downarrow} \\ a^+_{\mathbf{k}\uparrow} \end{pmatrix} \tag{6.33}$$

which gives

$$\begin{pmatrix} a_{\mathbf{k}\uparrow}(t) \\ a^+_{-\mathbf{k}\downarrow}(t) \end{pmatrix} = \begin{pmatrix} \cos t & -\sin t \\ \sin t & \cos t \end{pmatrix}\begin{pmatrix} a_{\mathbf{k}\uparrow} \\ a^+_{-\mathbf{k}\downarrow} \end{pmatrix} \tag{6.34}$$

$$\begin{pmatrix} a_{-\mathbf{k}\downarrow}(t) \\ a^+_{\mathbf{k}\uparrow}(t) \end{pmatrix} = \begin{pmatrix} \cos t & \sin t \\ -\sin t & \cos t \end{pmatrix}\begin{pmatrix} a_{-\mathbf{k}\downarrow} \\ a^+_{\mathbf{k}\uparrow} \end{pmatrix} \tag{6.35}$$

Thus, putting $\chi_{\mathbf{k}} := \chi(|e_{\mathbf{k}}| \leq \omega_D)$,

$$U_{\mathbf{k}}h_{\mathbf{k}}U^+_{\mathbf{k}} =$$

$$\frac{1}{L^d}\left(a^+_{\mathbf{k}\uparrow}a_{\mathbf{k}\uparrow} + a^+_{-\mathbf{k}\downarrow}a_{-\mathbf{k}\downarrow}\right)\{e_{\mathbf{k}}(\cos^2 t_{\mathbf{k}} - \sin^2 t_{\mathbf{k}}) - 2\chi_{\mathbf{k}}\Delta\sin t_{\mathbf{k}}\cos t_{\mathbf{k}}\}$$

$$+ \frac{1}{L^d}\left(a^+_{\mathbf{k}\uparrow}a^+_{-\mathbf{k}\downarrow} + a_{-\mathbf{k}\downarrow}a_{\mathbf{k}\uparrow}\right)\{-e_{\mathbf{k}}2\sin t_{\mathbf{k}}\cos t_{\mathbf{k}} - \chi_{\mathbf{k}}\Delta(\cos^2 t_{\mathbf{k}} - \sin^2 t_{\mathbf{k}})\}$$

$$+ 2e_{\mathbf{k}}\sin^2 t_{\mathbf{k}} + 2\chi_{\mathbf{k}}\Delta\sin t_{\mathbf{k}}\cos t_{\mathbf{k}} \tag{6.36}$$

Now we choose $t_{\mathbf{k}}$ such that the wavy brackets in the third line of (6.36) vanish. That is,

$$e_{\mathbf{k}}\sin 2t_{\mathbf{k}} + \chi_{\mathbf{k}}\Delta\cos 2t_{\mathbf{k}} = 0 \tag{6.37}$$

or, for $|e_{\mathbf{k}}| \leq \omega_D$,

$$t_{\mathbf{k}} = \frac{1}{2}\arctan\left(-\frac{\Delta}{e_{\mathbf{k}}}\right)$$

This gives

$$\cos 2t_{\mathbf{k}} = \frac{e_{\mathbf{k}}}{\sqrt{e_{\mathbf{k}}^2 + \Delta^2}}, \quad \sin 2t_{\mathbf{k}} = -\frac{\Delta}{\sqrt{e_{\mathbf{k}}^2 + \Delta^2}} \tag{6.38}$$

and

$$2e_{\mathbf{k}}\sin^2 t_{\mathbf{k}} + 2\Delta\sin t_{\mathbf{k}}\cos t_{\mathbf{k}} = e_{\mathbf{k}}(1 - \cos 2t_{\mathbf{k}}) + \Delta\sin 2t_{\mathbf{k}} = e_{\mathbf{k}} - E_{\mathbf{k}} \tag{6.39}$$

which proves part (b). Part **(c)** is an immediate consequence of (b).

d) Since the vacuum state $\mathbf{1}$ is the ground state of UHU^+, the ground state of H is given by $\Omega_0 = U^+\mathbf{1} = \Pi_{\mathbf{k}}\, e^{-t_{\mathbf{k}}S_{\mathbf{k}}}\mathbf{1}$. One has

$$e^{-t_{\mathbf{k}}S_{\mathbf{k}}}\mathbf{1} = \sum_{n=0}^{\infty} \frac{(-t_{\mathbf{k}})^n}{n!}\frac{1}{L^{nd}}\left(a_{\mathbf{k}\uparrow}^+ a_{-\mathbf{k}\downarrow}^+ - a_{-\mathbf{k}\downarrow}a_{\mathbf{k}\uparrow}\right)^n \mathbf{1}$$

and

$$\left(a_{\mathbf{k}\uparrow}^+ a_{-\mathbf{k}\downarrow}^+ - a_{-\mathbf{k}\downarrow}a_{\mathbf{k}\uparrow}\right)\mathbf{1} = a_{\mathbf{k}\uparrow}^+ a_{-\mathbf{k}\downarrow}^+\mathbf{1}$$

$$\left(a_{\mathbf{k}\uparrow}^+ a_{-\mathbf{k}\downarrow}^+ - a_{-\mathbf{k}\downarrow}a_{\mathbf{k}\uparrow}\right)^2 \mathbf{1} = -a_{-\mathbf{k}\downarrow}a_{\mathbf{k}\uparrow}a_{\mathbf{k}\uparrow}^+ a_{-\mathbf{k}\downarrow}^+\mathbf{1} = -L^{2d}\mathbf{1}$$

which gives

$$S_{\mathbf{k}}^{2n}\mathbf{1} = (-1)^n\mathbf{1}, \qquad S_{\mathbf{k}}^{2n+1}\mathbf{1} = (-1)^n \frac{1}{L^d} a_{\mathbf{k}\uparrow}^+ a_{-\mathbf{k}\downarrow}^+\mathbf{1} \qquad (6.40)$$

and part (d) follows.

e,f) The momentum distribution is given by

$$
\begin{aligned}
\langle a_{\mathbf{k}\uparrow}^+ a_{\mathbf{k}\uparrow}\rangle &= Tr[e^{-\beta H}a_{\mathbf{k}\uparrow}^+ a_{\mathbf{k}\uparrow}]/Tr\, e^{-\beta H}\\
&= Tr[U^+Ue^{-\beta H}U^+Ua_{\mathbf{k}\uparrow}^+ U^+Ua_{\mathbf{k}\uparrow}]/Tr\, U^+Ue^{-\beta H}\\
&= Tr[e^{-\beta UHU^+}Ua_{\mathbf{k}\uparrow}^+ U^+Ua_{\mathbf{k}\uparrow}U^+]/Tr\, e^{-\beta UHU^+}\\
&= Tr\left[e^{-\frac{\beta}{L^d}\sum_{\mathbf{k}\sigma}E_{\mathbf{k}}a_{\mathbf{k}\sigma}^+ a_{\mathbf{k}\sigma}}(a_{\mathbf{k}\uparrow}^+\cos t_{\mathbf{k}} - a_{-\mathbf{k}\downarrow}\sin t_{\mathbf{k}})\times\right.\\
&\qquad \left.(a_{\mathbf{k}\uparrow}\cos t_{\mathbf{k}} - a_{-\mathbf{k}\downarrow}^+\sin t_{\mathbf{k}})\right]\Big/ Tr\, e^{-\frac{\beta}{L^d}\sum_{\mathbf{k}\sigma}E_{\mathbf{k}}a_{\mathbf{k}\sigma}^+ a_{\mathbf{k}\sigma}}\\
&= \frac{Tr\left[e^{-\frac{\beta}{L^d}\sum_{\mathbf{k}\sigma}E_{\mathbf{k}}a_{\mathbf{k}\sigma}^+ a_{\mathbf{k}\sigma}}(a_{\mathbf{k}\uparrow}^+ a_{\mathbf{k}\uparrow}\cos^2 t_{\mathbf{k}} + a_{-\mathbf{k}\downarrow}a_{-\mathbf{k}\downarrow}^+\sin^2 t_{\mathbf{k}})\right]}{Tr\, e^{-\frac{\beta}{L^d}\sum_{\mathbf{k}\sigma}E_{\mathbf{k}}a_{\mathbf{k}\sigma}^+ a_{\mathbf{k}\sigma}}}\\
&= \cos 2t_{\mathbf{k}}\frac{Tr\left[e^{-\frac{\beta}{L^d}\sum_{\mathbf{k}\sigma}E_{\mathbf{k}}a_{\mathbf{k}\sigma}^+ a_{\mathbf{k}\sigma}}a_{\mathbf{k}\uparrow}^+ a_{\mathbf{k}\uparrow}\right]}{Tr\, e^{-\frac{\beta}{L^d}\sum_{\mathbf{k}\sigma}E_{\mathbf{k}}a_{\mathbf{k}\sigma}^+ a_{\mathbf{k}\sigma}}} + L^d\sin^2 t_{\mathbf{k}}\\
&= \cos 2t_{\mathbf{k}}\, L^d\frac{1}{1+e^{\beta E_{\mathbf{k}}}} + \frac{L^d}{2}(1-\cos 2t_{\mathbf{k}})\\
&= L^d\frac{1}{2}\left(1-\cos 2t_{\mathbf{k}}[1-\frac{2}{1+e^{\beta E_{\mathbf{k}}}}]\right)\\
&= L^d\frac{1}{2}\left(1-\frac{e_{\mathbf{k}}}{E_{\mathbf{k}}}\tanh(\beta E_{\mathbf{k}}/2)\right) \qquad (6.41)
\end{aligned}
$$

where we used (6.38) in the last line and in the sixth line we used $\langle a_{\mathbf{k}\uparrow}^+ a_{\mathbf{k}\uparrow}\rangle =$

$\langle a^+_{-\mathbf{k}\downarrow} a_{-\mathbf{k}\downarrow}\rangle$. This proves (6.26). Similarly

$$\langle a_{-\mathbf{k}\downarrow} a_{\mathbf{k}\uparrow}\rangle = Tr\left[e^{-\frac{\beta}{L^d}\sum_{\mathbf{k}\sigma} E_{\mathbf{k}} a^+_{\mathbf{k}\sigma} a_{\mathbf{k}\sigma}}(a_{-\mathbf{k}\downarrow}\cos t_{\mathbf{k}} + a^+_{\mathbf{k}\uparrow}\sin t_{\mathbf{k}}) \times\right.$$

$$\left.(a_{\mathbf{k}\uparrow}\cos t_{\mathbf{k}} - a^+_{-\mathbf{k}\downarrow}\sin t_{\mathbf{k}})\right]\Big/ Tr\, e^{-\frac{\beta}{L^d}\sum_{\mathbf{k}\sigma} E_{\mathbf{k}} a^+_{\mathbf{k}\sigma} a_{\mathbf{k}\sigma}}$$

$$= \sin 2t_{\mathbf{k}} \frac{Tr\left[e^{-\frac{\beta}{L^d}\sum_{\mathbf{k}\sigma} E_{\mathbf{k}} a^+_{\mathbf{k}\sigma} a_{\mathbf{k}\sigma}} a^+_{\mathbf{k}\uparrow} a_{\mathbf{k}\uparrow}\right]}{Tr\, e^{-\frac{\beta}{L^d}\sum_{\mathbf{k}\sigma} E_{\mathbf{k}} a^+_{\mathbf{k}\sigma} a_{\mathbf{k}\sigma}}} - \frac{1}{2} L^d \sin 2t_{\mathbf{k}}$$

$$= L^d \sin 2t_{\mathbf{k}} \left(\frac{1}{1+e^{\beta E_{\mathbf{k}}}} - \frac{1}{2}\right)$$

$$= L^d \frac{\Delta}{2E_{\mathbf{k}}} \tanh(\beta E_{\mathbf{k}}/2) \tag{6.42}$$

where we used (6.38) again. The BCS equation (6.15) becomes

$$\Delta = \frac{\lambda}{L^{2d}} \sum_{\substack{\mathbf{k}\\|e_{\mathbf{k}}|\le\omega_D}} \frac{Tr\left[a_{-\mathbf{k}\downarrow} a_{\mathbf{k}\uparrow}\, e^{-\beta(H_0+H_{MF})}\right]}{Tr\, e^{-\beta(H_0+H_{MF})}}$$

$$= \frac{\lambda}{L^d} \sum_{\substack{\mathbf{k}\\|e_{\mathbf{k}}|\le\omega_D}} \frac{\Delta}{2E_{\mathbf{k}}} \tanh(\beta E_{\mathbf{k}}/2) \overset{\beta\to\infty}{\to} \frac{\lambda}{L^d} \sum_{\substack{\mathbf{k}\\|e_{\mathbf{k}}|\le\omega_D}} \frac{\Delta}{2E_{\mathbf{k}}}$$

$$\overset{L\to\infty}{\to} \lambda\Delta \int \frac{d^d\mathbf{k}}{(2\pi)^d} \frac{\chi(|e_{\mathbf{k}}|\le\omega_D)}{2\sqrt{e_{\mathbf{k}}^2+\Delta^2}}$$

$$= \lambda\Delta \frac{\omega_{d-1}}{2(2\pi)^d} \int d\rho\, \rho^{d-1} \frac{\chi(|\rho^2/2m-\mu|\le\omega_D)}{\sqrt{(\rho-k_F)^2(\rho+k_F)^2/4m^2+\Delta^2}}$$

$$\approx \lambda\Delta \frac{\omega_{d-1}}{2(2\pi)^d} \int d\rho\, \rho^{d-1} \frac{\chi(\frac{k_F}{m}|\rho-k_F|\le\omega_D)}{\sqrt{\frac{k_F^2}{m^2}(\rho-k_F)^2+\Delta^2}}$$

$$\approx \lambda\Delta \frac{\omega_{d-1}}{2(2\pi)^d} k_F^{d-1} \frac{1}{\Delta} \int d\rho \frac{\chi(\frac{k_F}{\Delta m}|\rho-k_F|\le\frac{\omega_D}{\Delta})}{\sqrt{\frac{k_F^2}{\Delta^2 m^2}(\rho-k_F)^2+1}}$$

$$= \lambda\Delta \frac{\omega_{d-1}}{2(2\pi)^d} k_F^{d-1} \frac{m}{k_F} 2\int_0^{\frac{\omega_D}{\Delta}} \frac{dx}{\sqrt{x^2+1}}$$

$$= \lambda\Delta \frac{\omega_{d-1} k_F^{d-2} m}{(2\pi)^d} \log\left[\frac{\omega_D}{\Delta} + \sqrt{(\frac{\omega_D}{\Delta})^2+1}\right]$$

$$\approx \lambda\Delta \frac{\omega_{d-1} k_F^{d-2} m}{(2\pi)^d} \log\left[2\frac{\omega_D}{\Delta}\right] \tag{6.43}$$

which gives $\Delta = 2\omega_D\, e^{-\frac{const}{\lambda}}$ as stated in the theorem. ∎

6.2 The Quartic BCS Model

In this section we explicitly solve the model defined by the quartic Hamiltonian

$$H = H_0 + H_{\text{BCS}}^{\ell=0}$$
$$= \tfrac{1}{L^d} \sum_{\mathbf{k}\sigma} e_{\mathbf{k}} a_{\mathbf{k}\sigma}^+ a_{\mathbf{k}\sigma} + \tfrac{\lambda}{L^{3d}} \sum_{\substack{\mathbf{k},\mathbf{p} \\ |e_{\mathbf{k}}|,|e_{\mathbf{p}}| \le \omega_D}} a_{\mathbf{k}\uparrow}^+ a_{-\mathbf{k},\downarrow}^+ a_{-\mathbf{p},\downarrow} a_{\mathbf{p}\uparrow} \qquad (6.44)$$

The correlation functions are identical to those of the quadratic mean field model. In the next section we add higher ℓ-wave terms to the electron-electron interaction. While that model is still solvable, since the important point in the BCS approximation is 'putting $\mathbf{q} = 0$ without canceling a volume factor' which results in a volume dependence which force the integration variables in the bosonic functional integral representation to take values at the global minimum of the effective potential, the correlation functions of that model are no longer identical to those obtained by applying the quadratic Anderson Brinkmann mean field formalism. This is discussed in the next section. Here we prove

Theorem 6.2.1 *Let H be the quartic BCS Hamiltonian given by (6.44) and for some small 'external field' $r = |r|e^{i\alpha} \in \mathbb{C}$ let*

$$H_r = H + \tfrac{1}{L^d} \sum_{\mathbf{k}} \left(\bar{r} a_{\mathbf{k}\uparrow}^+ a_{-\mathbf{k}\downarrow}^+ + r a_{-\mathbf{k}\downarrow} a_{\mathbf{k}\uparrow} \right) \qquad (6.45)$$

For some operator A let $\langle A \rangle = Tr[e^{-\beta H} A]/Tr\, e^{-\beta H}$ and $\langle A \rangle_r$ the same expression with H replaced by H_r. Then

$$\lim_{r \to 0} \lim_{L \to \infty} \tfrac{1}{L^d} \langle a_{\mathbf{k}\sigma}^+ a_{\mathbf{k}\sigma} \rangle_r = \lim_{L \to \infty} \tfrac{1}{L^d} \langle a_{\mathbf{k}\sigma}^+ a_{\mathbf{k}\sigma} \rangle = \tfrac{1}{2} \left(1 - e_{\mathbf{k}} \frac{\tanh(\beta E_{\mathbf{k}}/2)}{E_{\mathbf{k}}} \right) \quad (6.46)$$

and there is symmetry breaking in the sense that

$$\lim_{r \to 0} \lim_{L \to \infty} \tfrac{1}{L^d} \langle a_{-\mathbf{k}\downarrow} a_{\mathbf{k}\uparrow} \rangle_r = \tfrac{\Delta}{2E_{\mathbf{k}}} \tanh(\beta E_{\mathbf{k}}/2) \ne 0 \qquad (6.47)$$

while

$$\lim_{L \to \infty} \lim_{r \to 0} \tfrac{1}{L^d} \langle a_{-\mathbf{k}\downarrow} a_{\mathbf{k}\uparrow} \rangle = 0 \qquad (6.48)$$

if $\Delta = |\Delta|e^{-i\alpha}$ and $|\Delta|$ is a solution of the BCS equation (6.23).

Proof: The Grassmann integral representation of $Z = Tr\, e^{-\beta H_r}/Tr\, e^{-\beta H_0}$, H_0 given by (6.1) (not to be confused with $H_{r=0} = H_0 + H_{BCS}$) is given by $Z = Z(s_{k\sigma} = 0, r_k = r)$ where (recall that $\kappa := \beta L^d$)

$$Z(\{s_{k\sigma}, r_k\}) = \int e^{-\frac{\lambda}{\kappa^3} \sum_{k,p,q_0} \bar\psi_{k_0,\mathbf{k},\uparrow} \bar\psi_{q_0-k_0,-\mathbf{k},\downarrow} \psi_{p_0,\mathbf{p},\uparrow} \psi_{q_0-p_0,-\mathbf{p},\downarrow}} \times \qquad (6.49)$$

$$e^{\frac{1}{\kappa} \sum_k \left(\bar r_k \bar\psi_{k\uparrow} \bar\psi_{-k\downarrow} + r_k \psi_{-k\downarrow} \psi_{k\uparrow} \right)} \times$$

$$\prod_{k\sigma} \frac{\kappa}{ik_0 - e_{\mathbf{k}}} e^{-\frac{1}{\kappa} \sum_{k\sigma} (ik_0 - e_{\mathbf{k}} - s_{k\sigma}) \bar\psi_{k\sigma} \psi_{k\sigma}} \prod_{k\sigma} d\psi_{k\sigma} d\bar\psi_{k\sigma}$$

Here $k = (k_0, \mathbf{k})$, $k_0, p_0 \in \frac{\pi}{\beta}(2\mathbb{Z} + 1)$, $q_0 \in \frac{2\pi}{\beta}\mathbb{Z}$. By making a Hubbard-Stratonovich transformation, we obtain as in Theorem 5.1.2

$$Z(\{s_{k\sigma}, r_k\}) = \int \frac{\det\left[\cdots \right]}{\det\left[\cdot \right]} e^{-\sum_{q_0} |\phi_{q_0}|^2} \prod_{q_0} \frac{d\phi_{q_0} d\bar\phi_{q_0}}{\pi} \qquad (6.50)$$

where the quotient of determinants is given by

$$\det \begin{bmatrix} (a_k - s_{k\uparrow})\delta_{k,p} & \left(\frac{ig}{\sqrt{\kappa}} \bar\phi_{p_0-k_0} - \bar r_k \delta_{p_0,k_0} \right)\delta_{\mathbf{p},\mathbf{k}} \\ \left(\frac{ig}{\sqrt{\kappa}} \phi_{k_0-p_0} + r_k \delta_{k_0,p_0} \right)\delta_{\mathbf{k},\mathbf{p}} & (a_{-k} - s_{-k\downarrow})\delta_{k,p} \end{bmatrix} \Big/ \det \begin{bmatrix} a_k \delta_{k,p} & 0 \\ 0 & a_{-k}\delta_{k,p} \end{bmatrix} = \qquad (6.51)$$

$$\prod_{\mathbf{k}} \frac{\det \begin{bmatrix} (a_{k_0,\mathbf{k}} - s_{k_0,\mathbf{k},\uparrow})\delta_{k_0,p_0} & \frac{ig}{\sqrt{\kappa}} \bar\phi_{p_0-k_0} - \bar r_{k_0,\mathbf{k}}\delta_{p_0,k_0} \\ \frac{ig}{\sqrt{\kappa}} \phi_{k_0-p_0} + r_{k_0,\mathbf{k}}\delta_{k_0,p_0} & (a_{-k_0,-\mathbf{k}} - s_{-k_0,-\mathbf{k},\downarrow})\delta_{k_0,p_0} \end{bmatrix}}{\det \begin{bmatrix} a_{k_0,\mathbf{k}}\delta_{k_0,p_0} & 0 \\ 0 & a_{-k_0,-\mathbf{k}}\delta_{k_0,p_0} \end{bmatrix}}$$

Observe that in the first line of (6.51) the matrices are labelled by $k, p = (k_0, \mathbf{k}), (p_0, \mathbf{p})$ whereas in the second line of (6.51) the matrices are only labelled by k_0, p_0 and there is a product over spatial momenta $\mathbf{k}$ since the matrices in the first line of (6.51) are diagonal in the spatial momenta $\mathbf{k}, \mathbf{p}$. Since we are dealing with the BCS approximation which is obtained from the full model by retaining only the $\mathbf{q} = 0$ term of the quartic part, the integration variables are no longer $\phi_{\mathbf{q},q_0}$ but only $\phi_{\mathbf{q}=0,q_0} \equiv \phi_{q_0}$. As in Theorem 5.1.2, the correlation functions are obtained from (6.50) by differentiating with respect to $s_{t\sigma}$ or r_t, $t = (t_0, \mathbf{t})$:

$$\frac{1}{\kappa} \langle \bar\psi_{t\sigma} \psi_{t\sigma} \rangle = \int \begin{bmatrix} a_{k_0,t}\delta_{k_0,p_0} & \frac{ig}{\sqrt{\beta}} \bar\phi_{p_0-k_0} - \bar r\delta_{p_0,k_0} \\ \frac{ig}{\sqrt{\beta}} \phi_{k_0-p_0} + r\delta_{k_0,p_0} & a_{-k_0,-t}\delta_{k_0,p_0} \end{bmatrix}^{-1}_{t_0\sigma,t_0\sigma} dP_r(\{\phi_{q_0}\})$$

$$\qquad (6.52)$$

$$\frac{1}{\kappa} \langle \psi_{t\uparrow} \psi_{-t\downarrow} \rangle = \int \begin{bmatrix} a_{k_0,t}\delta_{k_0,p_0} & \frac{ig}{\sqrt{\beta}} \bar\phi_{p_0-k_0} - \bar r\delta_{p_0,k_0} \\ \frac{ig}{\sqrt{\beta}} \phi_{k_0-p_0} + r\delta_{k_0,p_0} & a_{-k_0,-t}\delta_{k_0,p_0} \end{bmatrix}^{-1}_{t_0\downarrow,t_0\uparrow} dP_r(\{\phi_{q_0}\})$$

$$\qquad (6.53)$$

where

$$dP_r(\{\phi_{q0}\}) = \frac{e^{-L^d\,V_r(\{\phi_{q0}\})}\,\Pi_{q0}\,d\phi_{q0}\,d\bar{\phi}_{q0}}{\int e^{-L^d\,V_r(\{\phi_{q0}\})}\,\Pi_{q0}\,d\phi_{q0}\,d\bar{\phi}_{q0}} \tag{6.54}$$

with an effective potential

$$V_r(\{\phi_{q0}\}) = \tag{6.55}$$

$$\sum_{q0}|\phi_{q0}|^2 - \frac{1}{L^d}\sum_{\mathbf{k}}\log\frac{\det\begin{bmatrix} a_{k0,\mathbf{k}}\delta_{k0,p0} & \frac{ig}{\sqrt{\beta}}\bar{\phi}_{p0-k0}-\bar{r}\delta_{p0,k0} \\ \frac{ig}{\sqrt{\beta}}\phi_{k0-p0}+r\delta_{k0,p0} & a_{-k0,-\mathbf{k}}\delta_{k0,p0} \end{bmatrix}}{\det\begin{bmatrix} a_{k0,\mathbf{k}}\delta_{k0,p0} & 0 \\ 0 & a_{-k0,-\mathbf{k}}\delta_{k0,p0} \end{bmatrix}}$$

If we substitute the Riemannian sum $\frac{1}{L^d}\sum_{\mathbf{k}}$ in (6.55) by an integral $\int\frac{d^d\mathbf{k}}{(2\pi)^d}$, then the only place where the volume shows up in the above formulae is the prefactor in the exponent of (6.54). Thus, in the infinite volume limit we can evaluate the integrals by evaluating the integrands at the global minimum of the effective potential, or, if this is not unique, by averaging the integrands over all minimum configurations.

To this end observe that the external field r only appears in conjunction with the ϕ_0 variable through the combination $\phi_0/\sqrt{\beta}-ir/g$. Thus, by substitution of variables ($\phi_0 = u_0 + iv_0$ and $r = |r|\,e^{i\alpha}$)

$$\int_{\mathbb{R}^2}f\left(\frac{\phi_0}{\sqrt{\beta}}-i\frac{r}{g},\frac{\bar{\phi}_0}{\sqrt{\beta}}+i\frac{\bar{r}}{g}\right)e^{-[u_0^2+v_0^2]}du_0dv_0 = \tag{6.56}$$

$$\int_{\mathbb{R}^2}f\left(e^{i\alpha}\frac{\phi_0}{\sqrt{\beta}},e^{-i\alpha}\frac{\bar{\phi}_0}{\sqrt{\beta}}\right)e^{-[u_0^2+(v_0+\sqrt{\beta}\frac{|r|}{g})^2]}du_0dv_0$$

we obtain

$$\frac{1}{\kappa}\langle\bar{\psi}_{t\sigma}\psi_{t\sigma}\rangle = \int\begin{bmatrix} a_{k0,t}\delta_{k0,p0} & \frac{ig}{\sqrt{\beta}}\bar{\bar{\phi}}_{p0-k0} \\ \frac{ig}{\sqrt{\beta}}\tilde{\phi}_{k0-p0} & a_{-k0,-t}\delta_{k0,p0} \end{bmatrix}^{-1}_{t0\sigma,t0\sigma}dQ_r(\{\phi_{q0}\}) \tag{6.57}$$

$$\frac{1}{\kappa}\langle\psi_{t\uparrow}\psi_{-t\downarrow}\rangle = \int\begin{bmatrix} a_{k0,t}\delta_{k0,p0} & \frac{ig}{\sqrt{\beta}}\bar{\bar{\phi}}_{p0-k0} \\ \frac{ig}{\sqrt{\beta}}\tilde{\phi}_{k0-p0} & a_{-k0,-t}\delta_{k0,p0} \end{bmatrix}^{-1}_{t0\downarrow,t0\uparrow}dQ_r(\{\phi_{q0}\}) \tag{6.58}$$

where

$$dQ_r(\{\phi_{q0}\}) = \frac{e^{-L^d\,U_r(\{\phi_{q0}\})}\,\Pi_{q0}\,d\phi_{q0}\,d\bar{\phi}_{q0}}{\int e^{-L^d\,U_r(\{\phi_{q0}\})}\,\Pi_{q0}\,d\phi_{q0}\,d\bar{\phi}_{q0}} \tag{6.59}$$

with an effective potential

$$U_r(\{\phi_{q_0}\}) = u_0^2 + \left(v_0 + \sqrt{\beta}\,\frac{|r|}{g}\right)^2 + \sum_{q_0 \neq 0} |\phi_{q_0}|^2 \tag{6.60}$$

$$- \frac{1}{L^d} \sum_k \log \frac{\det \begin{bmatrix} a_{k_0,k}\delta_{k_0,p_0} & \frac{ig}{\sqrt{\beta}}\,\tilde{\phi}_{p_0-k_0} \\ \frac{ig}{\sqrt{\beta}}\,\tilde{\phi}_{k_0-p_0} & a_{-k_0,-k}\delta_{k_0,p_0} \end{bmatrix}}{\det \begin{bmatrix} a_{k_0,k}\delta_{k_0,p_0} & 0 \\ 0 & a_{-k_0,-k}\delta_{k_0,p_0} \end{bmatrix}}$$

and

$$\tilde{\phi}_{q_0} := \begin{cases} \phi_{q_0} & \text{for } q_0 \neq 0 \\ e^{i\alpha}\phi_0 & \text{for } q_0 = 0 \end{cases} \tag{6.61}$$

The effective potential (6.60) is discussed in Corollary 5.2.3. More precisely, if we denote by $\tilde{U}_r$ the effective potential considered in Corollary 5.2.3, then U_r of (6.60) is related to $\tilde{U}_r$ according to $U_r(\{\phi_{q_0}\}) = \tilde{U}_r(\{\phi_{q_0,\mathbf{q}} = 0 \text{ for } \mathbf{q} \neq 0, \phi_{q_0,0} \equiv \phi_{q_0}\})$. In particular, if the global minimum of $\tilde{U}_r$ fulfills $\phi_{q_0,\mathbf{q}} = 0$ if $\mathbf{q} \neq 0$, which is case, then the global minimum of $\tilde{U}_r$ is also the global minimum of U_r. Thus we have

$$\phi_{q_0}^{\min} = \delta_{q_0,0} \sqrt{\beta}\,iy_0 \tag{6.62}$$

where

$$\lim_{r \to 0} \lambda y_0^2 = |\Delta|^2 \tag{6.63}$$

and $|\Delta|$ the solution of the BCS equation. Hence, in the infinite volume limit,

$$\lim_{L \to \infty} \frac{1}{\beta L^d} \langle \bar{\psi}_{t\sigma}\psi_{t\sigma}\rangle = \begin{bmatrix} a_{k_0,t}\delta_{k_0,p_0} & ge^{-i\alpha}y_0\delta_{p_0,k_0} \\ -ge^{i\alpha}y_0\delta_{k_0,p_0} & a_{-k_0,-t}\delta_{k_0,p_0} \end{bmatrix}^{-1}_{t_0\sigma,t_0\sigma}$$

$$= \frac{a_{-t}}{|a_t|^2 + \lambda y_0^2} = \frac{-it_0 - e_t}{t_0^2 + e_t^2 + \lambda y_0^2} \tag{6.64}$$

$$\lim_{L \to \infty} \frac{1}{\beta L^d} \langle \psi_{t\uparrow}\psi_{-t\downarrow}\rangle = \begin{bmatrix} a_{k_0,t}\delta_{k_0,p_0} & ge^{-i\alpha}y_0\delta_{p_0,k_0} \\ -ge^{i\alpha}y_0\delta_{k_0,p_0} & a_{-k_0,-t}\delta_{k_0,p_0} \end{bmatrix}^{-1}_{t_0\downarrow,t_0\uparrow}$$

$$= \frac{ge^{i\alpha}y_0}{|a_t|^2 + \lambda y_0^2} = \frac{ge^{i\alpha}y_0}{t_0^2 + e_t^2 + \lambda y_0^2} \tag{6.65}$$

Performing the sums over the t_0 variables and using (6.63) one obtains (6.46) and (6.47). (6.48) follows from symmetry. ∎

6.3   BCS with Higher ℓ-Wave Interaction

We now consider the BCS model in two or three dimensions with an electron-electron interaction of the form (6.9). To have a uniform notation for $d = 2$ or $d = 3$ we write

$$V(\hat{\mathbf{k}} - \hat{\mathbf{p}}) = \sum_{l=0}^{J} \lambda_l \bar{y}_l(\hat{\mathbf{k}}) y_l(\hat{\mathbf{p}}) \qquad (6.66)$$

where we assume

$$y_l(-\hat{\mathbf{k}}) = (-1)^l y_l(\hat{\mathbf{k}}) \qquad (6.67)$$

Thus our Hamiltonian is

$$H = H_0 + H_{\text{BCS}}^{l \leq J} \qquad (6.68)$$

$$= \frac{1}{L^d} \sum_{\mathbf{k}\sigma} e_{\mathbf{k}} a^+_{\mathbf{k}\sigma} a_{\mathbf{k}\sigma} + \frac{1}{L^{3d}} \sum_{\sigma\tau} \sum_{\substack{\mathbf{k},\mathbf{p} \\ |e_{\mathbf{k}}|,|e_{\mathbf{p}}| \leq \omega_D}} \sum_{l=0}^{J} \lambda_l \bar{y}_l(\hat{\mathbf{k}}) y_l(\hat{\mathbf{p}}) a^+_{\mathbf{k}\sigma} a^+_{-\mathbf{k},\tau} a_{-\mathbf{p},\tau} a_{\mathbf{p}\sigma}$$

In the easiest case only even l terms contribute to (6.68). In that case only $a_\uparrow a_\downarrow$ pairs give a nonzero contribution since

$$\sum_{\mathbf{k}} \bar{y}_l(\hat{\mathbf{k}}) \, a^+_{\mathbf{k}\uparrow} a^+_{-\mathbf{k}\uparrow} = \sum_{\mathbf{k}'} \bar{y}_l(-\hat{\mathbf{k}'}) \, a^+_{-\mathbf{k}'\uparrow} a^+_{\mathbf{k}'\uparrow}$$

$$= -\sum_{\mathbf{k}'}(-1)^l \bar{y}_l(\hat{\mathbf{k}'}) \, a^+_{\mathbf{k}'\uparrow} a^+_{-\mathbf{k}'\uparrow}$$

$$\stackrel{l \text{ even}}{=} -\sum_{\mathbf{k}} \bar{y}_l(\hat{\mathbf{k}}) \, a^+_{\mathbf{k}\uparrow} a^+_{-\mathbf{k}\uparrow} \qquad (6.69)$$

Thus, for even l, the Hamiltonian becomes

$$H = \frac{1}{L^d} \sum_{\mathbf{k}\sigma} e_{\mathbf{k}} a^+_{\mathbf{k}\sigma} a_{\mathbf{k}\sigma} + \frac{1}{L^{3d}} \sum_{\substack{\mathbf{k},\mathbf{p} \\ |e_{\mathbf{k}}|,|e_{\mathbf{p}}| \leq \omega_D}} \sum_{\substack{l=0 \\ l \text{ even}}}^{J} 2\lambda_l \bar{y}_l(\hat{\mathbf{k}}) y_l(\hat{\mathbf{p}}) a^+_{\mathbf{k}\uparrow} a^+_{-\mathbf{k},\downarrow} a_{-\mathbf{p},\downarrow} a_{\mathbf{p}\uparrow}$$

$$(6.70)$$

In the next theorem we compute the $\langle a^+ a \rangle$ correlation function for the model defined by (6.70) in two and three dimensions. This model, although being quartic in the annihilation and creation operators and with a nontrivial interaction, is still exactly solvable. The important point, again, is that, compared to the full model, we put '$\mathbf{q} = 0$ without canceling a volume factor' which has the effect that, in the infinite volume limit, in the bosonic functional integral representation the integration variables are forced to take values at the global minimum of the effective potential which can be computed.

Then we compare our result to that one of the standard Anderson Brinkman mean field formalism [3, 5]. While in two dimensions the results are consistent, which has been shown by A. Schütte in his Diploma thesis [58], the quadratic mean field formalism in general gives the wrong answer in three dimensions, at least in the case where there is no explicit $SO(3)$ symmetry-breaking term present. We have the following

Theorem 6.3.1 a) *Let H be given by (6.70) with $2\lambda_l$ substituted by λ_l. Then*

$$\frac{1}{L^d}\langle a_{\mathbf{k}\sigma}^+ a_{\mathbf{k}\sigma}\rangle = \lim_{\epsilon\searrow 0}\frac{1}{\beta}\sum_{k_0\in\frac{\pi}{\beta}(2\mathbb{Z}+1)}\frac{1}{\beta L^d}\langle\bar{\psi}_{\mathbf{k}\sigma}\psi_{\mathbf{k}\sigma}\rangle\,e^{-ik_0\epsilon} \qquad (6.71)$$

where

$$\frac{1}{\beta L^d}\langle\bar{\psi}_{\mathbf{k}\sigma}\psi_{\mathbf{k}\sigma}\rangle = \frac{\int\frac{-ik_0-e_{\mathbf{k}}}{k_0^2+e_{\mathbf{k}}^2+\bar{\Phi}_{\mathbf{k}}\Phi_{\mathbf{k}}}\,e^{-\beta L^d V_{\mathrm{BCS}}(\{\phi_l\})}\prod_{\substack{l=0\\l\text{ even}}}^{J}du_l dv_l}{\int e^{-\beta L^d V_{\mathrm{BCS}}(\{\phi_l\})}\prod_{\substack{l=0\\l\text{ even}}}^{J}du_l dv_l} \qquad (6.72)$$

and $(\phi_l = u_l + iv_l, \ \bar{\phi}_l = u_l - iv_l, \ g_l = \sqrt{\lambda_l})$

$$\Phi_{\mathbf{k}} := \sum_{\substack{l=0\\l\text{ even}}}^{J} g_l\,\phi_l\,y_l(\mathbf{k}), \qquad \bar{\Phi}_{\mathbf{k}} := \sum_{\substack{l=0\\l\text{ even}}}^{J} g_l\,\bar{\phi}_l\,\bar{y}_l(\mathbf{k}) \qquad (6.73)$$

and the BCS effective potential is given by

$$V_{\mathrm{BCS}}(\{\phi_l\}) = \sum_{\substack{l=0\\l\text{ even}}}^{J}|\phi_l|^2 - \int\frac{d^dk}{(2\pi)^d}\frac{2}{\beta}\log\left[\frac{\cosh(\frac{\beta}{2}\sqrt{e_{\mathbf{k}}^2+\bar{\Phi}_{\mathbf{k}}\Phi_{\mathbf{k}}})}{\cosh\frac{\beta}{2}e_{\mathbf{k}}}\right] \qquad (6.74)$$

Observe that $\bar{\Phi}_{\mathbf{k}}$ is not necessarily the complex conjugate of $\Phi_{\mathbf{k}}$ if some of the λ_l are negative.

b) *Let $d = 2$ and suppose $\lambda_{l_0} > 0$ is attractive and $\lambda_{l_0} > \lambda_l$ for all $l \neq l_0$. Then*

$$\lim_{L\to\infty}\frac{1}{\beta L^d}\langle\bar{\psi}_{\mathbf{k}\sigma}\psi_{\mathbf{k}\sigma}\rangle = -\frac{ik_0+e_{\mathbf{k}}}{k_0^2+e_{\mathbf{k}}^2+|\Delta_{l_0}|^2} \qquad (6.75)$$

where $|\Delta_{l_0}|^2$ is a solution of the BCS equation

$$\frac{\lambda_{l_0}}{L^d}\sum_{\substack{\mathbf{k}\\|e_{\mathbf{k}}|\leq\omega_D}}\frac{\tanh(\frac{\beta}{2}\sqrt{e_{\mathbf{k}}^2+|\Delta_{l_0}|^2})}{2\sqrt{e_{\mathbf{k}}^2+|\Delta_{l_0}|^2}} = 1 \qquad (6.76)$$

c) Let $d = 3$ and suppose that the electron-electron interaction in (6.66) is given by a single, even ℓ term,

$$V(\hat{\mathbf{k}} - \hat{\mathbf{p}}) = \lambda_\ell \sum_{m=-\ell}^{\ell} \bar{Y}_{\ell m}(\hat{\mathbf{k}}) Y_{\ell m}(\hat{\mathbf{p}}) \qquad (6.77)$$

Then, if $e_{R\mathbf{k}} = e_{\mathbf{k}}$ for all $R \in SO(3)$, one has

$$\lim_{L\to\infty} \frac{1}{\beta L^d} \langle \bar{\psi}_{k\sigma} \psi_{k\sigma} \rangle_{\beta,L} = \int_{S^2} \frac{ik_0 + e_{\mathbf{k}}}{k_0^2 + e_{\mathbf{k}}^2 + \lambda_\ell \rho_0^2 |\Sigma_m \alpha_m^0 Y_{\ell m}(\mathbf{p})|^2} \frac{d\Omega(\mathbf{p})}{4\pi} \qquad (6.78)$$

where $\rho_0 \geq 0$ and $\alpha^0 \in \mathbb{C}^{2\ell+1}$, $\Sigma_m |\alpha_m^0|^2 = 1$, are values at the global minimum (which is degenerate) of

$$W(\rho, \alpha) = \rho^2 - \int_M \frac{d^d k}{(2\pi)^d} \frac{1}{\beta} \log \left[\frac{\cosh(\frac{\beta}{2} \sqrt{e_{\mathbf{k}}^2 + \lambda_\ell \rho^2 |\Sigma_m \alpha_m Y_{\ell m}(\hat{\mathbf{k}})|^2})}{\cosh \frac{\beta}{2} e_{\mathbf{k}}} \right]^2 \qquad (6.79)$$

In particular, the momentum distribution is given by

$$\lim_{L\to\infty} \frac{1}{L^d} \langle a_{\mathbf{k}\sigma}^+ a_{\mathbf{k}\sigma} \rangle_{\beta,L} = \int_{S^2} \frac{1}{2} \left(1 - e_{\mathbf{k}} \frac{\tanh(\frac{\beta}{2} \sqrt{e_{\mathbf{k}}^2 + |\Delta(\mathbf{p})|^2})}{\sqrt{e_{\mathbf{k}}^2 + |\Delta(\mathbf{p})|^2}} \right) \frac{d\Omega(\mathbf{p})}{4\pi} \qquad (6.80)$$

and has $SO(3)$ symmetry. Here $\Delta(\mathbf{p}) = \lambda_\ell^{\frac{1}{2}} \rho_0 \Sigma_m \alpha_m^0 Y_{\ell m}(\mathbf{p})$.

Proof: a) The Grassmann integral representation of $Z = Tr\, e^{-\beta H}/Tr\, e^{-\beta H_0}$, H given by (6.70), is given by $Z = Z(s_{k\sigma} = 0)$ where (recall that $\kappa := \beta L^d$)

$$Z(\{s_{k\sigma}\}) = \int \exp\left\{ -\frac{1}{\kappa^3} \sum_{k,p,q_0} \bar{\psi}_{k_0,\mathbf{k},\uparrow} \bar{\psi}_{q_0-k_0,-\mathbf{k},\downarrow} \sum_{l\ \text{even}} \lambda_l \bar{y}_l(\hat{\mathbf{k}}) y_l(\hat{\mathbf{p}}) \times \right.$$

$$\left. \psi_{p_0,\mathbf{p},\uparrow} \psi_{q_0-p_0,-\mathbf{p},\downarrow} \right\} \times$$

$$\prod_{k\sigma} \frac{\kappa}{ik_0 - e_{\mathbf{k}}} e^{-\frac{1}{\kappa} \Sigma_{k\sigma}(ik_0 - e_{\mathbf{k}} - s_{k\sigma}) \bar{\psi}_{k\sigma} \psi_{k\sigma}} \prod_{k\sigma} d\psi_{k\sigma} d\bar{\psi}_{k\sigma}$$

$$= \int \exp\left\{ -\frac{1}{\kappa^3} \sum_{l\ \text{even}} \sum_{q_0} \sqrt{\lambda_l} \sum_{k_0,\mathbf{k}} \bar{y}_l(\hat{\mathbf{k}}) \bar{\psi}_{k_0,\mathbf{k},\uparrow} \bar{\psi}_{q_0-k_0,-\mathbf{k},\downarrow} \times \right.$$

$$\left. \sqrt{\lambda_l} \sum_{p_0,\mathbf{p}} y_l(\hat{\mathbf{p}}) \psi_{p_0,\mathbf{p},\uparrow} \psi_{q_0-p_0,-\mathbf{p},\downarrow} \right\} \times$$

$$\prod_{k\sigma} \frac{\kappa}{ik_0 - e_{\mathbf{k}}} e^{-\frac{1}{\kappa} \Sigma_{k\sigma}(ik_0 - e_{\mathbf{k}} - s_{k\sigma}) \bar{\psi}_{k\sigma} \psi_{k\sigma}} \prod_{k\sigma} d\psi_{k\sigma} d\bar{\psi}_{k\sigma} \qquad (6.81)$$

Here $k = (k_0, \mathbf{k})$, $k_0, p_0 \in \frac{\pi}{\beta}(2\mathbb{Z} + 1)$, $q_0 \in \frac{2\pi}{\beta}\mathbb{Z}$. By making the Hubbard-Stratonovich transformation

$$e^{-\Sigma_{l,q_0} a_{l,q_0} b_{l,q_0}} = \int e^{i\Sigma_{l,q_0}(a_{l,q_0}\phi_{l,q_0} + b_{l,q_0}\bar{\phi}_{l,q_0})} e^{-\Sigma_{l,q_0} |\phi_{l,q_0}|^2} \prod_{l,q_0} \frac{d\phi_{l,q_0} d\bar{\phi}_{l,q_0}}{\pi}$$

with

$$a_{l,q_0} := \left(\tfrac{\lambda_l}{\kappa^3}\right)^{\frac{1}{2}} \sum_{p_0,\mathbf{p}} y_l(\hat{\mathbf{p}}) \psi_{p_0,\mathbf{p},\uparrow} \psi_{q_0-p_0,-\mathbf{p},\downarrow} \,,$$

$$b_{l,q_0} := \left(\tfrac{\lambda_l}{\kappa^3}\right)^{\frac{1}{2}} \sum_{k_0,\mathbf{k}} \bar{y}_l(\hat{\mathbf{k}}) \bar{\psi}_{k_0,\mathbf{k},\uparrow} \bar{\psi}_{q_0-k_0,-\mathbf{k},\downarrow}$$

we arrive at

$$Z(\{s_{k\sigma}\}) = \int \frac{\det\left[\,\cdots\,\right]}{\det\left[\,\cdot\,\right]} e^{-\sum_{l,q_0}|\phi_{l,q_0}|^2} \prod_{l,q_0} \frac{d\phi_{l,q_0} d\bar{\phi}_{l,q_0}}{\pi} \qquad (6.82)$$

where the quotient of determinants is given by

$$\frac{\det\begin{bmatrix} (a_k - s_{k\uparrow})\delta_{k,p} & \sum_l \frac{ig_l}{\sqrt{\kappa}} \bar{\phi}_{l,p_0-k_0} \bar{y}_l(\hat{\mathbf{k}})\delta_{\mathbf{p},\mathbf{k}} \\ \sum_l \frac{ig_l}{\sqrt{\kappa}} \phi_{l,k_0-p_0} y_\ell(\hat{\mathbf{k}})\delta_{\mathbf{k},\mathbf{p}} & (a_{-k} - s_{-k\downarrow})\delta_{k,p} \end{bmatrix}}{\det\begin{bmatrix} a_k \delta_{k,p} & 0 \\ 0 & a_{-k}\delta_{k,p} \end{bmatrix}} = \qquad (6.83)$$

$$\prod_{\mathbf{k}} \frac{\det\begin{bmatrix} (a_{k_0,\mathbf{k}} - s_{k_0,\mathbf{k},\uparrow})\delta_{k_0,p_0} & \frac{i}{\sqrt{\kappa}} \bar{\Phi}_{p_0-k_0}(\hat{\mathbf{k}}) \\ \frac{i}{\sqrt{\kappa}} \Phi_{k_0-p_0}(\hat{\mathbf{k}}) & (a_{-k_0,-\mathbf{k}} - s_{-k_0,-\mathbf{k},\downarrow})\delta_{k_0,p_0} \end{bmatrix}}{\det\begin{bmatrix} a_{k_0,\mathbf{k}}\delta_{k_0,p_0} & 0 \\ 0 & a_{-k_0,-\mathbf{k}}\delta_{k_0,p_0} \end{bmatrix}}$$

and we abbreviated

$$\Phi_{q_0}(\hat{\mathbf{k}}) := \sum_l g_l \,\phi_{l,q_0}\, y_l(\hat{\mathbf{k}}) \,, \qquad \bar{\Phi}_{q_0}(\hat{\mathbf{k}}) := \sum_l g_l \,\bar{\phi}_{l,q_0}\, \bar{y}_l(\hat{\mathbf{k}}) \qquad (6.84)$$

Observe that in the first line of (6.83) the matrices are labelled by $k, p = (k_0, \mathbf{k}), (p_0, \mathbf{p})$ whereas in the second line of (6.83) the matrices are only labelled by k_0, p_0 and there is a product over spatial momenta $\mathbf{k}$ since the matrices in the first line of (6.83) are diagonal in the spatial momenta $\mathbf{k}, \mathbf{p}$. As in Theorem 5.1.2, the correlation functions are obtained from (6.82) by differentiating with respect to $s_{t\sigma}$:

$$\tfrac{1}{\kappa}\langle \bar{\psi}_{t\sigma}\psi_{t\sigma}\rangle = \int \begin{bmatrix} a_{k_0,t}\delta_{k_0,p_0} & \frac{i}{\sqrt{\beta}} \bar{\Phi}_{p_0-k_0}(t) \\ \frac{i}{\sqrt{\beta}} \Phi_{k_0-p_0}(t) & a_{-k_0,-t}\delta_{k_0,p_0} \end{bmatrix}^{-1}_{t_0\sigma,t_0\sigma} dP(\{\phi_{l,q_0}\}) \qquad (6.85)$$

where, making the substitution of variables $\frac{1}{\sqrt{L^d}}\phi_{l,q_0} \to \phi_{l,q_0}$,

$$dP(\{\phi_{l,q_0}\}) = \frac{e^{-L^d V(\{\phi_{l,q_0}\})} \prod_{l,q_0} d\phi_{l,q_0} d\bar{\phi}_{l,q_0}}{\int e^{-L^d V(\{\phi_{l,q_0}\})} \prod_{l,q_0} d\phi_{l,q_0} d\bar{\phi}_{l,q_0}} \qquad (6.86)$$

with an effective potential

$$V(\{\phi_{l,q_0}\}) = \sum_{l,q_0} |\phi_{l,q_0}|^2 - \frac{1}{L^d} \sum_{\mathbf{k}} \log \frac{\det \begin{bmatrix} a_{k_0,\mathbf{k}}\delta_{k_0,p_0} & \frac{i}{\sqrt{\beta}}\bar{\Phi}_{p_0-k_0}(\mathbf{k}) \\ \frac{i}{\sqrt{\beta}}\Phi_{k_0-p_0}(\mathbf{k}) & a_{-k_0,-\mathbf{k}}\delta_{k_0,p_0} \end{bmatrix}}{\det \begin{bmatrix} a_{k_0,\mathbf{k}}\delta_{k_0,p_0} & 0 \\ 0 & a_{-k_0,-\mathbf{k}}\delta_{k_0,p_0} \end{bmatrix}}$$

(6.87)

If we substitute the Riemannian sum $\frac{1}{L^d}\sum_{\mathbf{k}}$ in (6.87) by an integral $\int \frac{d^d\mathbf{k}}{(2\pi)^d}$, then the only place where the volume shows up in the above formulae is the prefactor in the exponent of (6.86). Thus, in the infinite volume limit we can evaluate the integral by evaluating the integrand at the global minimum of the effective potential, or, if this is not unique, by averaging the integrand over all minimum configurations. Since the global minimum of (6.87) is at $\Phi_{q_0}(\mathbf{k}) = 0$ for $q_0 \neq 0$ (compare the proof of Theorem 6.2.1) we arrive at $(\Phi_{q_0=0}(\mathbf{k}) \equiv \Phi(\mathbf{k}))$

$$\frac{1}{\kappa}\langle \bar{\psi}_{t\uparrow}\psi_{t\uparrow}\rangle = \int \begin{bmatrix} a_{k_0,t}\delta_{k_0,p_0} & \frac{i}{\sqrt{\beta}}\bar{\Phi}(t)\delta_{k_0,p_0} \\ \frac{i}{\sqrt{\beta}}\Phi(t)\delta_{k_0,p_0} & a_{-k_0,-t}\delta_{k_0,p_0} \end{bmatrix}^{-1}_{t_0\uparrow,t_0\uparrow} dP(\{\phi_l\})$$

$$= \int \frac{a_{-t}}{|a_t|^2 + \Phi(t)\bar{\Phi}(t)} dP(\{\phi_l\})$$

(6.88)

where

$$dP(\{\phi_l\}) = \frac{e^{-\beta L^d V(\{\phi_l\})}\prod_l d\phi_l d\bar{\phi}_l}{\int e^{-L^d V(\{\phi_l\})}\prod_l d\phi_l d\bar{\phi}_l}$$

(6.89)

with an effective potential

$$V(\{\phi_l\}) = \sum_l |\phi_l|^2 - \frac{1}{\beta L^d}\sum_{k_0,\mathbf{k}} \log \left[\frac{|a_k|^2 + \Phi(\mathbf{k})\bar{\Phi}(\mathbf{k})}{|a_k|^2}\right]$$

(6.90)

which coincides with (6.72).

b) This part is due to A. Schütte [58]. We have to prove that the global minimum of the real part of the effective potential (6.74) is given by $\phi_l = 0$ for $l \neq l_0$. To this end we first prove that for $e \in \mathbb{R}$ and $z, w \in \mathbb{C}$

$$\left|\cosh\sqrt{e^2 + (z+iw)(\bar{z}+i\bar{w})}\right|^2 \leq \cosh^2\sqrt{e^2 + |z|^2}$$

(6.91)

and

$$\left|\cosh\sqrt{e^2 + (z+iw)(\bar{z}+i\bar{w})}\right|^2 = \cosh^2\sqrt{e^2 + |z|^2} \quad \Leftrightarrow \quad w = 0$$

(6.92)

Namely, for real a and b, $r = \sqrt{a^2 + b^2}$, $a + ib = r\, e^{i\varphi}$ one has

$$\sqrt{a + ib} = \sqrt{r}\left(\cos\tfrac{\varphi}{2} + i\sin\tfrac{\varphi}{2}\right) = \sqrt{r}\left(\sqrt{\tfrac{1+\cos\varphi}{2}} \pm i\sqrt{\tfrac{1-\cos\varphi}{2}}\right)$$

$$= \sqrt{\tfrac{r+a}{2}} \pm i\sqrt{\tfrac{r-a}{2}}$$

and

$$|\cosh(a + ib)|^2 = \cosh^2 a - \sin^2 b$$

which gives

$$\left|\cosh\sqrt{e^2 + (z + iw)(\bar z + i\bar w)}\right|^2 =$$

$$\cosh^2\sqrt{\tfrac{R}{2} + \tfrac{e^2+|z|^2-|w|^2}{2}} - \sin^2\sqrt{\tfrac{R}{2} - \tfrac{e^2+|z|^2-|w|^2}{2}}$$

where

$$R = \sqrt{(e^2 + |z|^2 - |w|^2)^2 + 4(\mathrm{Re}\, w\bar z)^2}$$

Since

$$\tfrac{R}{2} + \tfrac{e^2+|z|^2-|w|^2}{2} \le e^2 + |z|^2$$
$$\Leftrightarrow \quad \sqrt{(e^2 + |z|^2 - |w|^2)^2 + 4(\mathrm{Re}\, w\bar z)^2} \le e^2 + |z|^2 + |w|^2$$
$$\Leftrightarrow \quad (e^2 + |z|^2 - |w|^2)^2 + 4(\mathrm{Re}\, w\bar z)^2 \le (e^2 + |z|^2 + |w|^2)^2$$
$$\Leftrightarrow \quad (\mathrm{Re}\, w\bar z)^2 \le (e^2 + |z|^2)|w|^2$$

and because of $(\mathrm{Re}\, w\bar z)^2 \le |w|^2|z|^2$ (6.91) and (6.92) follow. Now suppose that $\{\lambda_l\} = \{\lambda_\ell\} \cup \{\lambda_m\}$ where the λ_ℓ are attractive couplings, $\lambda_\ell > 0$, and the λ_m are repulsive couplings, $\lambda_m \le 0$. Then

$$\Phi_{\mathbf{k}}\bar\Phi_{\mathbf{k}} = (z_{\mathbf{k}} + iw_{\mathbf{k}})(\bar z_{\mathbf{k}} + i\bar w_{\mathbf{k}})$$

where $z_{\mathbf{k}} = \sum_\ell \sqrt{\lambda_\ell}\,\phi_\ell\, y_\ell(\mathbf{k})$, $w_{\mathbf{k}} = \sum_m \sqrt{-\lambda_m}\,\phi_m\, y_m(\mathbf{k})$ and $\bar z_{\mathbf{k}}$ and $\bar w_{\mathbf{k}}$ are the complex conjugate of $z_{\mathbf{k}}$ and $w_{\mathbf{k}}$. Thus, using (6.91) and (6.92)

$$\min_{\phi_l} \mathrm{Re}\, V_{\mathrm{BCS}}(\{\phi_l\}) = \min_{\phi_l}\left\{\sum_l |\phi_l|^2 - \int \tfrac{d^2\mathbf{k}}{(2\pi)^2}\tfrac{1}{\beta}\log\left|\frac{\cosh(\tfrac{\beta}{2}\sqrt{e_{\mathbf{k}}^2 + \bar\Phi_{\mathbf{k}}\Phi_{\mathbf{k}}})}{\cosh\tfrac{\beta}{2}e_{\mathbf{k}}}\right|^2\right\}$$

$$= \min_{\phi_\ell}\left\{\sum_\ell |\phi_\ell|^2 - \int \tfrac{d^2\mathbf{k}}{(2\pi)^2}\tfrac{1}{\beta}\log\left|\frac{\cosh(\tfrac{\beta}{2}\sqrt{e_{\mathbf{k}}^2 + |z_{\mathbf{k}}|^2})}{\cosh\tfrac{\beta}{2}e_{\mathbf{k}}}\right|^2\right\} \qquad (6.93)$$

Using polar coordinates $\mathbf{k} = k(\cos\alpha, \sin\alpha)$, $\phi_l = \rho_l\, e^{i\varphi_l}$, we have

$$|z_{\mathbf{k}}|^2 = \sum_{\ell,\ell'} g_\ell g_{\ell'}\, \rho_\ell \rho_{\ell'}\, e^{i[\varphi_\ell - \varphi_{\ell'} + (\ell - \ell')\alpha]}$$

and $\int_0^{2\pi} \frac{d\alpha}{2\pi} |z_{\mathbf{k}}|^2 = \sum_\ell \lambda_\ell \rho_\ell^2$. Since $W(x) := \frac{1}{\beta} \log \left| \frac{\cosh(\frac{\beta}{2}\sqrt{e_{\mathbf{k}}^2+x})}{\cosh \frac{\beta}{2} e_k} \right|^2$ is a concave function, we can use Jensen's inequality to obtain

$$\int \frac{d^2\mathbf{k}}{(2\pi)^2} \frac{1}{\beta} \log \left| \frac{\cosh(\frac{\beta}{2}\sqrt{e_{\mathbf{k}}^2+|z_{\mathbf{k}}|^2})}{\cosh \frac{\beta}{2} e_k} \right|^2 \leq \int \frac{dk\, k}{2\pi} \frac{1}{\beta} \log \left| \frac{\cosh(\frac{\beta}{2}\sqrt{e_{\mathbf{k}}^2+\int_0^{2\pi} \frac{d\alpha}{2\pi}|z_{\mathbf{k}}|^2})}{\cosh \frac{\beta}{2} e_k} \right|^2$$

$$= \int \frac{dk\, k}{2\pi} \frac{1}{\beta} \log \left| \frac{\cosh(\frac{\beta}{2}\sqrt{e_{\mathbf{k}}^2+\sum_\ell \lambda_\ell \rho_\ell^2})}{\cosh \frac{\beta}{2} e_k} \right|^2$$

and

$$\min_{\phi_l} \mathrm{Re}\, V_{\mathrm{BCS}}(\{\phi_l\}) = \min_{\rho_\ell} \left\{ \sum_\ell \rho_\ell^2 - \int \frac{dk\, k}{2\pi} \frac{1}{\beta} \log \left| \frac{\cosh(\frac{\beta}{2}\sqrt{e_{\mathbf{k}}^2+\sum_\ell \lambda_\ell \rho_\ell^2})}{\cosh \frac{\beta}{2} e_k} \right|^2 \right\}$$

By assumption, we had $\lambda_{l_0} \equiv \lambda_{\ell_0} > \lambda_\ell$ for all $\ell \neq \ell_0$. Thus, if there is some $\rho_\ell > 0$ for $\ell \neq \ell_0$, we have $\sum_\ell \lambda_\ell \rho_\ell^2 < \lambda_{\ell_0} \sum_\ell \rho_\ell^2$ and, since $W(x)$ is strictly monotone increasing,

$$\int \frac{dk\, k}{2\pi} \frac{1}{\beta} \log \left| \frac{\cosh(\frac{\beta}{2}\sqrt{e_{\mathbf{k}}^2+\sum_\ell \lambda_\ell \rho_\ell^2})}{\cosh \frac{\beta}{2} e_k} \right|^2 < \int \frac{dk\, k}{2\pi} \frac{1}{\beta} \log \left| \frac{\cosh(\frac{\beta}{2}\sqrt{e_{\mathbf{k}}^2+\lambda_{\ell_0} \sum_\ell \rho_\ell^2})}{\cosh \frac{\beta}{2} e_k} \right|^2$$

Hence, we arrive at

$$\min_{\phi_l} \mathrm{Re}\, V_{\mathrm{BCS}}(\{\phi_l\}) = \min_{\rho_{\ell_0}} \left\{ \rho_{\ell_0}^2 - \int \frac{dk\, k}{2\pi} \frac{1}{\beta} \log \left| \frac{\cosh(\frac{\beta}{2}\sqrt{e_{\mathbf{k}}^2+\lambda_{\ell_0} \rho_{\ell_0}^2})}{\cosh \frac{\beta}{2} e_k} \right|^2 \right\} \quad (6.94)$$

which proves part (b).

c) We have to compute the infinite volume limit of

$$\frac{1}{\kappa} \langle \bar{\psi}_{p\sigma} \psi_{p\sigma} \rangle = - \frac{\int_{\mathbb{R}^{4\ell+2}} \frac{a_{-p}}{a_p a_{-p}+|\Phi_{\mathbf{p}}|^2} e^{-\kappa V(\phi)} \Pi_{m=-\ell}^{\ell} du_m dv_m}{\int_{\mathbb{R}^{4\ell+2}} e^{-\kappa V(\phi)} \Pi_{m=-\ell}^{\ell} du_m dv_m} \quad (6.95)$$

where

$$V(\phi) = \sum_{m=-\ell}^{\ell} |\phi_m|^2 - \int_M \frac{d^d\mathbf{k}}{(2\pi)^d} \frac{2}{\beta} \log \left[\frac{\cosh(\frac{\beta}{2}\sqrt{e_{\mathbf{k}}^2+|\Phi_{\mathbf{k}}|^2})}{\cosh \frac{\beta}{2} e_k} \right]$$

and $(v_m \to -v_m)$

$$\Phi_{\mathbf{k}} = \lambda_\ell^{\frac{1}{2}} \sum_{m=-\ell}^{\ell} \bar{\phi}_m Y_{\ell m}(\hat{\mathbf{k}})$$

Let $U(R)$ be the unitary representation of $SO(3)$ given by

$$Y_{\ell m}(R\hat{\mathbf{k}}) = \sum_{m'} U(R)_{mm'} Y_{\ell m'}(\hat{\mathbf{k}})$$

and let $\sum_m (\overline{U\phi})_m Y_{\ell m}(\hat{\mathbf{k}}) =: (U\Phi)_{\mathbf{k}}$. Then for all $R \in SO(3)$ one has $(U(R)\Phi)_{\mathbf{k}} = \Phi_{R^{-1}\mathbf{k}}$ and

$$
\begin{aligned}
V(U(R)\phi) &= \sum_m |[U(R)\phi]_m|^2 - \int_M \frac{d^d\mathbf{k}}{(2\pi)^d} \frac{2}{\beta} \log\left[\frac{\cosh(\frac{\beta}{2}\sqrt{e_{\mathbf{k}}^2 + |[U\Phi]_{\mathbf{k}}|^2})}{\cosh\frac{\beta}{2}e_{\mathbf{k}}}\right] \\
&= \sum_m |\phi_m|^2 - \int_M \frac{d^d\mathbf{k}}{(2\pi)^d} \frac{2}{\beta} \log\left[\frac{\cosh(\frac{\beta}{2}\sqrt{e_{\mathbf{k}}^2 + |\Phi_{R^{-1}\mathbf{k}}|^2})}{\cosh\frac{\beta}{2}e_{\mathbf{k}}}\right] \\
&= \sum_m |\phi_m|^2 - \int_M \frac{d^d\mathbf{k}}{(2\pi)^d} \frac{2}{\beta} \log\left[\frac{\cosh(\frac{\beta}{2}\sqrt{e_{\mathbf{k}}^2 + |\Phi_{\mathbf{k}}|^2})}{\cosh\frac{\beta}{2}e_{\mathbf{k}}}\right] \\
&= V(\phi)
\end{aligned}
\tag{6.96}
$$

Let $S^{4\ell+1} = \{\phi \in \mathbb{C}^{2\ell+1} \mid \sum_m |\phi_m|^2 = 1\}$. Since $U(R)$ leaves $S^{4\ell+1}$ invariant, $S^{4\ell+1}$ can be written as the union of disjoint orbits,

$$
S^{4\ell+1} = \cup_{[\alpha] \in O}[\alpha]
$$

where $[\alpha] = \{U(R)\alpha \mid R \in SO(3)\}$ is the orbit of $\alpha \in S^{4\ell+1}$ under the action of $U(R)$ and O is the set of all orbits. If one chooses a fixed representative α in each orbit $[\alpha]$, that is, if one chooses a fixed section $\sigma : O \to S^{4\ell+1}$, $[\alpha] \to \sigma_{[\alpha]}$ with $[\sigma_{[\alpha]}] = [\alpha]$, every $\phi \in \mathbb{C}^{2\ell+1}$ can be uniquely written as

$$
\phi = \rho U(R)\sigma_{[\alpha]}, \quad \rho = \|\phi\| \geq 0, \quad \alpha = \frac{\phi}{\|\phi\|}, \quad [\alpha] \in O \text{ and } R \in SO(3)/I_{[\alpha]}
$$

where $I_{[\alpha]} = I_{[\alpha]}^\sigma = \{S \in SO(3) \mid U(S)\sigma_{[\alpha]} = \sigma_{[\alpha]}\}$ is the isotropy subgroup of $\sigma_{[\alpha]}$. Let

$$
\int_{\mathbb{R}^{4\ell+2}} \Pi_m \, du_m dv_m \, f(\phi) = \int_{\mathbb{R}^+} D\rho \int_O D[\alpha] \int_{[\alpha]} DR \, f(\rho U(R)\sigma_{[\alpha]})
$$

be the integral in (6.95) over $\mathbb{R}^{4\ell+2}$ in the new coordinates. That is, for example, $D\rho = \rho^{4\ell+1} d\rho$. In the new coordinates

$$
|\Phi_{\mathbf{p}}|^2 = \lambda_\ell \rho^2 \left|\sum_m \left(\overline{U(R)\sigma_{[\alpha]}}\right)_m Y_{\ell m}(\hat{\mathbf{p}})\right|^2 = \lambda_\ell \rho^2 \left|\sum_m \bar{\sigma}_{[\alpha],m} Y_{\ell m}(R^{-1}\hat{\mathbf{p}})\right|^2
$$

such that

$$
\frac{-a_{-p}}{a_p a_{-p} + |\Phi_{\mathbf{p}}|^2} = \frac{-a_{-p}}{a_p a_{-p} + \lambda_\ell \rho^2 |\sum_m \bar{\sigma}_{[\alpha],m} Y_{\ell m}(R^{-1}\hat{\mathbf{p}})|^2} \equiv f(\rho, [\alpha], R^{-1}\mathbf{p})
$$

Since $V(\phi) = V(\rho, [\alpha])$ is independent of R, one obtains

$$\frac{1}{\kappa}\langle \bar{\psi}_{p\sigma}\psi_{p\sigma}\rangle = \frac{\int \frac{-a_{-p}}{a_p a_{-p} + |\Phi_{\mathbf{p}}|^2} e^{-\kappa V(\phi)} \prod\limits_{m=-\ell}^{\ell} du_m dv_m}{\int e^{-\kappa V(\phi)} \prod\limits_{m=-\ell}^{\ell} du_m dv_m} \tag{6.97}$$

$$= \frac{\int_{\mathbb{R}^+} D\rho \int_O D[\alpha] \int_{[\alpha]} DR\, f(\rho, [\alpha], R^{-1}\mathbf{p})\, e^{-\kappa V(\rho, [\alpha])}}{\int_{\mathbb{R}^+} D\rho \int_O D[\alpha] \int_{[\alpha]} DR\, e^{-\kappa V(\rho, [\alpha])}}$$

$$= \frac{\int_{\mathbb{R}^+} D\rho \int_O D[\alpha]\, \mathrm{vol}([\alpha])\, \frac{\int_{[\alpha]} DR f(\rho, [\alpha], R^{-1}\mathbf{p})}{\int_{[\alpha]} DR}\, e^{-\kappa V(\rho, [\alpha])}}{\int_{\mathbb{R}^+} D\rho \int_O D[\alpha]\, \mathrm{vol}([\alpha])\, e^{-\kappa V(\rho, [\alpha])}}$$

It is plausible to assume that at the global minimum of $V(\rho, [\alpha])$ ρ is uniquely determined, say ρ_0. Let $O_{\min} \subset O$ be the set of all orbits at which $V(\rho_0, [\alpha])$ takes its global minimum. Then in the infinite volume limit (6.97) becomes

$$\lim_{\kappa \to \infty} \frac{1}{\kappa}\langle \bar{\psi}_{p\sigma}\psi_{p\sigma}\rangle = \frac{\int_{O_{\min}} D[\alpha]\, \mathrm{vol}([\alpha])\, \frac{\int_{[\alpha]} DR\, f(\rho, [\alpha], R^{-1}\mathbf{p})}{\int_{[\alpha]} DR}}{\int_{O_{\min}} D[\alpha]\, \mathrm{vol}([\alpha])} \tag{6.98}$$

Consider the quotient of integrals in the numerator of (6.98). Since

$$f(\rho, [\alpha], R^{-1}\mathbf{p}) = f\left(\rho^2 \left|\Sigma_m \bar{\sigma}_{[\alpha],m} Y_{\ell m}\left(R^{-1}\hat{\mathbf{p}}\right)\right|^2\right)$$

$$= f\left(\rho^2 \left|\Sigma_m \left(\overline{U(S)\sigma_{[\alpha]}}\right)_m Y_{\ell m}\left(R^{-1}\mathbf{p}'\right)\right|^2\right)$$

$$= f\left(\rho^2 \left|\Sigma_m \bar{\sigma}_{[\alpha],m} Y_{\ell m}\left((RS)^{-1}\hat{\mathbf{p}}\right)\right|^2\right) = f(\rho, [\alpha], (RS)^{-1}\mathbf{p})$$

for all $S \in I_{[\alpha]}$, one has, since $[\alpha] \simeq SO(3)/I_{[\alpha]}$

$$\frac{\int_{[\alpha]} DR\, f(\rho, [\alpha], R^{-1}\mathbf{p})}{\int_{[\alpha]} DR} = \frac{\int_{SO(3)/I_{[\alpha]}} DR\, f(\rho, [\alpha], R^{-1}\mathbf{p}) \int_{I_{[\alpha]}} DS}{\int_{SO(3)/I_{[\alpha]}} DR \int_{I_{[\alpha]}} DS}$$

$$= \frac{\int_{SO(3)/I_{[\alpha]}} DR \int_{I_{[\alpha]}} DS\, f(\rho, [\alpha], (RS)^{-1}\mathbf{p})}{\int_{SO(3)/I_{[\alpha]}} \int_{I_{[\alpha]}} DR\, DS}$$

$$= \frac{\int_{SO(3)} DR\, f(\rho, [\alpha], R^{-1}\mathbf{p})}{\int_{SO(3)} DR}$$

$$= \frac{\int_{S^2} d\Omega(\mathbf{t}) \int_{SO(3)_{\mathbf{t}\to\mathbf{p}}} DR\, f(\rho, [\alpha], R^{-1}\mathbf{p})}{\int_{S^2} d\Omega(\mathbf{t}) \int_{SO(3)_{\mathbf{t}\to\mathbf{p}}} DR}$$

$$= \frac{\int_{S^2} d\Omega(\mathbf{t})\, f(\rho, [\alpha], \mathbf{t}) \int_{SO(3)_{\mathbf{t}\to\mathbf{p}}} DR}{\int_{S^2} d\Omega(\mathbf{t}) \int_{SO(3)_{\mathbf{t}\to\mathbf{p}}} DR} \tag{6.99}$$

where $SO(3)_{\mathbf{t}\to\mathbf{p}} = \{R \in SO(3) \mid R\mathbf{t} = \mathbf{p}\}$. If one assumes that DR has the usual invariance properties of the Haar measure, then $\int_{SO(3)_{\mathbf{t}\to\mathbf{p}}} DR$ does not depend on **t** such that it cancels out in (6.99). Then (6.98) gives

$$\lim_{\kappa\to\infty} \tfrac{1}{\kappa}\langle \bar\psi_{p\sigma}\psi_{p\sigma}\rangle = \frac{\int_{O_{\min}} D[\alpha]\ \mathrm{vol}([\alpha])\ \frac{\int_{S^2} d\Omega(\mathbf{t})\ f(\rho,[\alpha],\mathbf{t})}{\int_{S^2} d\Omega(\mathbf{t})}}{\int_{O_{\min}} D[\alpha]\ \mathrm{vol}([\alpha])} \tag{6.100}$$

Now, since the effective potential, which is constant on $O_{\min}$, may be written as

$$V(\rho,[\alpha]) = \int_{S^2} \frac{d\Omega(\mathbf{t})}{4\pi}\ G\Big(\rho^2 \big|\Sigma_m \bar\sigma_{[\alpha],m} Y_{\ell m}(\mathbf{t})\big|^2\Big)$$

with $G(X) = \rho^2 - \int \frac{dk\, k^2}{2\pi^2} \log\left[\frac{\cosh(\frac{\beta}{2}\sqrt{e_{\mathbf{k}}^2+\lambda_\ell X})}{\cosh\frac{\beta}{2}e_{\mathbf{k}}}\right]$, it is plausible to assume that also

$$\frac{\int_{S^2} d\Omega(\mathbf{t})\ f(\rho,[\alpha],\mathbf{t})}{\int_{S^2} d\Omega(\mathbf{t})} = \int_{S^2} \frac{d\Omega(\mathbf{t})}{4\pi}\ \frac{ip_0+e_{\mathbf{p}}}{p_0^2+e_{\mathbf{p}}^2+\lambda_\ell \rho^2 |\Sigma_m \bar\sigma_{[\alpha],m} Y_{\ell m}(\mathbf{t})|^2}$$

is constant on $O_{\min}$. In that case also the integrals over $O_{\min}$ in (6.100) cancel out and the theorem is proven. ■

We now compare the results of the above theorem to the predictions of the quadratic Anderson-Brinkman Balian-Werthamer [3, 5] mean field theory. This formalism gives

$$\lim_{L\to\infty} \tfrac{1}{L^d}\langle a_{\mathbf{k}\sigma}^+ a_{\mathbf{k}\sigma}\rangle = \tfrac{1}{2}\left(1 - e_{\mathbf{k}}\left[\frac{\tanh(\frac{\beta}{2}\sqrt{e_{\mathbf{k}}^2+\Delta_{\mathbf{k}}^*\Delta_{\mathbf{k}}})}{\sqrt{e_{\mathbf{k}}^2+\Delta_{\mathbf{k}}^*\Delta_{\mathbf{k}}}}\right]_{\sigma\sigma}\right) \tag{6.101}$$

where the 2×2 matrix $\Delta_{\mathbf{k}}$, $\Delta_{\mathbf{k}}^T = -\Delta_{-\mathbf{k}}$, is a solution of the gap equation

$$\Delta_{\mathbf{p}} = \int_{M_\omega} \frac{d^d\mathbf{k}}{(2\pi)^d}\ U(\mathbf{p}-\mathbf{k})\,\Delta_{\mathbf{k}} \frac{\tanh(\frac{\beta}{2}\sqrt{e_{\mathbf{k}}^2+\Delta_{\mathbf{k}}^*\Delta_{\mathbf{k}}})}{2\sqrt{e_{\mathbf{k}}^2+\Delta_{\mathbf{k}}^*\Delta_{\mathbf{k}}}} \tag{6.102}$$

Consider first the case $d=2$. The electron-electron interaction in (6.102) is given by

$$U(\mathbf{p}-\mathbf{k}) = \Sigma_l \lambda_l\, e^{il(\varphi_{\mathbf{p}}-\varphi_{\mathbf{k}})} \tag{6.103}$$

Usually it is argued that the interaction flows to an effective interaction which is dominated by a single attractive angular momentum sector $\lambda_{l_0} > 0$. In other words, one approximates

$$U(\mathbf{p}-\mathbf{k}) \approx \lambda_{l_0}\, e^{il_0(\varphi_{\mathbf{p}}-\varphi_{\mathbf{k}})} \tag{6.104}$$

where l_0 is chosen as in the theorem above. The proof of part (b) of the theorem gives a rigorous justification for that, the global minimum of the effective potential for an interaction of the form (6.103) is identical to the global minimum of the effective potential for the interaction (6.104) if $\lambda_{l_0} > \lambda_l$

for all $l \neq l_0$. Once $U(\mathbf{k} - \mathbf{p})$ is approximated by (6.104), one can solve the equation (6.102). In two dimensions there are the unitary isotropic solutions

$$\Delta(\mathbf{k}) = d \begin{pmatrix} \cos l\varphi_{\mathbf{k}} & \sin l\varphi_{\mathbf{k}} \\ \sin l\varphi_{\mathbf{k}} & -\cos l\varphi_{\mathbf{k}} \end{pmatrix} \tag{6.105}$$

for odd l and

$$\Delta(\mathbf{k}) = d \begin{pmatrix} 0 & e^{il\varphi_{\mathbf{k}}} \\ -e^{il\varphi_{\mathbf{k}}} & 0 \end{pmatrix} \tag{6.106}$$

for even l which gives $\Delta(\mathbf{k})^+\Delta(\mathbf{k}) = |d|^2 \, Id$ such that (6.101) is consistent with (6.75).

In three dimensions, for an interaction (6.77), it has been proven [21] that for all $\ell \geq 2$ (6.102) does not have unitary isotropic ($\Delta_{\mathbf{k}}^*\Delta_{\mathbf{k}} = const \, Id$) solutions. That is, the gap in (6.102) is angle dependent but part (c) of the above theorem states that $\langle a_{\mathbf{k}\sigma}^+ a_{\mathbf{k}\sigma}\rangle$ has $SO(3)$ symmetry. For $SO(3)$ symmetric $e_{\mathbf{k}}$ also the effective potential has $SO(3)$ symmetry which means that also the global minimum has $SO(3)$ symmetry. Since in the infinite volume limit the integration variables are forced to take values at the global minimum, the integral over the sphere in (6.80) is the averaging over all global minima.

However, it may very well be that in the physically relevant case there is $SO(3)$ symmetry breaking. That is, instead of the Hamiltonian H of (6.70) one should consider a Hamiltonian $H + H_B$ where H_B is an $SO(3)$-symmetry breaking term which vanishes if the external parameter B (say, a magnetic field) goes to zero. Then one has to compute the correlations

$$\lim_{B \to 0} \lim_{L \to \infty} \tfrac{1}{L^d} \langle a_{\mathbf{k}\sigma}^+ a_{\mathbf{k}\sigma}\rangle \tag{6.107}$$

which are likely to have no $SO(3)$ symmetry. However, the question to what extent the results of the quadratic mean field formalism for this model relate to the exact result for the quartic Hamiltonian $H + H_B$ (in the limit $B \to 0$) needs some further investigation.

Several authors [10, 12, 7, 36] have investigated the relation between the reduced quartic BCS Hamiltonian

$$H_{\mathrm{BCS}} = H_0 + \tfrac{1}{L^{3d}} \sum_{\sigma,\tau \in \{\uparrow,\downarrow\}} \sum_{\mathbf{k},\mathbf{p}} U(\mathbf{k} - \mathbf{p}) \, a_{\mathbf{k}\sigma}^+ a_{-\mathbf{k}\tau}^+ a_{\mathbf{p}\sigma} a_{-\mathbf{p}\tau} \tag{6.108}$$

and the quadratic mean field Hamiltonian

$$H_{\mathrm{MF}} = H_0 + \tfrac{1}{L^{3d}} \sum_{\sigma,\tau \in \{\uparrow,\downarrow\}} \sum_{\mathbf{k},\mathbf{p}} U(\mathbf{k} - \mathbf{p}) \Big(a_{\mathbf{k}\sigma}^+ a_{-\mathbf{k}\tau}^+ \langle a_{\mathbf{p}\sigma} a_{-\mathbf{p}\tau}\rangle \tag{6.109}$$

$$+ \langle a_{\mathbf{k}\sigma}^+ a_{-\mathbf{k}\tau}^+\rangle a_{\mathbf{p}\sigma} a_{-\mathbf{p}\tau} - \langle a_{\mathbf{k}\sigma}^+ a_{-\mathbf{k}\tau}^+\rangle \langle a_{\mathbf{p}\sigma} a_{-\mathbf{p}\tau}\rangle \Big)$$

where the numbers $\langle a_{\mathbf{p}\sigma} a_{-\mathbf{p}\tau} \rangle$ are to be determined according to the relation $\langle a_{\mathbf{p}\sigma} a_{-\mathbf{p}\tau} \rangle = Tr\, e^{-\beta H_{\mathrm{MF}}} a_{\mathbf{p}\sigma} a_{-\mathbf{p}\tau} / Tr\, e^{-\beta H_{\mathrm{MF}}}$. The idea is that

$$H' = H_{\mathrm{BCS}} - H_{\mathrm{MF}} \tag{6.110}$$

$$= \frac{1}{L^{3d}} \sum_{\sigma,\tau \in \{\uparrow,\downarrow\}} \sum_{\mathbf{k},\mathbf{p}} U(\mathbf{k}-\mathbf{p}) \left(a_{\mathbf{k}\sigma}^{+} a_{-\mathbf{k}\tau}^{+} - \langle a_{\mathbf{k}\sigma}^{+} a_{-\mathbf{k}\tau}^{+} \rangle \right) \times$$

$$\left(a_{\mathbf{p}\sigma} a_{-\mathbf{p}\tau} - \langle a_{\mathbf{p}\sigma} a_{-\mathbf{p}\tau} \rangle \right)$$

is only a small perturbation which should vanish in the infinite volume limit. It is argued that, in the infinite volume limit, the correlation functions of the models (6.108) and (6.110) should coincide. More precisely, it is claimed that

$$\lim_{L \to \infty} \frac{1}{L^d} \log \frac{Tr\, e^{-\beta H_{\mathrm{BCS}}}}{Tr\, e^{-\beta H_{\mathrm{MF}}}} \tag{6.111}$$

vanishes. To this end it is argued that each order of perturbation theory, with respect to H', of $Tr\, e^{-\beta(H_{\mathrm{MF}}+H')} / Tr\, e^{-\beta H_{\mathrm{MF}}}$ is finite as the volume goes to infinity. The Haag paper argues that spatial averages of field operators like $\frac{1}{L^d} \int_{[0,L]^d} d^d\mathbf{x}\, \psi_{\uparrow}^{+}(\mathbf{x}) \psi_{\downarrow}^{+}(\mathbf{x})$ may be substituted by numbers in the infinite volume limit, but there is no rigorous control of the error. However, in part (c) of Theorem 6.3.1 we have shown for the model (6.70), that the correlation functions of the quartic model do not necessarily have to coincide with those of the quadratic mean field model. Thus, in view of Theorem 6.3.1, it is questionable whether the above reasoning is actually correct.

Chapter 7

The Many-Electron System in a Magnetic Field

This chapter provides a nice application of the formalism of second quantization to the fractional quantum Hall effect. Throughout this chapter we will stay in the operator formalism and make no use of functional integrals. Whereas in the previous chapters we worked in the grand canonical ensemble, the fractional quantum Hall effect is usually discussed in the canonical ensemble which is mainly due to the huge degeneracy of the noninteracting system. For filling factors less than one, a suitable approximation is to consider only the contributions in the lowest Landau level. This projection of the many-body Hamiltonian onto the lowest Landau level can be done very easily with the help of annihilation and creation operators.

The model which will be discussed in this chapter contains a certain long range approximation. This approximation makes the model explicitly solvable but, as it is argued below, this approximation also has to be considered as unphysical. Nevertheless, we think it is worthwhile considering this model because, besides providing a nice illustration of the formalism, it has an, in finite volume, explicitly given eigenvalue spectrum, which, in the infinite volume limit, most likely has a gap for rational fillings and no one for irrational filling factors. This is interesting since a similar behavior one would like to prove for the unapproximated fractional quantum Hall Hamiltonian.

7.1 Solution of the Single Body Problem

In this section we solve the single body problem for one electron in a constant magnetic field in two dimensions. We consider a disc geometry and a rectangular geometry. The many-body problem is considered in the next section in a finite volume rectangular geometry.

7.1.1 Disk Geometry

We start by considering one electron in two dimensions in a not necessarily constant but radial symmetric magnetic field

$$\vec{B} = (0, 0, B(r)) \tag{7.1}$$

Let

$$H(r) = \frac{1}{r^2} \int_0^r B(s)\, s\, ds \tag{7.2}$$

then the vector potential

$$A(x, y) = H(r)\,(-y, x) = rH(r)\,\mathbf{e}_\varphi \tag{7.3}$$

satisfies

$$\mathrm{rot} A = (0, 0, B(r)) \tag{7.4}$$

and the Hamiltonian is given by

$$H = \left(\tfrac{\hbar}{i}\nabla - eA\right)^2$$
$$= \hbar^2 \left\{ -\Delta + 2ih(r)\tfrac{\partial}{\partial\varphi} + r^2 h(r)^2 \right\} \equiv \hbar^2 K \tag{7.5}$$

if we define

$$h(r) = \tfrac{e}{\hbar} H(r) \tag{7.6}$$

and

$$K = -\Delta + 2ih(r)\tfrac{\partial}{\partial\varphi} + r^2 h(r)^2 \tag{7.7}$$

As for the harmonic oscillator, we can solve the eigenvalue problem by introducing annihilation and creation operators. To this end we introduce complex variables

$$z = x + iy, \qquad \bar{z} = x - iy$$
$$\tfrac{\partial}{\partial z} = \tfrac{1}{2}\left(\tfrac{\partial}{\partial x} - i\tfrac{\partial}{\partial y}\right), \qquad \tfrac{\partial}{\partial \bar{z}} = \tfrac{1}{2}\left(\tfrac{\partial}{\partial x} + i\tfrac{\partial}{\partial y}\right)$$

such that

$$\tfrac{\partial}{\partial z} z = \tfrac{1}{2}\left(\tfrac{\partial x}{\partial x} - i^2\tfrac{\partial y}{\partial y}\right) = 1 = \tfrac{\partial}{\partial\bar{z}}\bar{z}, \qquad \tfrac{\partial}{\partial z}\bar{z} = 0 = \tfrac{\partial}{\partial\bar{z}}z$$

Now define the following operators $a = a(x, y, h)$ and $b = b(x, y, h)$:

$$a = \sqrt{2}\tfrac{\partial}{\partial\bar{z}} + \tfrac{1}{\sqrt{2}}h\,z, \qquad a^+ = -\sqrt{2}\tfrac{\partial}{\partial z} + \tfrac{1}{\sqrt{2}}h\,\bar{z} \tag{7.8}$$
$$b = -\sqrt{2}\tfrac{\partial}{\partial z} - \tfrac{1}{\sqrt{2}}h\,\bar{z}, \qquad b^+ = \sqrt{2}\tfrac{\partial}{\partial\bar{z}} - \tfrac{1}{\sqrt{2}}h\,z \tag{7.9}$$

Note that $b(h) = a^+(-h)$ and $b^+(h) = a(-h)$. There is the following

Lemma 7.1.1 a) *There are the following commutators:*

$$[a, a^+] = \tfrac{e}{\hbar} B(r) = [b, b^+] \tag{7.10}$$

and all other commutators $[a^{(+)}, b^{(+)}] = 0$.
b) *The Hamiltonian is given by* $H = \hbar^2 K$ *where*

$$K = a^+ a + a a^+ = 2a^+ a + \tfrac{e}{\hbar} B(r) \tag{7.11}$$

Proof: We have

$$[a, a^+] = \left[\tfrac{\partial}{\partial \bar{z}}, h\bar{z}\right] - \left[hz, \tfrac{\partial}{\partial z}\right] = \tfrac{\partial(h\bar{z})}{\partial \bar{z}} + \tfrac{\partial(hz)}{\partial z} = h + \bar{z}\tfrac{\partial h}{\partial \bar{z}} + h + z\tfrac{\partial h}{\partial z}$$

Since $h = h(r)$ depends only on r, one obtains with $\rho = r^2 = z\bar{z}$

$$z\tfrac{\partial h}{\partial z} = z\tfrac{\partial(z\bar{z})}{\partial z}\tfrac{dh}{d\rho} = \rho\tfrac{dh}{d\rho} = \bar{z}\tfrac{\partial h}{\partial \bar{z}} \tag{7.12}$$

which results in

$$[a, a^+] = 2\left(h + \rho\tfrac{dh}{d\rho}\right) = 2\tfrac{e}{\hbar}\left(H + \rho\tfrac{dH}{d\rho}\right) \tag{7.13}$$

Using $H(r) = \tfrac{1}{r^2}\int_0^r B(s)\, s\, ds$, one finds

$$
\begin{aligned}
H + \rho\tfrac{dH}{d\rho} &= H + r^2\left(-\tfrac{1}{r^4}\int_0^r B(s)s\, ds + \tfrac{1}{r^2}B(r)r\tfrac{dr}{d(r^2)}\right) \\
&= B(r)r\tfrac{1}{2r} = \tfrac{B(r)}{2}
\end{aligned} \tag{7.14}
$$

Thus one ends up with

$$[a, a^+] = \tfrac{e}{\hbar} B(r) = [b, b^+]$$

Furthermore

$$
\begin{aligned}
[b^+, a^+] = [a(-h), a^+(h)] &= \left[\tfrac{\partial}{\partial \bar{z}}, h\bar{z}\right] + \left[hz, \tfrac{\partial}{\partial z}\right] \\
&= \tfrac{\partial(h\bar{z})}{\partial \bar{z}} - \tfrac{\partial(hz)}{\partial z} = \bar{z}\tfrac{\partial h}{\partial \bar{z}} - z\tfrac{\partial h}{\partial z} = 0
\end{aligned}
$$

because of (7.12). This proves part (a). To obtain part (b), observe that

$$\tfrac{\partial}{\partial z}\tfrac{\partial}{\partial \bar{z}} = \tfrac{1}{4}\left(\tfrac{\partial^2}{\partial x^2} + \tfrac{\partial^2}{\partial y^2}\right) = \tfrac{1}{4}\Delta \tag{7.15}$$

and

$$
\begin{aligned}
\tfrac{\partial}{\partial \varphi} &= \tfrac{\partial x}{\partial \varphi}\tfrac{\partial}{\partial x} + \tfrac{\partial y}{\partial \varphi}\tfrac{\partial}{\partial y} = -y\tfrac{\partial}{\partial x} + x\tfrac{\partial}{\partial y} \\
&= -\tfrac{1}{2i}(z - \bar{z})\left(\tfrac{\partial}{\partial z} + \tfrac{\partial}{\partial \bar{z}}\right) + \tfrac{1}{2}(z + \bar{z})\tfrac{1}{i}\left(\tfrac{\partial}{\partial \bar{z}} - \tfrac{\partial}{\partial z}\right) \\
&= i\left(z\tfrac{\partial}{\partial z} - \bar{z}\tfrac{\partial}{\partial \bar{z}}\right)
\end{aligned} \tag{7.16}
$$

which gives

$$K = -\Delta + 2ih(r)\frac{\partial}{\partial\varphi} + r^2h(r)^2$$
$$= -4\frac{\partial}{\partial z}\frac{\partial}{\partial\bar{z}} - 2h\left(z\frac{\partial}{\partial z} - \bar{z}\frac{\partial}{\partial\bar{z}}\right) + h^2 z\bar{z} \tag{7.17}$$

Since

$$a^+a = -2\frac{\partial}{\partial z}\frac{\partial}{\partial\bar{z}} - \frac{\partial}{\partial z}hz + h\bar{z}\frac{\partial}{\partial\bar{z}} + \frac{1}{2}h^2 z\bar{z}$$
$$= -2\frac{\partial}{\partial z}\frac{\partial}{\partial\bar{z}} - h\left(z\frac{\partial}{\partial z} - \bar{z}\frac{\partial}{\partial\bar{z}}\right) + \frac{1}{2}h^2 z\bar{z} - \frac{\partial(hz)}{\partial z}$$
$$= \frac{1}{2}K - \frac{e}{\hbar}\frac{B(r)}{2}$$

the lemma follows. ∎

We now focus on the case of a constant magnetic field

$$B(r) = B = const \tag{7.18}$$

In that case

$$h(r) = \frac{eB}{2\hbar} = \frac{1}{2\ell_B^2} \tag{7.19}$$

where

$$\ell_B = \sqrt{\frac{\hbar}{eB}} \tag{7.20}$$

denotes the magnetic length. We change variables

$$w = \frac{1}{\sqrt{2}}\frac{z}{\ell_B} \tag{7.21}$$

and introduce the rescaled operators

$$c = \ell_B a = \frac{\partial}{\partial\bar{w}} + \frac{1}{2}w, \quad c^+ = \ell_B a^+ = -\frac{\partial}{\partial w} + \frac{1}{2}\bar{w} \tag{7.22}$$
$$d = \ell_B b = -\frac{\partial}{\partial w} - \frac{1}{2}\bar{w}, \quad d^+ = \ell_B b^+ = \frac{\partial}{\partial\bar{w}} - \frac{1}{2}w \tag{7.23}$$

which fulfill the commutation relations

$$[c, c^+] = [d, d^+] = 1, \quad [c^{(+)}, d^{(+)}] = 0 \tag{7.24}$$

Then the Hamiltonian becomes

$$\frac{1}{2m}\left(\frac{\hbar}{i}\nabla - eA\right)^2 = \hbar\frac{eB}{m}\left(c^+c + \frac{1}{2}\right) \tag{7.25}$$

and the orthonormalized eigenfunctions are given by

$$\phi_{nm}(w, \bar{w}) = \frac{1}{\sqrt{n!m!}}\left(c^+\right)^n\left(d^+\right)^m\phi_{00} \tag{7.26}$$

where

$$\phi_{00}(w,\bar{w}) = \frac{1}{(2\pi\ell_B^2)^{1/2}}\, e^{-\frac{w\bar{w}}{2}} = \frac{1}{(2\pi\ell_B^2)^{1/2}}\, e^{-\frac{|z|^2}{4\ell_B^2}} \tag{7.27}$$

The eigenfunctions can be expressed in terms of the generalized Laguerre polynomials L_n^α. One obtains

$$\phi_{nm}(w,\bar{w}) = \frac{1}{(2\pi\ell_B^2)^{1/2}}\left(\frac{n!}{m!}\right)^{\frac{1}{2}} w^{m-n} L_n^{m-n}(w\bar{w})\, e^{-\frac{w\bar{w}}{2}}$$

$$= \frac{1}{(2\pi\ell_B^2)^{1/2}}\left(\frac{n!}{m!}\right)^{\frac{1}{2}}\left(\frac{z}{\sqrt{2}\,\ell_B}\right)^{m-n} L_n^{m-n}\left(\frac{|z|^2}{2\ell_B^2}\right) e^{-\frac{|z|^2}{4\ell_B^2}} \tag{7.28}$$

We summarize some properties of the eigenfunctions in the following lemma. By a slight abuse of notation, we write $\phi_{nm}(z) = \phi_{nm}(z,\bar{z})$ in the lemma below instead of $\phi_{nm}\left(\frac{z}{\sqrt{2}\ell_B},\frac{\bar{z}}{\sqrt{2}\ell_B}\right)$.

Lemma 7.1.2 *Let* $\phi_{nm}(z)$ *be given by the second line of (7.28). Then*

(i)

$$\overline{\phi_{nm}(z)} = (-1)^{n-m}\phi_{mn}(z) \tag{7.29}$$

(ii)

$$\hat{\phi}_{nm}(k = k_1 + ik_2) = \int_{\mathbb{R}^2} dx\, dx\, e^{-i(k_1 x + k_2 y)}\phi_{nm}(x+iy)$$

$$= 4\pi\ell_B^2(-1)^n\phi_{nm}(-2i\ell_B^2 k) \tag{7.30}$$

(iii)

$$\sum_{m=0}^{\infty}\phi_{n_1 m}(z_1)\overline{\phi_{n_2 m}(z_2)} = \frac{1}{(2\pi\ell_B^2)^{1/2}}\phi_{n_1 n_2}(z_1 - z_2)\, e^{i\frac{\mathrm{Im}(z_1\bar{z}_2)}{2\ell_B^2}} \tag{7.31}$$

Proof: Part (i) is obtained by using the relation

$$L_n^{-l}(x) = \frac{(n-l)!}{n!}(-x)^l L_{n-l}^l(x)$$

for $l = n - m$. To obtain the second part, one can use the integral

$$\int_0^\infty x^{l+1} L_n^l(ax^2)\, e^{-\frac{ax^2}{2}} J_l(xy)\, dx = \frac{(-1)^n}{a^{l+1}}\, y^l L_n^l\left(\frac{y^2}{a}\right) e^{-\frac{y^2}{2a}}, \quad a > 0$$

where J_l is the l'th Bessel function. One may look in [43] for the details. The third part can be proven by using a generating function for the Laguerre polynomials,

$$\sum_{k=0}^{\infty} t^k L_k^{m-k}(x) L_n^{k-n}(y) = t^n (1+t)^{m-n} e^{-tx} L_n^{m-n}\left(\tfrac{1+t}{t}(tx+y)\right), \quad |t| < 1$$

but probably it is more instructive to give a proof in terms of annihilation and creation operators. To this end we reintroduce the rescaled variables $w = z/(\sqrt{2}\ell_B)$. Then we have to show

$$\sum_{m=0}^{\infty} \phi_{n_1 m}(w_1)\overline{\phi_{n_2 m}(w_2)} = \tfrac{1}{(2\pi\ell_B^2)^{1/2}} \phi_{n_1 n_2}(w_1 - w_2) e^{i\mathrm{Im}(w_1\bar{w}_2)}$$

Since $\bar{\phi}_{jm} = (-1)^{j-m}\phi_{mj}$ we have to compute (c_1^+ denotes the creation operator with respect to the w_1 variable)

$$\tfrac{(-1)^j}{(n!j!)^{1/2}}(c_1^+)^n(d_2^+)^j \sum_{m=0}^{\infty} \tfrac{(-d_1^+ c_2^+)^m}{m!} \phi_{00}(w_1)\phi_{00}(w_2)$$

$$= \tfrac{(-1)^j}{(n!j!)^{1/2}}(c_1^+)^n(d_2^+)^j \sum_{m=0}^{\infty} \tfrac{(w_1\bar{w}_2)^m}{m!} \phi_{00}(w_1)\phi_{00}(w_2)$$

$$= \tfrac{(-1)^j}{(n!j!)^{1/2}}(c_1^+)^n(d_2^+)^j e^{w_1\bar{w}_2} \tfrac{1}{2\pi\ell_B^2} e^{-\frac{1}{2}(w_1\bar{w}_1 + w_2\bar{w}_2)}$$

$$= \tfrac{1}{(2\pi\ell_B^2)^{1/2}} \tfrac{(-1)^j}{(n!j!)^{1/2}}(c_1^+)^n(d_2^+)^j e^{\frac{1}{2}(w_1\bar{w}_2 - \bar{w}_1 w_2)} \phi_{00}(w_1 - w_2)$$

Now, because of

$$e^A B e^{-A} = \sum_{k=0}^{\infty} \tfrac{1}{k!}\Big[A,\big[A,\cdots[A,B]\cdots\big]\Big]$$

and

$$\left[\tfrac{1}{2}(w_1\bar{w}_2 - \bar{w}_1 w_2), d_2^+\right] = \left[\tfrac{1}{2}(w_1\bar{w}_2 - \bar{w}_1 w_2), \tfrac{\partial}{\partial\bar{w}_2}\right] = -\tfrac{w_1}{2}$$

one gets

$$e^{-\frac{1}{2}(w_1\bar{w}_2 - \bar{w}_1 w_2)} d_2^+ e^{\frac{1}{2}(w_1\bar{w}_2 - \bar{w}_1 w_2)} = d_2^+ - \left[\tfrac{1}{2}(w_1\bar{w}_2 - \bar{w}_1 w_2), d_2^+\right]$$

$$= \tfrac{\partial}{\partial\bar{w}_2} - \tfrac{1}{2}w_2 + \tfrac{1}{2}w_1 = -d_{w_1 - w_2}^+$$

and

$$e^{-\frac{1}{2}(w_1\bar{w}_2 - \bar{w}_1 w_2)}(d_2^+)^j e^{\frac{1}{2}(w_1\bar{w}_2 - \bar{w}_1 w_2)} = (-d_{w_1 - w_2}^+)^j$$

Similarly one finds

$$e^{-\frac{1}{2}(w_1\bar{w}_2-\bar{w}_1 w_2)}\left(c_1^+\right)^n e^{\frac{1}{2}(w_1\bar{w}_2-\bar{w}_1 w_2)} = \left(c_{w_1-w_2}^+\right)^n$$

and we end up with

$$\frac{1}{(2\pi\ell_B^2)^{1/2}} e^{\frac{1}{2}(w_1\bar{w}_2-\bar{w}_1 w_2)} \frac{1}{(n!j!)^{1/2}} \left(c_{w_1-w_2}^+\right)^n \left(d_{w_1-w_2}^+\right)^j \phi_{00}(w_1-w_2)$$

$$= \frac{1}{(2\pi\ell_B^2)^{1/2}} e^{i\mathrm{Im}(w_1\bar{w}_2)} \phi_{nj}(w_1-w_2)$$

which proves the third part of the lemma. ∎

7.1.2 Rectangular Geometry

In this section we solve the eigenvalue problem for a single electron in a constant magnetic field in two dimensions for a rectangular geometry. We start with the half infinite case. That is, coordinate space is given by $[0, L_x] \times \mathbb{R}$. The following gauge is suitable

$$A(x,y) = (-By, 0, 0) \tag{7.32}$$

Then

$$\begin{aligned} H &= \left(\tfrac{\hbar}{i}\nabla + eA\right)^2 \\ &= \hbar^2\left\{-\Delta + 2i\tfrac{eB}{\hbar}y\tfrac{\partial}{\partial x} + \tfrac{e^2 B^2}{\hbar^2}y^2\right\} \equiv \hbar^2 K \end{aligned} \tag{7.33}$$

where, recalling that $\ell_B = \{\hbar/(eB)\}^{1/2}$ denotes the magnetic length,

$$K = -\Delta + \tfrac{2i}{\ell_B^2}y\tfrac{\partial}{\partial x} + \tfrac{1}{\ell_B^4}y^2 \tag{7.34}$$

We make the ansatz

$$\psi(x,y) = e^{ikx}\varphi(y) \tag{7.35}$$

Imposing periodic boundary conditions on $[0, L_x]$ gives $k = \frac{2\pi}{L_x}m$ with $m \in \mathbb{Z}$. The eigenvalue problem $H\psi = \varepsilon\psi$ is equivalent to

$$\hbar^2\left\{-\tfrac{\partial^2}{\partial y^2} + k^2 - \tfrac{2ky}{\ell_B^2} + \tfrac{y^2}{\ell_B^4}\right\}\varphi(y) = \varepsilon\,\varphi(y)$$

$$\Leftrightarrow \quad \hbar^2\left\{-\tfrac{\partial^2}{\partial(y-\ell_B^2 k)^2} + \tfrac{1}{\ell_B^4}(y-\ell_B^2 k)^2\right\}\varphi(y) = \varepsilon\,\varphi(y)$$

This is the eigenvalue equation for the harmonic oscillator shifted by $\ell_B^2 k$. Thus, if h_n denotes the normalized Hermite function, then the normalized eigenfunctions of H read

$$\psi_{nk}(x,y) = \frac{1}{\sqrt{L_x \ell_B}} e^{ikx} h_n\big((y-\ell_B^2 k)/\ell_B\big) \tag{7.36}$$

with eigenvalues $1/(2m)H\psi_{nk} = \varepsilon_n\psi_{nk}$, $\varepsilon_n = \hbar\frac{eB}{m}(n+1/2)$.

As in the last section, the eigenvalues have infinite degeneracy. This is due to the fact that, as in the last section, we computed in infinite volume. If one turns to finite volume, the degeneracy is reduced to a finite value. Namely, the degeneracy is equal to the number of flux quanta flowing through the sample. A flux quantum is given by

$$\phi_0 = 2\pi\frac{\hbar}{e} = 4,14 \cdot 10^{-11}\text{T cm}^2 \tag{7.37}$$

and there is flux quantization which means that the magnetic flux through a given sample has to be an integer multiple of ϕ_0. Thus, for finite volume $[0, L_x] \times [0, L_y]$ the number of flux quanta is equal to

$$M := \frac{\Phi}{\phi_0} = \frac{BL_xL_y}{2\pi\hbar/e} = \frac{L_xL_y}{2\pi\ell_B^2} \overset{!}{\in} \mathbb{N} \tag{7.38}$$

This is the case we consider next. So let coordinate space be $[0, L_x] \times [0, L_y]$. The following boundary conditions are suitable ('magnetic boundary conditions'):

$$\psi(x + L_x, y) = \psi(x, y), \qquad \psi(x, y + L_y) = e^{ixL_y/\ell_B^2}\psi(x, y) \tag{7.39}$$

The finite volume eigenvalue problem can be solved by periodizing the states (7.36). This in turn can be obtained by a suitable superposition. To this end observe that in the y-direction the wavefunctions (7.36) are centered at $y_k = \ell_B^2 k$. Now, if we shift $k = \frac{2\pi}{L_x}m$ by $K := \frac{2\pi}{L_x}M$, then the center y_{k+K} is shifted by $\ell_B^2\frac{2\pi}{L_x}M = \ell_B^2\frac{2\pi}{L_x}\frac{L_xL_y}{2\pi\ell_B^2} = L_y$. As a result, one finds the following eigenfunctions:

Lemma 7.1.3 *Let H be given by (7.33) on finite volume $[0, L_x] \times [0, L_y]$ with the Landau gauge (7.34) and with magnetic boundary conditions (7.37). Then a complete orthonormal set of eigenfunctions of H is given by*

$$\psi_{n,k}(x, y) = \frac{1}{\sqrt{L_x\ell_B}}\sum_{j=-\infty}^{\infty} e^{i(k+jK)x}h_{n,k}(y - jL_y) \tag{7.40}$$

$$n = 0, 1, 2, \dots \qquad k = \frac{2\pi}{L_x}m, \quad m = 1, 2, \dots, M$$

where $K := \frac{2\pi}{L_x}M = \frac{L_y}{\ell_B^2}$ and $h_{n,k}(y) = h_n\left((y-\ell_B^2 k)/\ell_B\right)$, $h_n(y) = c_n H_n(y)e^{-\frac{y^2}{2}}$ the normalized Hermite function, $c_n = \pi^{-\frac{1}{4}}(2^n n!)^{-\frac{1}{2}}$. The eigenvalues are given by

$$\frac{1}{2m}H\psi_{n,k} = \varepsilon_n\psi_{n,k}, \qquad \varepsilon_n = \hbar\frac{eB}{m}\left(n + \frac{1}{2}\right) \tag{7.41}$$

For the details of the computation one may look, for example, in [42].

7.2 Diagonalization of the Fractional Quantum Hall Hamiltonian in a Long Range Limit

Having discussed the single body problem in the previous two sections, we now turn to the N-body problem. We consider the many-electron system in two dimensions in a finite, rectangular volume $[0, L_x] \times [0, L_y]$ in a constant magnetic field $\vec{B} = (0, 0, B)$. The noninteracting Hamiltonian is

$$H_{0,N} = \sum_{i=1}^{N} \left(\tfrac{\hbar}{i} \nabla_i - eA(\mathbf{x}_i) \right)^2 \tag{7.42}$$

with A given by the Landau gauge (7.32). Apparently the eigenstates of (7.42) are given by wedge products of single body eigenfunctions (7.40),

$$\Psi_{n_1 k_1, \cdots, n_N k_N} = \psi_{n_1 k_1} \wedge \cdots \wedge \psi_{n_N k_N} \tag{7.43}$$

with eigenvalues $E_{n_1 \cdots n_N} = \varepsilon_{n_1} + \cdots + \varepsilon_{n_N}$. Recall that N denotes the number of electrons and M (see (7.38) and (7.40)) is the degeneracy per Landau level which is equal to the number of flux quanta flowing through the sample. The quotient

$$\nu = \frac{N}{M} \tag{7.44}$$

is the filling factor of the system. For integer filling, there is a unique ground state, namely the wedge product of the $N = \nu M$ single body states of the lowest $\nu \in \mathbb{N}$ Landau levels. This state is separated by a gap of size $\hbar \frac{eB}{m}$ from the other states. The existence of this gap leads to the integer quantum Hall effect. If the filling is not an integer, the ground state of the noninteracting N-body system is highly degenerate. Suppose that $\nu = 1/3$. Then $M = 3N$ and, identifying $k = \frac{2\pi}{L_x} m$ with m, the ground states are given by $\Psi_{0m_1, \cdots, 0m_N}$ where $m_1, ..., m_N \in \{0, 1, ...M - 1\}$ and $m_1 < \cdots < m_N$. Apparently, there are

$$\binom{M}{N} = \binom{3N}{N} = \frac{(3N)!}{(2N)! \, N!} \approx \frac{\sqrt{3}}{\sqrt{4\pi N}} \left(\frac{3^3}{2^2} \right)^N \tag{7.45}$$

such choices. Here we used Stirling's formula, $n! \approx \sqrt{2\pi n} \, (n/e)^n$ to evaluate the factorials. Since $3^3/2^2 = 6\frac{3}{4}$, this is an extremely large number for macroscopic values of N like 10^{11} or 10^{12} which is the number of conduction electrons in Ga-As samples where the fractional quantum Hall effect is observed. Now, if the electron-electron interaction is turned on, some of these states are energetically more favorable and others less, but, due to the huge degeneracy of about $(6.75)^{(10^{11})}$, one would expect a continuum of energies. However, the discovery of the fractional quantum Hall effect in 1982 demonstrated that this is not true. For certain rational values of ν with odd denominators, mainly $\nu = \frac{n}{2pn\pm1}$, $p, n \in \mathbb{N}$, the interacting Hamiltonian

$$H_N = \sum_{i=1}^{N} \left(\tfrac{\hbar}{i}\nabla_i - eA(\mathbf{x}_i)\right)^2 + \sum_{\substack{i,j=1 \\ i\neq j}} V(\mathbf{x}_i - \mathbf{x}_j) \tag{7.46}$$

with V being the Coulomb interaction, should have a gap. Since then, a lot of work has been done on the system (7.46) (see [34, 39, 13] for an overview). Numerical data for small system size are available which give evidence for a gap. However, so far it has not been possible to give a rigorous mathematical proof that the Hamiltonian (7.46) has a gap for certain rational values of the filling factor.

In this section we consider the Hamiltonian (7.46) in a certain approximation in which it becomes explicitly solvable. We will argue below that this approximation has to be considered as unphysical. However it is interesting since it gives a model with an, in finite volume, explicitly given energy spectrum which, in the infinite volume limit, most likely has a gap for $\nu \in \mathbb{Q}$ and no gap for $\nu \notin \mathbb{Q}$. For that reason we think it is worth discussing that model.

We consider the complete spin polarized case and neglect the Zeemann energy. We take a Gaussian as the electron-electron interaction,

$$V(x, y) = \lambda e^{-\frac{x^2+y^2}{2r^2}} \tag{7.47}$$

Opposite to Coulomb, this has no singularity at small distances. We assume that it is long range in the sense that $r >> \ell_B$, ℓ_B being the magnetic length. This length is of the order 10^{-8}m for typical FQH magnetic fields which are about $B \approx 10T$. The long range condition is used to make the approximation (see (7.68, 7.70) below)

$$\int ds\, ds'\, h_n(s)\, h_{n'}(s')\, e^{-\frac{\ell_B^2}{2r^2}(s-s')^2} \approx \int ds\, ds'\, h_n(s)\, h_{n'}(s') \tag{7.48}$$

where $h_n(s) = c_n H_n(s)\, e^{-\frac{s^2}{2}}$ denotes the normalized Hermite function. With this approximation, the Hamiltonian $P_{LL}H_N P_{LL}$, P_{LL} being the projection onto the lowest Landau level, can be explicitly diagonalized. There is the following

Theorem 7.2.1 *Let H_N be the Hamiltonian (7.46) in finite volume $[0, L_x] \times [0, L_y]$ with magnetic boundary conditions (7.39), let $A(x,y) = (-By, 0, 0)$ and let the interaction be Gaussian with long range,*

$$V(x, y) = \lambda e^{-\frac{x^2+y^2}{2r^2}}, \qquad r >> \ell_B \tag{7.49}$$

Let $P_{LL} : \mathcal{F}_N \to \mathcal{F}_N^{LL}$ be the projection onto the lowest Landau level, where $\mathcal{F}_N$ is the antisymmetric N-particle Fock space and $\mathcal{F}_N^{LL}$ is the antisymmetric Fock space spanned by the eigenfunctions of the lowest Landau level. Then, with the approximation (7.48), the Hamiltonian $H_{N,LL} = P_{LL}H_N P_{LL}$ becomes exactly diagonalizable. Let M be the number of flux quanta flowing through $[0, L_x] \times [0, L_y]$ such that $\nu = N/M$ is the filling factor. Then the eigenstates and eigenvalues are labelled by N-tuples $(n_1, \cdots, n_N)$, $n_1 < \cdots < n_N$ and $n_i \in \{1, 2, \cdots, M\}$ for all i,

$$H_{N,LL}\Psi_{n_1\cdots n_N} = (\varepsilon_0 N + E_{n_1\cdots n_N})\Psi_{n_1\cdots n_N} \tag{7.50}$$

where $\varepsilon_0 = \hbar eB/(2m)$ and

$$E_{n_1\cdots n_N} = \sum_{\substack{i,j=1 \\ i\neq j}}^{N} W(n_i - n_j), \qquad W(n) = \lambda\sum_{j\in\mathbb{Z}} e^{-\frac{1}{2r^2}\left(L_x\frac{n}{M}-jL_x\right)^2} \tag{7.51}$$

and the normalized eigenstates are given by $\Psi_{n_1\cdots n_N} = \phi_{n_1} \wedge \cdots \wedge \phi_{n_N}$ where

$$\phi_n(x,y) = \frac{\pi^{-\frac{1}{4}}}{\sqrt{\ell_B L_y}} \sum_{s=-\infty}^{\infty} e^{-\frac{1}{2\ell_B^2}\left(x-\frac{n}{M}L_x-sL_x\right)^2} e^{i(x-\frac{n}{M}L_x-sL_x)y/\ell_B^2} \tag{7.52}$$

$$= \frac{1}{\sqrt{M}} \frac{\pi^{-\frac{1}{4}}}{\sqrt{\ell_B L_x}} \sum_{r=-\infty}^{\infty} e^{-\frac{1}{2\ell_B^2}\left(y-\frac{r}{M}L_y\right)^2} e^{i\left(x-\frac{n}{M}L_x\right)\frac{r}{M}L_y/\ell_B^2} \tag{7.53}$$

Before we start with the proof we make some comments. The approximation (7.48) looks quite innocent. However, by reviewing the computations in the proof one finds that it is actually equivalent to the approximation $V(x,y) = \lambda e^{-(x^2+y^2)/(2r^2)} \approx \lambda e^{-x^2/(2r^2)}$. That is, if we write in (7.49) $V(x,y) = \lambda e^{-x^2/(2r^2)}$ then the above Theorem is an exact statement. Recall that the single body eigenfunctions are localized in the y-direction and are given by plane waves in the x-direction. Since the eigenstates are given by wedge products, we can explicitly compute the expectation value of the energy. One may speculate that for fillings $\nu = 1/q$ the ground states are labelled by the N-tuples $(n_1, \cdots, n_N) = (j, j+q, j+2q, \cdots, j+(N-1)q)$ which have a q-fold degeneracy, $1 \le j \le q$. A q-fold degeneracy for fillings $\nu = p/q$, p, q without common divisor, follows already from general symmetry considerations (see [42] or [61]). In particular, for $\nu = 1/3$, one may speculate that there are three ground states labelled by $(3, 6, 9, ..., 3N)$, $(2, 5, 8, ..., 3N-1)$ and $(1, 4, 7, ..., 3N-2)$. Below Lemma 7.2.3 we compute the expectation value of the energy of these states with respect to the Coulomb interaction. It is about as twice as big as the energy of the Laughlin wavefunction. This is not too surprising since a wedge product vanishes only linearly in the differences $x_i - x_j$ whereas the Laughlin wavefunction, given by

$$\psi(w_1, ... w_N) = \prod_{i<j} (w_i - w_j)^3 \, e^{-\frac{1}{2}\sum_j |w_j|^2}, \qquad w_i = z_i/(\sqrt{2}\ell_B) \tag{7.54}$$

vanishes like $|x_i - x_j|^3$ which results in a lower contribution of the Coulomb energy $1/|x_i - x_j|$. Thus we have to conclude that the approximation (7.48), implemented in (7.68) and (7.70) below, has to be considered as unphysical. Nevertheless, the energy spectrum (7.51) seems to have the property that it has a gap for rational fillings in the infinite volume limit. That is, we expect

$$\Delta(\nu) := \lim_{\substack{N,M \to \infty \\ N/M = \nu}} \left(E_1(N,M) - E_0(N,M)\right) \begin{cases} > 0 \text{ if } \nu \in \mathbb{Q} \\ = 0 \text{ if } \nu \notin \mathbb{Q} \end{cases} \quad (7.55)$$

E_0 being the lowest and E_1 the second lowest eigenvalue in finite volume. This is interesting since a similar behavior one would like to prove for the original Hamiltonian (7.46). However, it seems that the energy (7.51) does not distinguish between even and odd denominators q. That is, it looks like the approximate model does not select the observed fractional quantum Hall fillings (see [34, 39, 13] for an overview). This should be due to the unphysical nature of the approximation (7.48).

Proof of Theorem 7.2.1: We proceed in three steps: Projection onto the lowest Landau level using fermionic annihilation and creation operators, implementation of the approximation (7.48) and finally diagonalization.

(i) Projection onto the Lowest Landau Level

To project H_N onto the lowest Landau level, we rewrite H_N in terms of fermionic annihilation and creation operators

$$H_N = \left\{ \int d^2x\, \psi^+(\mathbf{x}) \left(\tfrac{\hbar}{i}\nabla - eA(\mathbf{x})\right)^2 \psi(\mathbf{x}) \right.$$
$$\left. + \int d^2x\, d^2x'\, \psi^+(\mathbf{x})\psi^+(\mathbf{x}')V(\mathbf{x}-\mathbf{x}')\psi(\mathbf{x}')\psi(x) \right\}\Big|_{\mathcal{F}_N} \quad (7.56)$$

where $\mathcal{F}_N$ is the antisymmetric N-particle Fock space. We consider the complete spin polarized case in which only one spin direction (say $\psi = \psi_\uparrow$) contributes and we neglect the Zeeman energy. Let ψ and ψ^+ be the fermionic annihilation and creation operators in coordinate space. In order to avoid confusion with the single body eigenfunctions $\psi_{n,k}$ (7.40), we denote the latter ones as $\varphi_{n,k}$. Introducing $a_{n,k}$, $a_{n,k}^+$ according to

$$\psi(\mathbf{x}) = \sum_{n,k} \varphi_{n,k}(\mathbf{x})a_{n,k}, \qquad \psi^+(\mathbf{x}) = \sum_{n,k} \bar{\varphi}_{n,k}(\mathbf{x})a_{n,k}^+ \quad (7.57)$$

$$a_{n,k} = \int d^2x\, \bar{\varphi}_{n,k}(\mathbf{x})\psi(\mathbf{x}), \qquad a_{n,k}^+ = \int d^2x\, \varphi_{n,k}(\mathbf{x})\psi^+(\mathbf{x}) \quad (7.58)$$

the $a_{n,k}$ obey the canonical anticommutation relations

$$\{a_{n,k}, a_{n',k'}^+\} = \delta_{n,n'}\delta_{k,k'} \quad (7.59)$$

and (7.56) becomes, if $H = \oplus_N H_N$,

$$H = H_{\text{kin}} + H_{\text{int}} \tag{7.60}$$

where

$$H_{\text{kin}} = \sum_{n,k} \varepsilon_n \, a_{n,k}^+ a_{n,k} \tag{7.61}$$

The interacting part becomes

$$
\begin{aligned}
H_{\text{int}} &= \sum_{\substack{n,k \\ n',k'}} \int d^2x \, d^2x' \, \psi^+(\mathbf{x}) \psi^+(\mathbf{x}') \bar{\varphi}_{n,k}(\mathbf{x}) \, \langle nk|V|\overline{n'k'}\rangle \, \varphi_{n',k'}(\mathbf{x}') \psi(\mathbf{x}') \psi(\mathbf{x}) \\
&= \sum_{\substack{n,k \\ n',k'}} \sum_{\substack{n_1,\cdots,n_4 \\ l_1,\cdots,l_4}} (\overline{n_1 l_1}; n_2 l_2; \overline{nk}) \, \langle nk|V|\overline{n'k'}\rangle \, \times \\
&\qquad\qquad (n'k'; \overline{n_3 l_3}; n_4 l_4) \, a_{n_1,l_1}^+ a_{n_3,l_3}^+ a_{n_4,l_4} a_{n_2,l_2}
\end{aligned}
\tag{7.62}
$$

where we used the notation

$$\langle nk|V|\overline{n'k'}\rangle := \int d^2x \int d^2x' \varphi_{n,k}(\mathbf{x}) \, V(\mathbf{x}-\mathbf{x}') \, \bar{\varphi}_{n',k'}(\mathbf{x}') \tag{7.63}$$

$$(\overline{n_1 l_1}; n_2 l_2; \overline{nk}) := \int d^2x \, \bar{\varphi}_{n_1,l_1}(\mathbf{x}) \, \varphi_{n_2,l_2}(\mathbf{x}) \, \bar{\varphi}_{n,k}(\mathbf{x}) \tag{7.64}$$

Now we consider systems with fillings

$$\nu = \frac{N}{M} < 1 \tag{7.65}$$

and restrict the electrons to the lowest Landau level. Since the kinetic energy is constant, we consider only the interacting part,

$$H_{\text{LL}} := P_{\text{LL}} H_{\text{int}} P_{\text{LL}} \tag{7.66}$$

$$= \sum_{\substack{n,k \\ n',k'}} \sum_{l_1,\cdots,l_4} (\overline{0l_1}; 0l_2; \overline{nk}) \, \langle nk|V|\overline{n'k'}\rangle \, (n'k'; \overline{0l_3}; 0l_4) \, a_{l_1}^+ a_{l_3}^+ a_{l_4} a_{l_2}$$

where we abbreviated

$$a_l := a_{0,l}, \qquad a_l^+ := a_{0,l}^+ \tag{7.67}$$

(ii) The Approximation

The matrix element $\langle nk|V|\overline{n'k'}\rangle$ is computed in part (a) of Lemma 7.2.2 below. For a gaussian interaction (7.47) the exact result is

$$\langle n, k|V|\overline{n', k'}\rangle = \tag{7.68}$$

$$\sqrt{2\pi}\ell_B \, \delta_{k,k'} \, \lambda \, r \, [e^{-\frac{r^2}{2}k^2}]_M \int ds \int ds' h_n(s) \, h_{n'}(s') \, e^{-\frac{\ell_B^2}{2r^2}(s-s')^2}$$

where, if $k = 2\pi m/L_x$,

$$[e^{-\frac{r^2}{2}k^2}]_M := \sum_{j=-\infty}^{\infty} e^{-\frac{r^2}{2}(k-jK)^2} = \sum_{j=-\infty}^{\infty} e^{-\frac{r^2}{2}[\frac{2\pi}{L_x}(m-jM)]^2} \qquad (7.69)$$

is an M-periodic function (as a function of m). For a long range interaction $r \gg \ell_B$, we may approximate this by

$$\langle n, k|V|\overline{n', k'}\rangle \approx \sqrt{2\pi}\ell_B\, \delta_{k,k'}\, \lambda r\, [e^{-\frac{r^2}{2}k^2}]_M \int ds\, h_n(s) \int ds'\, h_{n'}(s')$$
$$=: \delta_{k,k'}\, v_k \int ds\, h_n(s) \int ds'\, h_{n'}(s') \qquad (7.70)$$

Then H_{LL} becomes

$$H_{LL} =$$
$$\sum_{\substack{n,n' \\ k}} \sum_{l_1,\cdots,l_4} (\overline{0l_1}; 0l_2; \overline{nk})\, v_k \int h_n(s)ds \int h_{n'}(s')ds'\, (n'k; \overline{0l_3}; 0l_4)\, a_{l_1}^+ a_{l_3}^+ a_{l_4} a_{l_2}$$
$$= \sum_{k} \sum_{l_1,\cdots,l_4} (\overline{0l_1}; 0l_2; \overline{1_y k})\, v_k\, (1_y k; \overline{0l_3}; 0l_4)\, a_{l_1}^+ a_{l_3}^+ a_{l_4} a_{l_2} \qquad (7.71)$$

Here we used that

$$\sum_{n=0}^{\infty} h_n(y) \int h_n(s)ds = 1 \qquad (7.72)$$

which is a consequence of $\sum_{n=0}^{\infty} h_n(y)\, h_n(s) = \delta(y - s)$. Thus

$$\sum_{n=0}^{\infty} \bar{\varphi}_{n,k}(x,y) \int h_n(s)ds$$
$$= \frac{1}{\sqrt{L_x \ell_B}} \sum_{j=-\infty}^{\infty} e^{-i(k+jK)x} \sum_{n=0}^{\infty} h_n\big((y - y_k - jL_y)/\ell_B\big) \int h_n(s)ds$$
$$= \frac{1}{\sqrt{L_x \ell_B}} \sum_{j=-\infty}^{\infty} e^{-i(k+jK)x} \qquad (7.73)$$

and (7.71) follows if we define

$$(\overline{0l_1}; 0l_2; \overline{1_y k}) := \int dx\, dy\, \bar{\varphi}_{0,l_1}(x,y)\, \varphi_{0,l_2}(x,y) \frac{1}{\sqrt{L_x \ell_B}} \sum_{j=-\infty}^{\infty} e^{-i(k+jK)x} \qquad (7.74)$$

These matrix elements are computed in part (b) of Lemma 7.2.2 below and the result is

$$(\overline{0l_1}; 0l_2; \overline{1_y k}) = \delta_{m,m_2-m_1}^M \frac{1}{\sqrt{L_x \ell_B}} [e^{-\frac{\ell_B^2}{4}(l_1-l_2)^2}]_M \qquad (7.75)$$

if $k = \frac{2\pi}{L_x}m$, $l_j = \frac{2\pi}{L_x}m_j$ and $\delta^M_{m_1,m_2} = 1$ iff $m_1 = m_2$ mod M. In the following we write, by a slight abuse of notation, also $\delta^M_{l,l'}$ if $l = \frac{2\pi}{L_x}m$. Then the Hamiltonian (7.71) becomes

$$H_{LL} = \frac{1}{L_x \ell_B} \sum_k \sum_{l_1,\cdots,l_4} \delta^M_{k,l_2-l_1} \, [e^{-\frac{\ell_B^2}{4}(l_1-l_2)^2}]_M \, v_k \, \delta^M_{k,l_3-l_4} \times$$

$$[e^{-\frac{\ell_B^2}{4}(l_3-l_4)^2}]_M \, a^+_{l_1} a^+_{l_3} a^+_{l_4} a_{l_2}$$

$$= \frac{1}{L_x} \sum_{l_1,\cdots,l_4} \delta^M_{l_2-l_1,l_3-l_4} \, w_{l_2-l_1} \, a^+_{l_1} a^+_{l_3} a_{l_4} a_{l_2} \qquad (7.76)$$

where the interaction is given by

$$w_k := \sqrt{2\pi} \, \lambda \, r \, [e^{-\frac{r^2}{2}k^2}]_M [e^{-\frac{\ell_B^2}{4}k^2}]^2_M \qquad (7.77)$$

(iii) Diagonalization

Apparently (7.76) looks like a usual one-dimensional many-body Hamiltonian in momentum space. Thus, since the kinetic energy is constant, we can easily diagonalize it by taking the discrete Fourier transform. For $1 \le n \le M$ let

$$\psi_n := \frac{1}{\sqrt{M}} \sum_{m=1}^{M} e^{2\pi i \frac{nm}{M}} a_m, \qquad \psi^+_n = \frac{1}{\sqrt{M}} \sum_{m=1}^{M} e^{-2\pi i \frac{nm}{M}} a^+_m \qquad (7.78)$$

or

$$a_m = \frac{1}{\sqrt{M}} \sum_{n=1}^{M} e^{-2\pi i \frac{nm}{M}} \psi_n, \qquad a^+_m = \frac{1}{\sqrt{M}} \sum_{n=1}^{M} e^{2\pi i \frac{nm}{M}} \psi^+_n \qquad (7.79)$$

Substituting this in (7.76), we get

$$H_{LL} = \sum_{n,n'} \psi^+_n \psi^+_{n'} \, W(n-n') \, \psi_{n'} \psi_n \qquad (7.80)$$

with an interaction

$$W(n) = \frac{1}{L_x} \sum_{m=1}^{M} e^{2\pi i \frac{nm}{M}} w_m \qquad (7.81)$$

where $w_m \equiv w_k$ is given by (7.77), $k = 2\pi m/L_x$. The N-particle eigenstates of (7.80) are labelled by N-tuples $(n_1, \cdots, n_N)$ where $1 \le n_j \le M$ and

$n_1 < n_2 < \cdots < n_N$ and are given by

$$
\begin{aligned}
\Psi_{n_1 \cdots n_N} &= \psi^+_{n_1} \psi^+_{n_2} \cdots \psi^+_{n_N} |1\rangle \\
&= \frac{1}{M^{N/2}} \sum_{j_1 \cdots j_N} e^{-\frac{2\pi i}{M}(n_1 j_1 + \cdots n_N j_N)} a^+_{j_1} \cdots a^+_{j_N} |1\rangle \\
&= \frac{1}{M^{N/2}} \sum_{j_1 \cdots j_N} e^{-\frac{2\pi i}{M}(n_1 j_1 + \cdots n_N j_N)} \varphi_{0j_1} \wedge \cdots \wedge \varphi_{0j_N} \\
&= \phi_{n_1} \wedge \cdots \wedge \phi_{n_N}
\end{aligned}
\tag{7.82}
$$

if we define

$$
\phi_n(x,y) := \frac{1}{\sqrt{M}} \sum_{j=1}^{M} e^{-2\pi i \frac{nj}{M}} \varphi_{0j}(x,y)
\tag{7.83}
$$

The energy eigenvalues are

$$
H_{\text{LL}} \Psi_{n_1 \cdots n_N} = E_{n_1 \cdots n_N} \Psi_{n_1 \cdots n_N}
\tag{7.84}
$$

where

$$
E_{n_1 \cdots n_N} = \sum_{\substack{i,j=1 \\ i \neq j}}^{N} W(n_i - n_j)
\tag{7.85}
$$

The Fourier sums in (7.81) and (7.83) can be performed with the Poisson summation formula. This is done in part (c) and (d) of Lemma 7.2.2. If we approximate $w_k \approx \sqrt{2\pi}\,\lambda\, r\, [e^{-\frac{r^2}{2} k^2}]_M$, since by assumption $r >> \ell_B$, we find for this w_k

$$
\begin{aligned}
W(n) &= \lambda \sum_{j \in \mathbb{Z}} e^{-\frac{1}{2r^2}\left(L_x \frac{n}{M} - jL_x\right)^2} \\
&= \lambda \sum_{j \in \mathbb{Z}} e^{-\frac{1}{2r^2}\left(\ell_B^2 \frac{2\pi n}{L_y} - jL_x\right)^2}
\end{aligned}
\tag{7.86}
$$

Thus the theorem is proven. ∎

Lemma 7.2.2 a) *For the matrix element in (7.63) one has*

$$
\langle nk|V|\overline{n'k'}\rangle =
\tag{7.87}
$$

$$
\sqrt{2\pi}\ell_B\, \delta_{k,k'}\, \lambda\, r\, [e^{-\frac{r^2}{2} k^2}]_M \int ds \int ds'\, h_n(s)\, h_{n'}(s')\, e^{-\frac{\ell_B^2}{2r^2}(s-s')^2}
$$

where, if $k = 2\pi m/L_x$,

$$[e^{-\frac{r^2}{2}k^2}]_M := \sum_{j=-\infty}^{\infty} e^{-\frac{r^2}{2}(k-jK)^2} = \sum_{j=-\infty}^{\infty} e^{-\frac{r^2}{2}\left[\frac{2\pi}{L_x}(m-jM)\right]^2} \qquad (7.88)$$

is an M-periodic function (as a function of m).

b) The matrix elements (7.74) are given by

$$(0, l_1; 0, l_2; \overline{1_y}, k) = \delta^M_{m, m_2 - m_1} \frac{1}{\sqrt{L_x \ell_B}} \, [e^{-\frac{\ell_B^2}{4}(l_1 - l_2)^2}]_M \qquad (7.89)$$

if $k = \frac{2\pi}{L_x} m$, $l_j = \frac{2\pi}{L_x} m_j$ and $\delta^M_{m_1, m_2} = 1$ iff $m_1 = m_2 \bmod M$.

c) For $m \in \mathbb{Z}$ let

$$v_m = [e^{-\frac{r^2}{2}k^2}]_M = \sum_{j\in\mathbb{Z}} e^{-\frac{r^2}{2}\left(\frac{2\pi}{L_x}\right)^2 (m-jM)^2} \qquad (7.90)$$

and let $V(n) = \frac{1}{L_x} \sum_{m=1}^{M} e^{2\pi i \frac{nm}{M}} v_m$. Then

$$V(n) = \frac{1}{\sqrt{2\pi}\,r} \sum_{j\in\mathbb{Z}} e^{-\frac{1}{2r^2}\left(L_x \frac{n}{M} - jL_x\right)^2} \qquad (7.91)$$

d) Let $k = 2\pi m/L_x$ and let $\varphi_{0,m} \equiv \varphi_{0,k}$ be the single-body eigenfunction (7.40). Then

$$\frac{1}{\sqrt{M}} \sum_{m=1}^{M} e^{-2\pi i \frac{nm}{M}} \varphi_{0,m}(x, y)$$

$$= \frac{\pi^{-\frac{1}{4}}}{\sqrt{\ell_B L_y}} \sum_{s=-\infty}^{\infty} e^{-\frac{1}{2\ell_B^2}\left(x - \frac{n}{M}L_x - sL_x\right)^2} e^{i\frac{(x - \frac{n}{M}L_x - sL_x)y}{\ell_B^2}} \qquad (7.92)$$

$$= \frac{1}{\sqrt{M}} \frac{\pi^{-\frac{1}{4}}}{\sqrt{\ell_B L_x}} \sum_{r=-\infty}^{\infty} e^{-\frac{1}{2\ell_B^2}\left(y - \frac{r}{M}L_y\right)^2} e^{i\left(x - \frac{n}{M}L_x\right)\frac{r}{M}L_y/\ell_B^2} \qquad (7.93)$$

Proof: a) We have

$$\langle n, k|V|\overline{n', k'}\rangle = \int d^2x \, d^2x' \, \varphi_{n,k}(\mathbf{x}) \, V(\mathbf{x} - \mathbf{x}') \, \bar{\varphi}_{n',k'}(\mathbf{x}') \qquad (7.94)$$

$$= \frac{1}{L_x \ell_B} \sum_{j,j'} \int dx\,dx'\,dy\,dy' \, e^{i(k-jK)x - i(k'-j'K)x'} \times$$

$$h_{n,k}(y - jL_y) \, h_{n',k'}(y' - j'L_y) \, V(\mathbf{x} - \mathbf{x}')$$

$$= \frac{1}{L_x \ell_B} \sum_{j,j'} \int dx\,dx'\,dy\,dy' \, e^{i(k-jK)(x-x')} e^{i(k-jK-k'+j'K)x'} \times$$

$$h_n\big((y - y_k - jL_y)/\ell_B\big) \, h_{n'}\big((y' - y_{k'} - j'L_y)/\ell_B\big) \, \lambda e^{-\frac{(x-x')^2}{2r^2}} e^{-\frac{(y-y')^2}{2r^2}}$$

The x'-integral gives $L_x \, \delta_{m-jM,m'-j'M} = L_x \, \delta_{m,m'} \delta_{j,j'}$ if $k = 2\pi m/L_x$, $k' = 2\pi m'/L_x$, $0 \le m, m' \le M - 1$. Thus we get

$$\langle n, k|V|n', k'\rangle =$$

$$\tfrac{\lambda}{\ell_B} \, \delta_{k,k'} \sum_j \int dx \, e^{i(k-jK)x} e^{-\frac{x^2}{2r^2}} \int dy dy' h_n\big((y - y_k - jL_y)/\ell_B\big) \times$$

$$h_{n'}\big((y' - y_k - jL_y)/\ell_B\big) \, e^{-\frac{(y-y')^2}{2r^2}}$$

$$= \sqrt{2\pi} \ell_B \lambda r \, \delta_{k,k'} \sum_j e^{-\frac{r^2}{2}(k-jK)^2} \int ds ds' h_n(s) \, h_{n'}(s') \, e^{-\frac{\ell_B^2}{2r^2}(s-s')^2}$$

$$= \sqrt{2\pi} \ell_B \lambda r \, \delta_{k,k'} \, \big[e^{-\frac{r^2}{2}k^2}\big]_M \int ds ds' h_n(s) \, h_{n'}(s') \, e^{-\frac{\ell_B^2}{2r^2}(s-s')^2} \qquad (7.95)$$

and part (a) follows.

b) One has

$$(\overline{0, l_1}; 0, l_2; \overline{1_y, k}) = \qquad (7.96)$$

$$\tfrac{1}{\sqrt{L_x \ell_B}^3} \sum_{j_1, j_2, j} \int dx dy \, e^{-i(l_1 + j_1 K)x} e^{i(l_2 + j_2 K)x} e^{-i(k+jK)x} h_{0,l_1}(y) \, h_{0,l_2}(y)$$

The plane waves combine to

$$\exp\left[i\tfrac{2\pi}{L_x}(m_2 + j_2 M - m_1 - j_1 M - m - jM)x\right] \qquad (7.97)$$

and the x-integral gives a volume factor L_x times a Kroenecker delta which is one iff

$$m_2 + j_2 M - m_1 - j_1 M - m - jM = 0$$

or

$$m = m_2 - m_1 \ \wedge \ j = j_2 - j_1 \quad \text{if } m_2 \ge m_1$$
$$m = m_2 - m_1 + M \ \wedge \ j = j_2 - j_1 - 1 \quad \text{if } m_2 < m_1 \qquad (7.98)$$

Thus (7.96) becomes

$$(\overline{0, l_1}; 0, l_2; \overline{1_y, k}) =$$

$$\delta^M_{m, m_2 - m_1} \tfrac{1}{\sqrt{\ell_B}^3} \tfrac{1}{\sqrt{L_x}} \sum_{j_1, j_2} \int_0^{L_y} dy \, h_0\big((y - y_{l_1} - j_1 L_y)/\ell_B\big) \times$$

$$h_0\big((y - y_{l_2} - j_2 L_y)/\ell_B\big)$$

$$= \delta^M_{m, m_2 - m_1} \tfrac{1}{\sqrt{\ell_B}^3} \tfrac{1}{\sqrt{L_x}} \sum_{j_1, j_2} \int_0^{L_y} dy \, h_0\big((y - y_{l_1} - j_1 L_y)/\ell_B\big) \times$$

$$h_0\big((y - y_{l_2} - j_1 L_y + (j_1 - j_2)L_y)/\ell_B\big)$$

$$= \delta^M_{m,m_2-m_1} \frac{1}{\sqrt{\ell_B{}^3}} \frac{1}{\sqrt{L_x}} \sum_{j_1,j} \int_0^{L_y} dy\, h_0\left((y-y_{l_1}-j_1 L_y)/\ell_B\right) \times$$

$$h_0\left((y-y_{l_2}-j_1 L_y + j L_y)/\ell_B\right)$$

$$= \delta^M_{m,m_2-m_1} \frac{1}{\sqrt{\ell_B{}^3}} \frac{1}{\sqrt{L_x}} \sum_{j} \int_{-\infty}^{\infty} dy\, h_0\left((y-y_{l_1})/\ell_B\right) h_0\left((y-y_{l_2}+jL_y)/\ell_B\right)$$

$$= \delta^M_{m,m_2-m_1} \frac{1}{\sqrt{L_x \ell_B}} \sum_{j} e^{-\frac{1}{4\ell_B^2}(y_{l_1}-y_{l_2}+jL_y)^2}$$

$$= \delta^M_{m,m_2-m_1} \frac{1}{\sqrt{L_x \ell_B}} \sum_{j} e^{-\frac{\ell_B^2}{4}(l_1-l_2+jK)^2}$$

$$= \delta^M_{m,m_2-m_1} \frac{1}{\sqrt{L_x \ell_B}} \left[e^{-\frac{\ell_B^2}{4}(l_1-l_2)^2}\right]_M \tag{7.99}$$

where $\delta^M_{m,m'}$ equals one iff $m = m'$ mod M and equals zero otherwise.

c) It is

$$\left[e^{-\frac{r^2}{2}k^2}\right]_M = \sum_{j\in\mathbb{Z}} e^{-\frac{r^2}{2}\frac{M^2}{L_x^2}\left(2\pi\frac{m}{M}-2\pi j\right)^2} \tag{7.100}$$

We use the following formula which is obtained from the Poisson summation theorem

$$\sum_{j\in\mathbb{Z}} e^{-\frac{1}{2t}(x-2\pi j)^2} = \sqrt{\frac{t}{2\pi}} \sum_{j\in\mathbb{Z}} e^{-\frac{t}{2}j^2} e^{ijx} \tag{7.101}$$

with

$$x = 2\pi\frac{m}{M}, \qquad t = \frac{L_x^2}{r^2 M^2} \tag{7.102}$$

Then

$$v_m = \frac{1}{\sqrt{2\pi}\, r} \frac{L_x}{M} \sum_{j\in\mathbb{Z}} e^{-\frac{1}{2r^2}\frac{L_x^2}{M^2}j^2} e^{-2\pi i \frac{jm}{M}} \tag{7.103}$$

and $V(n)$ becomes

$$V(n) = \frac{1}{L_x} \sum_{m=1}^{M} e^{2\pi i \frac{nm}{M}} \frac{1}{\sqrt{2\pi}\, r} \frac{L_x}{M} \sum_{j\in\mathbb{Z}} e^{-\frac{1}{2r^2}\frac{L_x^2}{M^2}j^2} e^{-2\pi i \frac{jm}{M}}$$

$$= \frac{1}{\sqrt{2\pi}\, r} \frac{1}{M} \sum_{j\in\mathbb{Z}} e^{-\frac{1}{2r^2}\frac{L_x^2}{M^2}j^2} \sum_{m=1}^{M} e^{2\pi i \frac{(n-j)m}{M}}$$

$$= \frac{1}{\sqrt{2\pi}\, r} \frac{1}{M} \sum_{j\in\mathbb{Z}} e^{-\frac{1}{2r^2}\frac{L_x^2}{M^2}j^2} M \delta^M_{n,j}$$

$$= \frac{1}{\sqrt{2\pi}\, r} \sum_{s\in\mathbb{Z}} e^{-\frac{1}{2r^2}\frac{L_x^2}{M^2}(n-sM)^2} \tag{7.104}$$

which proves part (c).

d) According to (7.40) we have

$$\frac{1}{\sqrt{M}} \sum_{m=1}^{M} e^{-2\pi i \frac{nm}{M}} \varphi_{0,m}(x,y) =$$

$$\frac{1}{\sqrt{M}} \sum_{m=1}^{M} e^{-2\pi i \frac{nm}{M}} \frac{1}{\sqrt{L_x \ell_B}} \sum_{j=-\infty}^{\infty} e^{i \frac{2\pi}{L_x}(m+jM)x} h_{0,k}(y-jL_y)$$

$$= \frac{\pi^{-\frac{1}{4}}}{\sqrt{M L_x \ell_B}} \sum_{j=-\infty}^{\infty} \sum_{m=1}^{M} e^{i\frac{2\pi}{L_x}(x-\frac{n}{M}L_x)m} e^{i\frac{2\pi}{L_x}jMx} e^{-\frac{1}{2\ell_B^2}\left(y-\ell_B^2\frac{2\pi}{L_x}m-jL_y\right)^2}$$

$$= \frac{\pi^{-\frac{1}{4}}}{\sqrt{M L_x \ell_B}} \sum_{m=1}^{M} \sum_{j=-\infty}^{\infty} e^{i\frac{2\pi}{L_x}(x-\frac{n}{M}L_x)m} e^{i\frac{L_y}{\ell_B^2}jx} \times$$

$$\frac{1}{\sqrt{2\pi}} \int dq\, e^{-\frac{q^2}{2}} e^{iq\left(\frac{y}{\ell_B}-\ell_B\frac{2\pi}{L_x}m-j\frac{L_y}{\ell_B}\right)}$$

$$= \frac{\pi^{-\frac{1}{4}}}{\sqrt{M L_x \ell_B}} \frac{1}{\sqrt{2\pi}} \int dq\, e^{-\frac{q^2}{2}} e^{iq\frac{y}{\ell_B}} \sum_{m=1}^{M} e^{i\frac{2\pi}{L_x}(x-\frac{n}{M}L_x-q\ell_B)m} \times$$

$$\sum_{j=-\infty}^{\infty} e^{i\left(\frac{x}{\ell_B}-q\right)\frac{L_y}{\ell_B}j} \qquad (7.105)$$

$$= \frac{\pi^{-\frac{1}{4}}}{\sqrt{M L_x \ell_B}} \frac{1}{\sqrt{2\pi}} \int dq\, e^{-\frac{q^2}{2}} e^{iq\frac{y}{\ell_B}} \sum_{m=1}^{M} e^{i\frac{2\pi}{L_x}(x-\frac{n}{M}L_x-q\ell_B)m} \times$$

$$\sum_{r=-\infty}^{\infty} 2\pi\, \delta\left(\left(\frac{x}{\ell_B}-q\right)\frac{L_y}{\ell_B}-2\pi r\right)$$

The delta function forces q to take values

$$q = \frac{x}{\ell_B} - 2\pi r \frac{\ell_B}{L_y}$$

which gives

$$x - \frac{n}{M}L_x - q\ell_B = -\frac{n}{M}L_x + 2\pi r \frac{\ell_B^2}{L_y} = \frac{r-n}{M}L_x$$

Therefore the m-sum in (7.105) becomes

$$\sum_{m=1}^{M} e^{i\frac{2\pi}{L_x}(x-\frac{n}{M}L_x-q\ell_B)m} = \sum_{m=1}^{M} e^{2\pi i \frac{(r-n)m}{M}} = M\, \delta_{r,n}^{M} \qquad (7.106)$$

and we get

$$\frac{1}{\sqrt{M}} \sum_{m=1}^{M} e^{-2\pi i \frac{nm}{M}} \varphi_{0,m}(x,y) =$$

$$\pi^{-\frac{1}{4}} \frac{\sqrt{M}}{\sqrt{L_x \ell_B}} \sqrt{2\pi} \frac{\ell_B}{L_y} \sum_{r=-\infty}^{\infty} e^{-\frac{1}{2\ell_B^2}\left(x - r\frac{2\pi \ell_B^2}{L_y}\right)^2} e^{i\left(\frac{x}{\ell_B} - 2\pi r \frac{\ell_B}{L_y}\right)\frac{y}{\ell_B}} \delta_{r,n}^{M}$$

$$= \frac{\pi^{-\frac{1}{4}}}{\sqrt{\ell_B L_y}} \sum_{s=-\infty}^{\infty} e^{-\frac{1}{2\ell_B^2}\left(x - (n+sM)\frac{2\pi \ell_B^2}{L_y}\right)^2} e^{i\left(\frac{x}{\ell_B} - 2\pi(n+sM)\frac{\ell_B}{L_y}\right)\frac{y}{\ell_B}}$$

$$= \frac{\pi^{-\frac{1}{4}}}{\sqrt{\ell_B L_y}} \sum_{s=-\infty}^{\infty} e^{-\frac{1}{2\ell_B^2}\left(x - \frac{n}{M}L_x - sL_x\right)^2} e^{i\frac{(x - \frac{n}{M}L_x - sL_x)y}{\ell_B^2}} \tag{7.107}$$

This proves (7.92). (7.93) is obtained directly from (7.40) by putting $r = m + jM$. ∎

Since the eigenstates of the approximate model are given by pure wedge products, we cannot expect that their energies, for a Coulomb interaction, are close to those of the Laughlin or Jain wavefunctions. The reason is that a wedge product only vanishes linearly in $x_i - x_j$ while the Laughlin wavefunction vanishes like $(x_i - x_j)^3$ if x_i goes to x_j. This gives a lower contribution to the Coulomb repulsion $1/|x_i - x_j|$.

In general it is not possible to make an exact analytical computation of the expectation value of the energy if the wavefunction is not given by a pure wedge product, like the Laughlin or composite fermion wavefunction. One has to rely on numerical and analytical approximations or exact numerical results for small system size. For a pure wedge product, the exact result can be written down and it looks as follows.

Lemma 7.2.3 a) *Let $\psi(x_1, ..., x_N)$ be a normalized antisymmetric wavefunction and let $W(x_1, ..., x_N) = \sum_{i<j}^{N} V(x_i - x_j)$. Let $A = \int d^2x$ be the sample size, let $n_\nu = N/A = \nu/(2\pi \ell_B^2)$ be the density and let ρ be the density of the constant N-particle wavefunction such that $\int dx_1 \cdots dx_N \rho = 1$, that is, $\rho = 1/A^N$. Let $\langle W \rangle = \int d^{2N}x\, W(|\psi|^2 - \rho)$. Then*

$$\frac{\langle W \rangle}{N} = \frac{1}{2} n_\nu \frac{1}{A} \int dx_1\, dx_2\, V(x_1 - x_2)\big[g(x_1, x_2) - 1\big] \tag{7.108}$$

where

$$g(x_1, x_2) = \frac{N(N-1)}{n_\nu^2} \int dx_3 \cdots dx_N\, |\psi(x_1, ..., x_N)|^2 \tag{7.109}$$

b) *Suppose that*

$$\psi(x_1, ..., x_N) = \phi_{m_1} \wedge \cdots \wedge \phi_{m_N}(x_1, ..., x_N)$$
$$= \frac{1}{\sqrt{N!}} \det [\phi_{m_i}(x_j)_{1 \le i,j \le N}] \tag{7.110}$$

Then

$$g(x_1, x_2) = \frac{1}{n_\nu^2} \left\{ P(x_1, x_1) P(x_2, x_2) - |P(x_1, x_2)|^2 \right\} \qquad (7.111)$$

where P is the kernel of the projector onto the space spanned by $\{\phi_{m_1}, ..., \phi_{m_N}\}$. That is, $P(x, x') = \sum_{j=1}^{N} \phi_{m_j}(x) \bar{\phi}_{m_j}(x')$.

Proof: a) One has

$$\langle W \rangle = \int d^{2N}x \sum_{\substack{i,j=1 \\ i<j}}^{N} V(x_i - x_j) \left\{ |\psi(x_1, ..., x_n)|^2 - \rho \right\}$$

$$= \frac{N(N-1)}{2} \int dx_1 dx_2 \, V(x_1 - x_2) \int dx_3 \cdots dx_N \left\{ |\psi(x_1, ..., x_n)|^2 - \rho \right\}$$

$$= \frac{1}{2} \int dx_1 dx_2 \, V(x_1 - x_2) \times$$

$$\left\{ N(N-1) \int dx_3 \cdots dx_N |\psi(x_1, ..., x_n)|^2 - \frac{N(N-1)}{A^2} \right\}$$

$$= N \frac{N-1}{A} \frac{1}{2} \frac{1}{A} \int dx_1 dx_2 \, V(x_1 - x_2) \times$$

$$\left\{ \frac{N(N-1)}{n_\nu^2} \int dx_3 \cdots dx_N |\psi(x_1, ..., x_n)|^2 - 1 \right\}$$

where we used $n_\nu = N/A \approx (N-1)/A$. This proves part (a).

b) Let $\psi = \phi_{m_1} \wedge \cdots \wedge \phi_{m_N}$. Then

$$\int dx_3 \cdots dx_N |\psi(x_1, ..., x_N)|^2$$

$$= \frac{1}{N!} \sum_{\pi, \sigma \in S_N} \varepsilon_\pi \varepsilon_\sigma \int dx_3 \cdots dx_N \phi_{m_{\pi 1}}(x_1) \bar{\phi}_{m_{\sigma 1}}(x_1) \cdots \phi_{m_{\pi N}}(x_N) \bar{\phi}_{m_{\sigma N}}(x_N)$$

$$= \frac{1}{N!} \sum_{\pi \in S_N} \left\{ |\phi_{m_{\pi 1}}(x_1)|^2 |\phi_{m_{\pi 2}}(x_2)|^2 - \phi_{m_{\pi 1}}(x_1) \bar{\phi}_{m_{\pi 1}}(x_2) \phi_{m_{\pi 2}}(x_2) \bar{\phi}_{m_{\pi 2}}(x_1) \right\}$$

$$= \frac{1}{N(N-1)} \sum_{\substack{i,j=1 \\ i \neq j}}^{N} \left\{ |\phi_{m_i}(x_1)|^2 |\phi_{m_j}(x_2)|^2 - \phi_{m_i}(x_1) \bar{\phi}_{m_i}(x_2) \phi_{m_j}(x_2) \bar{\phi}_{m_j}(x_1) \right\}$$

$$= \frac{1}{N(N-1)} \sum_{i,j=1}^{N} \left\{ \cdots \right\}$$

$$= \frac{1}{N(N-1)} \left\{ P_{\underline{m}}(x_1, x_1) P_{\underline{m}}(x_2, x_2) - P_{\underline{m}}(x_1, x_2) P_{\underline{m}}(x_2, x_1) \right\} \qquad (7.112)$$

where $P_{\underline{m}}(x_1, x_2) = \sum_{j=1}^{N} \phi_{m_j}(x_1) \bar{\phi}_{m_j}(x_2)$. ∎

Now consider the eigenvalues $E_{n_1...n_N}$ given by (7.51). For $-M/2 \leq n \leq M/2$, the dominant contribution from the periodizing j-sum for $W(n)$ is the $j = 0$ term which is $\exp\{-\frac{L^2}{2r^2}(\frac{n}{M})^2\}$. This is small if $n = n_i - n_j$ is large. Thus, it seems that those configurations $(n_1, ..., n_N)$ have low energy for which $n_i - n_j$ in average is large. Hence, one may speculate that, for $\nu = 1/3$, the minimizing configurations are $(3, 6, 9, ..., 3N)$, $(2, 5, 8, ..., 3N-1)$ and $(1, 4, 7, ..., 3N-2)$ and the lowest excited states should be obtained from these states by just changing one n_i or a whole group of neighboring n_i's by one each. The projection $P_{\{3k\}}$ onto the space spanned by $\phi_3, \phi_6, ..., \phi_{3N}$ can be explicitly computed. In the infinite volume limit, one simply obtains $P_{\{3k\}} = \frac{1}{3}P_{\nu=1}$ where $P_{\nu=1}$ is the projection onto the whole lowest Landau level, spanned by all the φ_n's. Thus, the energy per particle $U_{\{3k\}}$ for the wavefunction $\phi_3 \wedge \phi_6 \wedge \cdots \wedge \phi_{3N}$ is $\frac{1}{3}U_{\nu=1} = -\frac{1}{3}\sqrt{\frac{\pi}{8}}\frac{e^2}{\ell_B} \approx -0,21\frac{e^2}{\ell_B}$ which is much bigger than the energy of the Laughlin wavefunction which is about $U_{\nu=1/3} = -0,42\frac{e^2}{\ell_B}$.

Chapter 8

Feynman Diagrams

8.1 The Typical Behavior of Field Theoretical Perturbation Series

In chapters 3 and 4 we wrote down the perturbation series for the partition function and for some correlation functions. We found that the coefficients of λ^n were given by a sum $(d+1)n$-dimensional integrals if the space dimension is d. Typically, some of these integrals diverge if the cutoffs of the theory are removed. This does not mean that something is wrong with the model, but merely means first of all that the function which has been expanded is not analytic if the cutoffs are removed. To this end we consider a small example. Let

$$G_\delta(\lambda) := \int_0^\infty dx \int_0^1 dk \frac{1}{\sqrt{k+\lambda x+\delta}} e^{-x} \tag{8.1}$$

where $\delta > 0$ is some cutoff and the coupling λ is positive. One may think of $\delta = T$, the temperature, or $\delta = 1/L$ if L^d is the volume of the system. By explicit computation, using Lebesgue's theorem of dominated convergence to interchange the limit with the integrals,

$$G_0(\lambda) = \lim_{\delta\to 0} G_\delta(\lambda) = \int_0^\infty dx\, 2(\sqrt{1+\lambda x} - \sqrt{\lambda x})\, e^{-x}$$

$$= 2 + O(\lambda) - O(\sqrt{\lambda}) \tag{8.2}$$

Thus, the $\delta \to 0$ limit is well defined but it is not analytic. This fact has to show up in the Taylor expansion. It reads

$$G_\delta(\lambda) = \sum_{j=0}^n \binom{-\frac{1}{2}}{j} \int_0^\infty dx \int_0^1 dk \frac{x^j e^{-x}}{(k+\delta)^{j+\frac{1}{2}}} \lambda^j + r_{n+1} \tag{8.3}$$

Apparently, all integrals over k diverge for $j \geq 1$ in the limit $\delta \to 0$. Now, very roughly speaking, renormalization is the passage from the expansion (8.3) to the expansion

$$G_0(\lambda) = \sum_{\ell=0}^n \binom{\frac{1}{2}}{\ell} \int_0^\infty dx\, 2x^\ell\, e^{-x}\, \lambda^\ell - c\sqrt{\lambda} + R_{n+1} \tag{8.4}$$

where the last one is obtained from (8.2) by expanding the $\sqrt{1 + \lambda x}$ term. One would say 'the diverging integrals have been resumed to the nonanalytic term $c\sqrt{\lambda}$'. In the final expansion (8.4) all coefficients are finite and, for small λ, the lowest order terms are a good approximation since $(\theta_\lambda \in [0, \lambda])$

$$
\begin{aligned}
|R_{n+1}| &= \left| 2 \int_0^\infty dx \binom{\frac{1}{2}}{n+1} \frac{1}{(1+\theta_\lambda x)^{n+\frac{1}{2}}} x^{n+1} \lambda^{n+1} e^{-x} \right| \\
&\leq 2 \left| \binom{\frac{1}{2}}{n+1} \right| \int_0^\infty dx\, x^{n+1} e^{-x} \, \lambda^{n+1} \\
&= \frac{1}{2^{2n}} \frac{(2n)!}{n!} \lambda^{n+1} \overset{n \to \infty}{\sim} \sqrt{2} \left(\frac{n}{e} \right)^n \lambda^{n+1}
\end{aligned}
\tag{8.5}
$$

Here we used the Lagrange representation of the $n+1$'st Taylor remainder in the first line and Stirling's formula, $n! \sim \sqrt{2\pi n}(n/e)^n$, in the last line. An estimate of the form (8.5) is typical for renormalized field theoretic perturbation series. The lowest order terms are a good approximation for weak coupling, but the renormalized expansion is only asymptotic, the radius of convergence of the whole series is zero. The approximation becomes more accurate if n approaches $1/\lambda$, but then quickly diverges if $n > e/\lambda$. Or, for fixed n, the n lowest order terms are a good approximation as long as $\lambda < 1/n$.

In this small example we went from 'the unrenormalized' or 'naive' perturbation expansion (8.3) to the 'renormalized' perturbation expansion (8.4) by going through the exact answer (8.2). Of course, for the models we are interested in, we do not know the exact answer. Thus, for weak coupling, the whole problem is to find this rearrangement which transforms a naive perturbation expansion into a renormalized expansion which is (at least) asymptotic.

In the next section we prove a combinatorial formula which rewrites the perturbation series in terms of n'th order diagrams whose connected components are at least of order m, which, for $m = n$, results in a proof of the linked cluster theorem. This reordering has nothing to do with the rearrangements considered above, it simply states that the logarithm of the partition function is still given by a sum of diagrams which is not obvious. In section 8.3 we start with estimates on Feynman diagrams. That section is basic for an understanding of renormalization, since it identifies the divergent contributions in a sum of diagrams.

8.2 Connected Diagrams and the Linked Cluster Theorem

The perturbation series for the partition function reads

$$Z(\lambda) = \sum_{n=0}^{\infty} \frac{(-\lambda)^n}{n!} \int d\xi_1 \cdots d\xi_{2n} \, U(\xi_1 - \xi_2) \times \cdots \qquad (8.6)$$

$$\cdots \times U(\xi_{2n-1} - \xi_{2n}) \, \det\left[C(\xi_i, \xi_j)\right]_{1 \le i,j \le 2n}$$

If we expand the $2n \times 2n$ determinant and interchange the sum over permutations with the ξ-integrals, we obtain the expansion into Feynman diagrams:

$$Z(\lambda) = \sum_{n=0}^{\infty} \frac{(-\lambda)^n}{n!} \sum_{\pi \in S_{2n}} \operatorname{sign}\pi \, G(\pi) \qquad (8.7)$$

where the graph or the value of the graph defined by the permutation π is given by

$$G(\pi) = \int d\xi_1 \cdots d\xi_{2n} \prod_{i=1}^{n} U(\xi_{2i-1} - \xi_{2i}) \, C(\xi_1, \xi_{\pi 1}) \cdots C(\xi_{2n}, \xi_{\pi 2n}) \qquad (8.8)$$

In general the above integral factorizes into several connected components. The number of U's in each component defines the order of that component. The goal of this subsection is to prove the linked cluster theorem which states that the logarithm of the partition function is given by the sum of all connected diagrams. A standard proof of this fact can be found in many books on field theory or statistical mechanics [40, 56]. In the following we give a slightly more general proof which reorders the perturbation series in terms of n'th order diagrams whose connected components are at least of order m where $1 \le m \le n$ is an arbitrary given number. See (8.17) below.

Linked Cluster Theorem: *The logarithm of the partition function is given by the sum of all connected diagrams,*

$$\log Z(\lambda) = \sum_{n=0}^{\infty} \frac{(-\lambda)^n}{n!} \sum_{\substack{\pi \in S_{2n} \\ G(\pi) \text{ connected}}} \operatorname{sign}\pi \, G(\pi) \qquad (8.9)$$

We use the following

Lemma 8.2.1 *Let $\{w_n\}_{n\in\mathbb{N}}$ be a sequence with $w_i w_j = w_j w_i \ \forall i,j \in \mathbb{N}$ and let a be given with $a w_i = w_i a \ \forall i \in \mathbb{N}$ (for example $w_i, a \in \mathbb{C}$ or even elements of a Grassmann algebra). For fixed $m \in \mathbb{N}$ define the sequence $\{v_n\}_{n\in\mathbb{N}}$ by $(n = mk + l, \ 0 \leq l \leq m-1, \ k \in \mathbb{N})$*

$$\frac{v_{mk+l}}{(mk+l)!} = \sum_{j=0}^{k}(-1)^j \frac{a^j}{j!} \frac{w_{mk-mj+l}}{(mk-mj+l)!} \ . \tag{8.10}$$

Then the w_n's can be computed from the v_n's by

$$\frac{w_{mk+l}}{(mk+l)!} = \sum_{j=0}^{k} \frac{a^j}{j!} \frac{v_{mk-mj+l}}{(mk-mj+l)!} \tag{8.11}$$

and one has

$$\sum_{n=0}^{\infty} \frac{w_n}{n!} = \sum_{n=0}^{\infty} \frac{v_n}{n!} e^a \ . \tag{8.12}$$

Proof: One has

$$\sum_{j=0}^{k} \frac{a^j}{j!} \frac{v_{mk-mj+l}}{(mk-mj+l)!} = \sum_{j=0}^{k} \frac{a^j}{j!} \sum_{i=0}^{k-j}(-1)^i \frac{a^i}{i!} \frac{w_{m(k-j)-mi+l}}{(m(k-j)-mi+l)!}$$

$$\stackrel{r=i+j}{=} \sum_{r=0}^{k} \frac{a^r}{r!} \frac{w_{mk-mr+l}}{(mk-mr+l)!} \sum_{i=0}^{r} \binom{r}{i}(-1)^i$$

$$= \sum_{r=0}^{k} \frac{a^r}{r!} \frac{w_{mk-mr+l}}{(mk-mr+l)!} \delta_{r,0} = \frac{w_{mk+l}}{(mk+l)!}$$

which proves the first formula. The second formula is obtained as follows

$$\sum_{n=0}^{\infty} \frac{w_n}{n!} = \sum_{k=0}^{\infty}\sum_{l=0}^{m-1} \frac{w_{mk+l}}{(mk+l)!} = \sum_{k=0}^{\infty}\sum_{l=0}^{m-1}\sum_{j=0}^{k} \frac{a^j}{j!} \frac{v_{mk-mj+l}}{(mk-mj+l)!}$$

$$= \sum_{j=0}^{\infty} \frac{a^j}{j!} \sum_{k=j}^{\infty}\sum_{l=0}^{m-1} \frac{v_{mk-mj+l}}{(mk-mj+l)!} \stackrel{r=k-j}{=} \sum_{j=0}^{\infty} \frac{a^j}{j!} \sum_{r=0}^{\infty}\sum_{l=0}^{m-1} \frac{v_{mr+l}}{(mr+l)!}$$

$$= e^a \sum_{n=0}^{\infty} \frac{v_n}{n!} \tag{8.13}$$

which proves the lemma. ∎

Now we define the sum $\mathrm{Det}_n^{(m)}$ of n'th order diagrams whose connected components are at least of order m inductively by

$$\mathrm{Det}_n^{(1)} = \mathrm{Det}_n^{(1)}(C,U)$$

$$:= \int d\xi_1 \cdots d\xi_{2n} \prod_{i=1}^{n} U(\xi_{2i-1} - \xi_{2i}) \, \det\,[C(\xi_i,\xi_j)]_{1\leq i,j\leq 2n} \qquad (8.14)$$

and for $m \geq 1$, $n = mk + l$, $0 \leq l \leq m-1$

$$\frac{\mathrm{Det}_{mk+l}^{(m+1)}}{(mk+l)!} = \sum_{j=0}^{k}(-1)^j \frac{\left(\frac{1}{m!}\mathrm{Det}_m^{(m)}\right)^j}{j!} \frac{\mathrm{Det}_{mk-mj+l}^{(m)}}{(mk-mj+l)!} \qquad (8.15)$$

Then the fact that the logarithm gives only the connected diagrams may be formulated as follows.

Theorem 8.2.2 *The logarithm of the partition function is given by*

$$\log Z(\lambda) = \sum_{n=1}^{\infty} \frac{\lambda^n}{n!} \, \mathrm{Det}_n^{(n)}(C,U) \qquad (8.16)$$

and $\mathrm{Det}_n^{(n)}(C,U)$ *is the sum of all connected n'th order diagrams.*

Proof: We claim that for arbitrary m

$$Z(\lambda) = \sum_{n=0}^{\infty} \frac{\lambda^n}{n!}\mathrm{Det}_n^{(m+1)} \, \exp\left\{ \sum_{s=1}^{m} \frac{\lambda^s}{s!}\mathrm{Det}_s^{(s)} \right\} \qquad (8.17)$$

For $m = 0$, (8.17) is obviously correct. Suppose (8.17) is true for $m-1$. Then, because of

$$\lambda^n \frac{\mathrm{Det}_n^{(m+1)}}{n!} = \lambda^{mk+l} \frac{\mathrm{Det}_{mk+l}^{(m+1)}}{(mk+l)!}$$

$$= \sum_{j=0}^{k}(-1)^j \frac{\left(\frac{\lambda^m}{m!}\mathrm{Det}_m^{(m)}\right)^j}{j!} \lambda^{mk-mj+l} \frac{\mathrm{Det}_{mk-mj+l}^{(m)}}{(mk-mj+l)!} \qquad (8.18)$$

and the lemma, one obtains

$$Z(\lambda) = \sum_{n=0}^{\infty} \frac{\lambda^n}{n!}\mathrm{Det}_n^{(m)} \, \exp\left\{ \sum_{s=1}^{m-1} \frac{\lambda^s}{s!}\mathrm{Det}_s^{(s)} \right\}$$

$$= \sum_{n=0}^{\infty} \frac{\lambda^n}{n!}\mathrm{Det}_n^{(m+1)} \, e^{\frac{\lambda^m}{m!}\mathrm{Det}_m^{(m)}} \, \exp\left\{ \sum_{s=1}^{m-1} \frac{\lambda^s}{s!}\mathrm{Det}_s^{(s)} \right\}$$

$$= \sum_{n=0}^{\infty} \frac{\lambda^n}{n!}\mathrm{Det}_n^{(m+1)} \, \exp\left\{ \sum_{s=1}^{m} \frac{\lambda^s}{s!}\mathrm{Det}_s^{(s)} \right\} \qquad (8.19)$$

Furthermore we claim that for a given $m \in \mathbb{N}$

$$\text{Det}_1^{(m+1)} = \text{Det}_2^{(m+1)} = \cdots = \text{Det}_m^{(m+1)} = 0 \tag{8.20}$$

Namely, for $n = mk + l < m$ one has $k = 0$, $n = l$ such that

$$\frac{\text{Det}_n^{(m+1)}}{n!} = \sum_{j=0}^{0}(-1)^j \frac{\left(\frac{1}{m!}\text{Det}_m^{(m)}\right)^j}{j!} \frac{\text{Det}_n^{(m)}}{n!} = \frac{\text{Det}_n^{(m)}}{n!} \tag{8.21}$$

and it follows

$$\text{Det}_n^{(m+1)} = \text{Det}_n^{(m)} = \cdots = \text{Det}_n^{(n+1)} \tag{8.22}$$

But, for $n = m = 1m + 0$, by definition

$$\frac{\text{Det}_m^{(m+1)}}{m!} = \sum_{j=0}^{1}(-1)^j \frac{\left(\frac{1}{m!}\text{Det}_m^{(m)}\right)^j}{j!} \frac{\text{Det}_{m-mj}^{(m)}}{(m-mj)!}$$

$$= \frac{\text{Det}_m^{(m)}}{m!} - \frac{\left(\frac{1}{m!}\text{Det}_m^{(m)}\right)^1}{1!} = 0 \tag{8.23}$$

which proves (8.20). Thus one obtains

$$Z(\lambda) = \left(1 + \sum_{n=m+1}^{\infty} \frac{\lambda^n}{n!}\text{Det}_n^{(m+1)}\right) \exp\left\{\sum_{s=1}^{m} \frac{\lambda^s}{s!}\text{Det}_s^{(s)}\right\} \tag{8.24}$$

By taking the limit $m \to \infty$ one gets

$$\log Z(\lambda) = \sum_{s=1}^{\infty} \frac{\lambda^s}{s!}\text{Det}_s^{(s)} \tag{8.25}$$

which proves (8.16). It remains to prove that $\text{Det}_n^{(n)}$ is the sum of all connected n'th order diagrams. To this end we make the following definitions. Let $\pi \in S_{2n}$ be given. We say that π is of type

$$t(\pi) = 1^{b_1} 2^{b_2} \cdots n^{b_n} \tag{8.26}$$

iff G_π consists of precisely b_1 first order connected components, b_2 second order connected components, $\cdots$, b_n n'th order connected components, where G_π is the graph produced by the permutation π. Observe that, contrary to section (1.3.1), the b_r are not the number of r-cycles of the permutation π, but, as defined above, the number of r'th order connected components of the diagram (8.8) given by the permutation π. Let

$$S_{2n}^{(b_1,\cdots,b_n)} = \{\pi \in S_{2n} \mid t(\pi) = 1^{b_1} 2^{b_2} \cdots n^{b_n}\} \tag{8.27}$$

Then S_{2n} is the disjoint union

$$S_{2n} = \bigcup_{\substack{0 \le b_1, \cdots, b_n \le n \\ 1b_1 + \cdots + nb_n = n}} S_{2n}^{(b_1, \cdots, b_n)} \tag{8.28}$$

Let

$$S_{2n}^{(m)} = \bigcup_{\substack{0 \le b_m, \cdots, b_n \le n \\ mb_m + \cdots + nb_n = n}} S_{2n}^{(0, \cdots, 0, b_m, \cdots, b_n)} \tag{8.29}$$

and

$$S_{2n}^c = S_{2n}^{(n)} = S_{2n}^{(0, \cdots, 0, 1)} \tag{8.30}$$

One has

$$\left| S_{2n}^{(b_1, \cdots, b_n)} \right| = \frac{n!}{(1!)^{b_1} \cdots (n!)^{b_n} b_1! \cdots b_n!} |S_2^c|^{b_1} \cdots |S_{2n}^c|^{b_n} , \tag{8.31}$$

in particular

$$\left| S_{2n}^{(0, \cdots, 0, b_m, \cdots, b_n)} \right| = \frac{n!}{(n - mb_m)! \, (m!)^{b_m} \, b_m!} |S_{2m}^c|^{b_m} \left| S_{2n-2mb_m}^{(0, \cdots, 0, b_{m+1}, \cdots, b_{n-mb_m})} \right| \tag{8.32}$$

We now prove by induction on m that

$$\mathrm{Det}_n^{(m)} = \int d\xi_1 d\xi_2 \cdots d\xi_{2n} \prod_{i=1}^{n} U(\xi_{2i-1} - \xi_{2i}) \det{}^{(m)} [C(\xi_j, \xi_k)_{j,k=1,\cdots,2n}] \tag{8.33}$$

where

$$\det{}^{(m)} [C(\xi_j, \xi_k)_{j,k=1,\cdots,2n}] = \sum_{\pi \in S_{2n}^{(m)}} C(\xi_1, \xi_{\pi 1}) \cdots C(\xi_{2n}, \xi_{\pi(2n)}) \tag{8.34}$$

It follows from (8.33) that $\mathrm{Det}_n^{(n)}$ is the sum of all connected diagrams. Obviously (8.33) is correct for $m = 1$. Suppose (8.33) is true for m. Let

$$\widetilde{\mathrm{Det}}_n^{(m+1)} = \int d\xi_1 d\xi_2 \cdots d\xi_{2n} \prod_{i=1}^{n} U(\xi_{2i-1} - \xi_{2i}) \det{}^{(m+1)} [C(\xi_j, \xi_k)_{j,k=1,\cdots,2n}] \tag{8.35}$$

Then, for $n = mk + l$ with $k \in \mathbb{N}$ and $0 \leq l \leq m - 1$

$$\mathrm{Det}_n^{(m)} = \sum_{\pi \in S_{2n}^{(m)}} \int \prod_{i=1}^{2n} d\xi_i \prod_{i=1}^{n} U(\xi_{2i-1} - \xi_{2i}) \prod_{r=1}^{2n} C(\xi_r, \xi_{\pi r})$$

$$= \sum_{\substack{b_m, b_{m+1}, \cdots, b_n = 0 \\ mb_m + \cdots + nb_n = n}} \sum_{\pi \in S_{2n}^{(0, \cdots, 0, b_m, b_{m+1}, \cdots, b_n)}} \int \prod_{i=1}^{2n} d\xi_i \cdots$$

$$= \sum_{b_m = 0}^{k} \frac{n!}{(n - mb_m)! \, (m!)^{b_m} b_m!} \times$$

$$\left\{ \sum_{\pi \in S_{2m}^c} \int \prod_{i=1}^{2m} d\xi_i \prod_{i=1}^{m} U(\xi_{2i-1} - \xi_{2i}) \prod_{r=1}^{2m} C(\xi_r, \xi_{\pi r}) \right\}^{b_m} \times$$

$$\sum_{\pi \in S_{2n-2mb_m}^{(m+1)}} \int \prod_{i=1}^{2(n-mb_m)} d\xi_i \prod_{i=1}^{n-mb_m} U(\xi_{2i-1} - \xi_{2i}) \prod_{r=1}^{2(n-mb_m)} C(\xi_r, \xi_{\pi r})$$

$$= \sum_{b_m = 0}^{k} \frac{n!}{(n - mb_m)! \, (m!)^{b_m} b_m!} \left[\mathrm{Det}_m^{(m)} \right]^{b_m} \widetilde{\mathrm{Det}}_{n-mb_m}^{(m+1)} \tag{8.36}$$

or

$$\frac{\mathrm{Det}_{mk+l}^{(m)}}{(mk+l)!} = \sum_{b_m = 0}^{k} \frac{\left[\frac{1}{m!} \mathrm{Det}_m^{(m)} \right]^{b_m}}{b_m!} \frac{\widetilde{\mathrm{Det}}_{mk-mb_m+l}^{(m+1)}}{(mk - mb_m + l)!} \tag{8.37}$$

Then, by the lemma and the definition of $\mathrm{Det}_{mk+l}^{(m+1)}$,

$$\frac{\widetilde{\mathrm{Det}}_{mk+l}^{(m+1)}}{(mk+l)!} = \sum_{b_m = 0}^{k} (-1)^{b_m} \frac{\left[\frac{1}{m!} \mathrm{Det}_m^{(m)} \right]^{b_m}}{b_m!} \frac{\mathrm{Det}_{mk-mb_m+l}^{(m)}}{(mk - mb_m + l)!} = \frac{\mathrm{Det}_{mk+l}^{(m+1)}}{(mk+l)!}$$

which proves (8.33). ∎

8.3 Estimates on Feynman Diagrams

8.3.1 Elementary Bounds

In this section we identify the large contributions which are typically contained in a sum of Feynman diagrams. These large contributions have the effect that the lowest order terms of the naive perturbation expansion are not a good approximation. The elimination of these large contributions is called renormalization. We find that the size of a graph is determined by its subgraph structure. For the many-electron system with short range interaction, that is, for $V(\mathbf{x}) \in L^1$, the dangerous subgraphs are the two- and four-legged ones. Indeed, in Theorem 8.3.4 below we show that an n'th order diagram without two- and four-legged subgraphs is bounded by $const^n$ which is basically the best case which can happen. A sum of diagrams where each diagram is bounded by $const^n$ can be expected to be asymptotic. That is, the lowest order terms in this expansion would be indeed a good approximation if the coupling is not too big.

However usually there are certain subgraphs which produce anomalously large contributions which prevent the lowest order terms in the perturbation series from being a good approximation. For the many-electron system with short range interaction these are two- and four-legged subdiagrams. Four-legged subgraphs produce factorials, that is, an n'th order bound of the form $const^n n!$, the constant being independent of the cutoffs. Diagrams which contain two-legged subgraphs in general diverge if the cutoffs are removed. The goal of this section is to prove these assertions.

We start with a lemma which sets up all the graph theoretical notation and gives the basic bound in coordinate space. The diagrams in Lemma 8.3.1 consist of generalized vertices or subgraphs which are represented by some functions $I_{2q_v}(x_1, \cdots, x_{2q_v})$ and lines to which are assigned propagators $C(x - x')$. A picture may be helpful. In figure 8.1 below, $G = G(x_1, x_2)$ and there is an integral over the remaining variables $x_3, \cdots, x_{10}$.

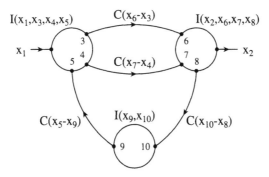

Figure 8.1

138

Lemma 8.3.1 (Coordinate Space Bound) *Let* $I_{2q} = I_{2q}(x_1, \cdots, x_{2q})$, $x_i \in \mathbb{R}^d$, *be a generalized vertex (or subgraph) with $2q$ legs obeying*

$$\|I_{2q}\|_\emptyset := \sup_i \sup_{x_i} \left\{ \int \prod_{\substack{j=1 \\ j \neq i}}^{2q} dx_j \, |I_{2q}(x_1, \cdots, x_{2q})| \right\} < \infty \qquad (8.38)$$

Let G be a connected graph built up from vertices I_{2q_v}, $v \in V_G$, the set of all vertices of G, by pairing some of their legs. Two paired legs are by definition a line $\ell \in L_G$, the set of all lines of G. To each line, assign a propagator $C_\ell(x_{i_\ell}^{v_\ell}, x_{i'_\ell}^{v'_\ell}) = C_\ell(x_{i_\ell}^{v_\ell} - x_{i'_\ell}^{v'_\ell})$. Suppose $2q$ legs remain unpaired. The value of G is by definition

$$G(x_1, \cdots, x_{2q}) = \int \prod_{\substack{v \in V_G \\ x_i^v \text{ int.}}}^{2q_v} dx_i^v \prod_{v \in V_G} I_{2q_v}(x_1^v, \cdots, x_{2q_v}^v) \prod_{\ell \in L_G} C_\ell(x_{i_\ell}^{v_\ell}, x_{i'_\ell}^{v'_\ell})$$

$$(8.39)$$

where $x_1, \cdots, x_{2q} \in \cup_{v \in V_G} \cup_{i=1}^{2q_v} \{x_i^v\}$ are by definition the variables of the unpaired legs and $x_{i_\ell}^{v_\ell}, x_{i'_\ell}^{v'_\ell} \in \cup_{v \in V_G} \cup_{i=1}^{2q_v} \{x_i^v\}$ are the variables of the legs connected by the line ℓ.

Let $f_1, \cdots, f_{2q}$ be some test functions. Fix $s = |S|$ legs of G where $S \subset \{1, \cdots, 2q\}$. These s legs, which will be integrated against test functions, are by definition the external legs of G, and the other legs, which are integrated over $\mathbb{R}^d$, are called internal. For $S \neq \emptyset$, define the norm

$$\|G\|_S := \left(\prod_{k \in S} \int dx_k \, |f_k(x_k)| \right) \left(\prod_{i \in S^c} \int dx_i \right) |G(x_1, \cdots, x_{2p})| \qquad (8.40)$$

Then there are the following bounds:

a)

$$\|G\|_\emptyset \leq \prod_{\ell \in T} \|C_\ell\|_{L^1} \prod_{\ell \in L \setminus T} \|C_\ell\|_\infty \prod_{v \in V} \|I_{2q_v}\|_\emptyset \qquad (8.41)$$

where T is a spanning tree for G which is a collection of lines which connects all vertices of G such that no loops are formed.

b)

$$\|G\|_S \leq \prod_{\ell \in \bar{T}} \|C_\ell\|_{L^1} \prod_{\ell \in L \setminus \bar{T}} \|C_\ell\|_\infty \prod_{v \in V} \|I_{2q_v}\|_{S_v} \qquad (8.42)$$

where now $\bar{T} = \cup_{i=1}^w T_i$ is a union of w trees which spans G and w is the number of vertices to which at least one external leg is hooked which, by definition, is the number of external vertices. Each T_i contains precisely one external vertex. Finally, S_v is the set of external legs at I_{2q_v}.

c) *Let G be a vacuum diagram, that is, a diagram without unpaired legs* $(q = 0)$. *Then*

$$|G| \leq L^d \prod_{\ell \in T} \|C_\ell\|_{L^1} \prod_{\ell \in L \setminus T} \|C_\ell\|_\infty \prod_{v \in V} \|I_{2q_v}\|_\emptyset \qquad (8.43)$$

where $|G|$ is the usual modulus of G and $L^d = \int dx\, 1$.

Proof: a) Choose a spanning tree T for G, that is, choose a set of lines $T \subset L = L_G$ which connects all vertices such that no loops are formed. For all lines not in T take the L^∞-norm. Let $x_i \in \{x_1, \cdots, x_{2q}\}$ be the variable where the supremum is taken over. We get

$$\|G\|_\emptyset \leq \int \prod_{v \in V_G} \prod_{\substack{i=1 \\ x^v_i \neq x_i}}^{2q_v} dx^v_i \prod_{v \in V_G} |I_{2q_v}(x^v_1, \cdots, x^v_{2q_v})| \prod_{\ell \in L_G} |C_\ell(x^{v_\ell}_{i_\ell} - x^{v'_\ell}_{i'_\ell})|$$

$$\leq \int \prod_{v \in V_G} \prod_{\substack{i=1 \\ x^v_i \neq x_i}}^{2q_v} dx^v_i \prod_{v \in V_G} |I_{2q_v}(x^v_1, \cdots, x^v_{2q_v})| \prod_{\ell \in T} |C_\ell(x^{v_\ell}_{i_\ell} - x^{v'_\ell}_{i'_\ell})| \times$$

$$\prod_{\ell \in L_G \setminus T} \|C_\ell\|_\infty$$

The vertex to which the variable x_i belongs we define as the root of the tree. To perform the integrations, we start at the extremities of the tree, that is, at those vertices I_{2q_v} which are not the root and which are connected to the tree only by one line. To be specific, choose one of these vertices, $I_{2q_{v_1}}(x^{v_1}_1, \cdots, x^{v_1}_{2q_{v_1}})$. Let $x^{v_1}_r$ be the variable which belongs to the tree. This variable also shows up in the propagator for the corresponding line ℓ, $C_\ell(x^{v_1}_r - x^{v'}_{r'})$ where v' is a vertex necessarily different from v since ℓ is on the tree. Now we bound as follows

$$\int dx^{v_1}_1 \cdots dx^{v_1}_{2q_{v_1}} |I_{2q_{v_1}}(x^{v_1}_1, \cdots, x^{v_1}_{2q_{v_1}})| |C_\ell(x^{v_1}_r - x^{v'}_{r'})| |I_{2q_{v'}}(\{x^{v'}_j\})|$$

$$= \int dx^{v_1}_r \int \prod_{\substack{i=1 \\ i \neq r}}^{2q_{v_1}} dx^{v_1}_i |I_{2q_{v_1}}(x^{v_1}_1, \cdots, x^{v_1}_{2q_v})| |C_\ell(x^{v_1}_r - x^{v'}_{r'})| |I_{2q_{v'}}(\{x^{v'}_j\})|$$

$$\leq \int dx^{v_1}_r \left\{ \sup_{x^{v_1}_r} \left[\int \prod_{\substack{i=1 \\ i \neq r}}^{2q_{v_1}} dx^{v_1}_i |I_{2q_{v_1}}(x^{v_1}_1, \cdots, x^{v_1}_{2q_v})| \right] |C_\ell(x^{v_1}_r - x^{v'}_{r'})| \times \right.$$

$$\left. |I_{2q_{v'}}(\{x^{v'}_j\})| \right\}$$

$$= \sup_{x^{v_1}_r} \left[\int \prod_{\substack{i=1 \\ i \neq r}}^{2q_{v_1}} dx^{v_1}_i |I_{2q_{v_1}}(x^{v_1}_1, \cdots, x^{v_1}_{2q_v})| \right] \int dx^{v_1}_r |C_\ell(x^{v_1}_r - x^{v'}_{r'})| |I_{2q_{v'}}(\{x^{v'}_j\})|$$

$$\leq \|I_{2q_{v_1}}\|_\emptyset \|C_\ell\|_{L^1} |I_{2q_{v'}}(\{x^{v'}_j\})| \qquad (8.44)$$

Now we repeat this step until we have reached the root of the tree. To obtain an estimate in this way we refer in the following to as 'we apply the tree identity'. For each line on the tree we get the L^1 norm, lines not on the tree give the L^∞ norm and each vertex is bounded by the $\|\cdot\|_\emptyset$ norm which results in (8.41).

b) Choose w trees $T_1, \cdots, T_w$ with the properties stated in the lemma. For each line not in $\bar{T}$ take the L^∞-norm. Then for each T_i apply the tree identity with the external vertex as root. Suppose this vertex is $I_{2q_v}(y_1, \cdots, y_{2q_v})$ and $y_1, \cdots, y_{p_v}$ are the external variables. Then, instead of $\|I_{2q_v}\|_\emptyset$ as in case (a) one ends up with

$$\left(\prod_{k=1}^{p_v} \int dy_k\, f_{j_k}(y_k) \right) \left(\prod_{i=p_v+1}^{2q_v} \int dy_i \right) |I_{2q_v}(y_1, \cdots, y_{2q_v})| \qquad (8.45)$$

which by definition is $\|I_{2q_v}\|_{S_v}$.

c) Here we proceed as in (a), however we can choose an arbitrary vertex I_{2q_v} to be the root of the tree, since we do not have to take a supremum over some x_i. We apply the tree identity, and the integrations at the last vertex, the root I_{2q_v}, are bounded by

$$\int \prod_{j=1}^{2q_v} dx_j^v\, |I_{2q_v}(x_1^v, \cdots, x_{2q_v}^v)| \leq \int dx_i^v \sup_{x_i^v} \prod_{\substack{j=1 \\ j \neq i}}^{2q_v} dx_j^v\, |I_{2q_v}(x_1^v, \cdots, x_{2q_v}^v)|$$

$$\leq L^d \|I_{2q_v}\|_\emptyset$$

which proves the lemma ∎

For an interacting many body system the basic vertex is

$$\int d\xi d\xi'\, \bar{\psi}(\xi)\psi(\xi)U(\xi - \xi')\bar{\psi}(\xi')\psi(\xi') \qquad (8.46)$$

which corresponds to the diagram

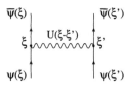

In order to represent this by some generalized vertex $I_4(\xi_1, \xi_2, \xi_3, \xi_4)$ which corresponds to the diagram

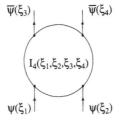

we rewrite (8.46) as

$$\int d\xi_1 d\xi_2 d\xi_3 d\xi_4 \, \bar\psi(\xi_3)\psi(\xi_1)\bar\psi(\xi_4)\psi(\xi_2)I_4(\xi_1,\xi_2,\xi_3,\xi_4) \qquad (8.47)$$

which coincides with (8.46) if we choose

$$I_4(\xi_1,\xi_2,\xi_3,\xi_4) = \delta(\xi_3 - \xi_1)\delta(\xi_4 - \xi_2)U(\xi_1 - \xi_2) \qquad (8.48)$$

Then

$$\|I_4\|_\emptyset = \|U\|_{L^1(\mathbb{R}^{d+1})} = \|V\|_{L^1(\mathbb{R}^d)} < \infty \qquad (8.49)$$

if we assume a short range potential.

The next lemma is the momentum space version of Lemma 8.3.1. We use the same letters for the Fourier transformed quantities, hats will be omitted. For translation invariant $I_4(\xi_1,\xi_2,\xi_3,\xi_4) = I_4(\xi_1 + \xi',\xi_2 + \xi',\xi_3 + \xi',\xi_4 + \xi')$ we have

$$\int d\xi_1 d\xi_2 d\xi_3 d\xi_4 \bar\psi(\xi_3)\psi(\xi_1)\bar\psi(\xi_4)\psi(\xi_2)I_4(\xi_1,\xi_2,\xi_3,\xi_4) =$$

$$\int dk_1 dk_2 dk_3 dk_4 \, (2\pi)^d \delta(k_1 + k_2 - k_3 - k_4) \, I_4(k_1,k_2,k_3,k_4)\bar\psi_{k_3}\psi_{k_1}\bar\psi_{k_4}\psi_{k_2}$$

where we abbreviated $dk := \frac{d^d k}{(2\pi)^d}$. For example,

$$\int dx_1 dx_2 dx_3 dx_4 \, e^{i(k_1 x_1 + k_2 x_2 - k_3 x_3 - k_4 x_4)} \, \delta(x_3 - x_1)\delta(x_4 - x_2)V(x_1 - x_2)$$

$$= \int dx_1 dx_2 \, e^{i(k_1 - k_3)x_1 + i(k_2 - k_4)x_2} \, V(x_1 - x_2)$$

$$= \int dx_1 dx_2 \, e^{i(k_1 - k_3)(x_1 - x_2)} \, V(x_1 - x_2) \, e^{i(k_1 + k_2 - k_3 - k_4)x_2}$$

$$= (2\pi)^d \delta(k_1 + k_2 - k_3 - k_4) \, V(k_1 - k_3) \qquad (8.50)$$

The value of the graph G defined in the above Lemma 8.3.1 reads in momentum space

$$(2\pi)^d \delta(k_1 + \cdots + k_{2q}) \, G(k_1, \cdots, k_{2q}) = \qquad (8.51)$$

$$\int \prod_{\ell \in L_G} dk_\ell \prod_{\ell \in L_G} C_\ell(k_\ell) \prod_{v \in V_G} (2\pi)^d \delta(k_1^v + \cdots + k_{2q_v}^v) I_{2q_v}(k_1^v, \cdots, k_{2q_v}^v)$$

where $k_i^v \in \cup_{\ell \in L} \{k_\ell\}$ is the momentum $\pm k_\ell$ if ℓ is the line to which the i'th leg of I_{2q_v} is paired. Observe that $G(k_1, \cdots, k_{2q})$ is the value of the Fourier transformed diagram after the conservation of momentum delta function has been removed. However, for a vacuum diagram, we do not explicitly remove the factor $\delta(0) = L^d$.

In (8.51) we have $|L_G| =: |L|$ integrals and $|V_G| =: |V|$ constraints. The $|V|$ delta functions produce one overall delta function which is explicitly written in (8.51) and $|V| - 1$ constraints on the $|L|$ integration variables. Let T be a tree for G, that is, a collection of lines which connect all the vertices such that no loops are formed. Since $|T| = |V| - 1$, we can use the momenta k_ℓ for lines $\ell \in L \setminus T$ which are not on the tree as independent integration variables. We get

$$G(k_1, \cdots, k_{2q}) = \int \prod_{\ell \in L \setminus T} dk_\ell \prod_{\ell \in L} C_\ell(K_\ell) \prod_{v \in V} I_{2q_v}(K_1^v, \cdots, K_{2q_v}^v) \quad (8.52)$$

Here the sum of all momenta K_ℓ flowing through the line ℓ is defined as follows. Each line ℓ not on the tree defines a unique loop which contains only lines on the tree with the exception of ℓ itself. To each line in this loop assign the loop momentum k_ℓ. The sum of the external variables $k_1 + \cdots + k_{2q}$ vanishes, thus we may write

$$G(k_1, \cdots, k_{2q}) = G(k_1, \cdots, k_{2q-1}, -k_1 - \cdots - k_{2q-1}) \quad (8.53)$$

There are $2q - 1$ unique paths γ_i on G, containing only lines on the tree, which connect the i'th leg, to which the external momentum k_i is assigned, to the $2q$'th leg, to which the momentum $k_{2q} = -k_1 - \cdots - k_{2q-1}$ is assigned. To each line on γ_i assign the momentum k_i. Then K_ℓ is the sum of all assigned momenta. In particular, $K_\ell = k_\ell$ for all $\ell \in L \setminus T$. The K_i appearing in $I_{2q_v}(K_1, \cdots, K_{2q_v})$ are the sum of all momenta flowing through the leg of I_{2q_v} labelled by i.

Lemma 8.3.2 (Momentum Space Bound) *Define the following norms*

$$\|I_{2q}\|_\infty = \sup_{\substack{k_1, \cdots, k_{2q} \in \mathbb{R}^d \\ k_1 + \cdots + k_{2q} = 0}} |I_{2q}(k_1, \cdots, k_{2q})| \quad (8.54)$$

and for $S \subset \{1, \cdots, 2q\}$, $S \neq \emptyset$,

$$\|I\|_S = \sup_{\substack{k_s \in \mathbb{R}^d \\ s \notin S}} \int \prod_{s \in S} dk_s \, \delta(k_1 + \cdots + k_{2q}) |I_{2q}(k_1, \cdots, k_{2q})| \quad (8.55)$$

Recall the notation of Lemma 8.3.1. Then there are the following bounds

a)

$$\|G\|_\infty \;\le\; \prod_{\ell\in T}\|C_\ell\|_\infty \;\prod_{\ell\in L\backslash T}\|C_\ell\|_1 \;\prod_{v\in V}\|I_{2q_v}\|_\infty \tag{8.56}$$

b)

$$\|G\|_S \;\le\; \prod_{\ell\in\bar T}\|C_\ell\|_\infty \;\prod_{\ell\in L\backslash\bar T}\|C_\ell\|_1 \;\prod_{\substack{v\in V\\ S_v\ne\emptyset}}\|I_{2q_v}\|_{S_v} \;\prod_{\substack{v\in V\\ S_v=\emptyset}}\|I_{2q_v}\|_\infty \tag{8.57}$$

c) *Let G be a vacuum diagram, that is, G has no unpaired legs ($q = 0$). Then*

$$|G| \;\le\; L^d \prod_{\ell\in T}\|C_\ell\|_\infty \;\prod_{\ell\in L\backslash T}\|C_\ell\|_1 \;\prod_{v\in V}\|I_{2q_v}\|_\emptyset \tag{8.58}$$

where L^d is the volume of the system.

Remark: Observe that the $\|\cdot\|_S$-norms in coordinate space depend on the choice of testfunctions whereas in momentum space the $\|\cdot\|_S$-norms are defined independent of testfunctions.

Proof: a) We have

$$|G(k_1,\cdots,k_{2q})| \;\le$$

$$\int \prod_{\ell\in L\backslash T} dk_\ell \prod_{\ell\in T}|C_\ell(K_\ell)| \prod_{\ell\in L\backslash T}|C_\ell(K_\ell)| \prod_{v\in V}|I_{2q_v}(K_1,\cdots,K_{2q_v})|$$

$$\le \prod_{\ell\in T}\|C_\ell\|_\infty \prod_{v\in V}\|I_{2q_v}\|_\infty \int \prod_{\ell\in L\backslash T} dk_\ell \prod_{\ell\in L\backslash T}|C_\ell(K_\ell)|$$

$$= \prod_{\ell\in T}\|C_\ell\|_\infty \prod_{v\in V}\|I_{2q_v}\|_\infty \int \prod_{\ell\in L\backslash T} dk_\ell \prod_{\ell\in L\backslash T}|C_\ell(k_\ell)|$$

$$= \prod_{\ell\in T}\|C_\ell\|_\infty \prod_{v\in V}\|I_{2q_v}\|_\infty \prod_{\ell\in L\backslash T}\|C_\ell\|_1 \tag{8.59}$$

b) Let $k_{i_1},\cdots,k_{i_{|S|}}$ be the external momenta of G with respect to S. Observe that exactly $|S_{v_i}|$, $i=1,\cdots,w$, of these momenta are external to $I_{2q_{v_i}}$ and

$$|S_{v_1}| + \cdots + |S_{v_w}| \;=\; |S| \tag{8.60}$$

For each $j=1,\cdots,w$, let k_j^*, be one of the momenta external to I_{v_j}. That is, $k_j^* = k_{i_{j^*}}$ for some $1\le j^*\le|S|$. Let $\tilde k_1,\cdots,\tilde k_{w'}$ be the complementary external momenta. Here, $w' = |S| - w$. We have

$$\{k_1^*,\cdots,k_w^*,\tilde k_1,\cdots,\tilde k_{w'}\} \;=\; \{k_{i_1},\cdots,k_{i_{|S|}}\} \tag{8.61}$$

By definition,

$$\|G\|_S = \sup_{\substack{k_s \in \mathbb{R}^d \\ s \notin S}} \int \prod_{s \in S} dk_s \, \delta(k_1 + \cdots + k_{2q}) \, |G(k_1, \cdots, k_{2q})| \qquad (8.62)$$

$$= \sup_{\substack{k_s \in \mathbb{R}^d \\ s \notin S}} \int \prod_{j=1}^{w'} d\tilde{k}_j \prod_{j=1}^{w} dk_j^* \, \delta(k_1 + \cdots + k_{2q}) \, |G(k_1, \cdots, k_{2q})|$$

$$= \sup_{\substack{k_s \in \mathbb{R}^d \\ s \notin S}} \int dk_0 \prod_{j=1}^{w'} d\tilde{k}_j \prod_{j=1}^{w} dk_j^* \, \delta(k_0 + k_1^* + \cdots + k_w^*) \times$$

$$\delta(k_1 + \cdots + k_{2q}) \, |G(k_1, \cdots, k_{2q})|$$

Observe that

$$\delta(k_1 + \cdots + k_{2q}) \, |G(k_1, \cdots, k_{2q})| \le$$

$$\int \prod_{\ell \in L_G} dk_\ell \, |C_\ell(k_\ell)| \prod_{v \in V} |I_{2q_v}(k_{\ell_1}, \cdots, k_{\ell_{2q_v}})| \, \delta(k_{\ell_1} + \cdots + k_{\ell_{2q_v}}) \quad (8.63)$$

Thus,

$$\|G\|_S \le \sup_{\substack{k_s \in \mathbb{R}^d \\ s \notin S}} \int dk_0 \prod_{j=1}^{w'} d\tilde{k}_j \, H(k_0, \tilde{k}_1, \cdots, \tilde{k}_{w'}, k_s; s \notin S) \qquad (8.64)$$

where

$$H(k_0, \tilde{k}_1, \cdots, \tilde{k}_{w'}, k_s; s \notin S) =$$

$$\int \prod_{j=1}^{w} dk_j^* \prod_{\ell \in L_G} dk_\ell \, |C_\ell(k_\ell)| \, \delta(k_0 + k_1^* + \cdots + k_w^*) \qquad (8.65)$$

$$\times \prod_{v \in V} |I_{2q_v}(k_{\ell_1}, \cdots, k_{\ell_{2q_v}})| \, \delta(k_{\ell_1} + \cdots + k_{\ell_{2q_v}})$$

Let G^* be the graph obtained from G and the vertex $I_* = \delta(k_0 + k_1^* + \cdots + k_w^*)$ with $w+1$ legs by joining the leg of I_{v_j} with momentum k_j^* to one leg of I_* other than the leg labeled by k_0. Let T_i^* be the tree obtained from T_i by adjoining the line that connects the vertex at the end of T_i to I_*. Observe that $T^* = T_1^* \cup \cdots \cup T_w^*$ is a spanning tree for G^*. For each $\ell \in L_{G^*} \setminus T^*$ there is a momentum cycle obtained by attaching ℓ to the unique path in T^* that joins the ends of ℓ. Let p_ℓ^* be the momentum flowing around that cycle. The external legs of G^* are labeled by the momenta $\tilde{k}_1, \cdots, \tilde{k}_{w'}, k_s; s \notin S$, and k_0. Connect each of the external legs $\tilde{k}_1, \cdots, \tilde{k}_{w'}, k_s; s \notin S$, to k_0 by the unique path in T^* that joins them and let the external momenta flow along these paths. We have

$$H(k_0, \tilde{k}_1, \cdots, \tilde{k}_{w'}, k_s; s \notin S) = \delta(\cdots) \, G^*(k_0, \tilde{k}_1, \cdots, \tilde{k}_{w'}, k_s; s \notin S) \quad (8.66)$$

where

$$\delta(\cdots) = \delta\left(k_0 + \tilde{k}_1 + \cdots + \tilde{k}_{w'} + \sum_{s \notin S} k_s\right) \tag{8.67}$$

and

$$G^*(k_0, \tilde{k}_1, \cdots, \tilde{k}_{w'}, k_s; s \notin S) =$$
$$\int \prod_{\ell \in L_{G^*} \setminus T^*} dp_\ell^* \prod_{\ell \in L_G} |C_\ell(K_\ell^*)| \prod_{v \in V} |I_{2q_v}(K_1^*, \cdots, K_{q_v}^*)|$$

Here, K_ℓ^* is the sum of all momenta flowing through ℓ and K_i^* appearing in I_{2q_v} is the sum of all momenta flowing through the leg of I_{2q_v} labelled by i. Observe that, by construction, $G^*(k_0, \tilde{k}_1, \cdots, \tilde{k}_{w'}, k_s; s \notin S)$ is independent of k_0. Now split $L_G = T_1 \cup \cdots \cup T_w \cup \left(L_G \setminus (T_1 \cup \cdots \cup T_w)\right)$. Then

$$G^* = \int \prod_{\ell \in L_{G^*} \setminus T^*} dp_\ell^* \prod_{\ell \in T_1 \cup \cdots \cup T_w} |C_\ell(K_\ell^*)| \prod_{\ell \in L \setminus (T_1 \cup \cdots \cup T_w)} |C_\ell(K_\ell^*)| \times$$
$$\prod_{v \in V} |I_{2q_v}(K_1^*, \cdots, K_{q_v}^*)|$$

$$= \int \prod_{\ell \in L_{G^*} \setminus T^*} dp_\ell^* \prod_{\ell \in T_1 \cup \cdots \cup T_w} |C_\ell(K_\ell^*)| \prod_{\ell \in L \setminus (T_1 \cup \cdots \cup T_w)} |C_\ell(p_\ell^*)| \times$$
$$\prod_{v \in V} |I_{2q_v}(K_1^*, \cdots, K_{2q_v}^*)|$$

$$= \int \prod_{\ell \in L_{G^*}) \setminus T^*} dp_\ell^* \prod_{\ell \in T_1 \cup \cdots \cup T_w} |C_\ell(K_\ell^*)| \prod_{\ell \in L \setminus (T_1 \cup \cdots \cup T_w)} |C_\ell(p_\ell^*)| \times$$
$$\prod_{i=1}^{w} |I_{2q_{v_i}}(K_1^*, \cdots, K_{2q_{v_i}}^*)| \prod_{\substack{v \neq v_i \\ i=1, \cdots, w}} |I_{2q_v}(K_1^*, \cdots, K_{2q_v}^*)|$$

Thus we get

$$\|G\|_S \leq \sup_{\substack{k_s \in \mathbb{R}^d \\ s \notin S}} \int dk_0 \prod_{j=1}^{w'} d\tilde{k}_j \, \delta\left(k_0 + \tilde{k}_1 + \cdots + \tilde{k}_{w'} + \sum_{s \notin S} k_s\right) \times$$
$$G^*(k_0, \tilde{k}_1, \cdots, \tilde{k}_{w'}, k_s; s \notin S)$$

$$= \sup_{\substack{k_s \in \mathbb{R}^d \\ s \notin S}} \int \prod_{j=1}^{w'} d\tilde{k}_j \, G^*(\tilde{k}_1, \cdots, \tilde{k}_{w'}, k_s; s \notin S)$$

$$\leq \sup_{\substack{k_s \in \mathbf{R}^d \\ s \notin S}} \int \prod_{j=1}^{w'} d\tilde{k}_j \int \prod_{\ell \in L_{G^*}\setminus T^*} dp_\ell^* \prod_{\ell \in T_1 \cup \cdots \cup T_w} |C_\ell(K_\ell^*)| \prod_{\ell \in L \setminus (T_1 \cup \cdots \cup T_w)} |C_\ell(p_\ell^*)|$$

$$\prod_{i=1}^{w} |I_{2q_{v_i}}(K_1^*, \cdots, K_{2q_{v_i}}^*)| \prod_{\substack{v \neq v_i \\ i=1,\cdots,w}} |I_{2q_v}(K_1^*, \cdots, K_{2q_v}^*)|$$

$$\leq \prod_{\ell \in T_1 \cup \cdots \cup T_w} \|C_\ell\|_\infty \prod_{\substack{v \neq v_i \\ i=1,\cdots,w}} \|I_{2q_v}\|_\infty$$

$$\times \sup_{\substack{k_s \in \mathbf{R}^d \\ s \notin S}} \int \prod_{j=1}^{w'} d\tilde{k}_j \int \prod_{\ell \in L_{G^*}\setminus T^*} dp_\ell^* \prod_{\ell \in L \setminus (T_1 \cup \cdots \cup T_w)} |C_\ell(p_\ell^*)| \times$$

$$\prod_{i=1}^{w} |I_{2q_{v_i}}(K_1^*, \cdots, K_{2q_{v_i}}^*)| \tag{8.68}$$

Exchanging integrals, we obtain

$$\int \prod_{j=1}^{w'} d\tilde{k}_j \int \prod_{\ell \in L_{G^*}\setminus T^*} dp_\ell^* \prod_{\ell \in L \setminus (T_1 \cup \cdots \cup T_w)} |C_\ell(p_\ell^*)| \times$$

$$\prod_{i=1}^{w} |I_{2q_{v_i}}(K_1^*, \cdots, K_{2q_{v_i}}^*)|$$

$$= \int \prod_{\ell \in L_{G^*}\setminus T^*} dp_\ell^* \prod_{\ell \in L \setminus (T_1 \cup \cdots \cup T_w)} |C_\ell(p_\ell^*)| \times$$

$$\int \prod_{j=1}^{w'} d\tilde{k}_j \prod_{i=1}^{w} |I_{2q_{v_i}}(K_1^*, \cdots, K_{2q_{v_i}}^*)|$$

$$= \int \prod_{\ell \in L_{G^*}\setminus T^*} dp_\ell^* \prod_{\ell \in L \setminus (T_1 \cup \cdots \cup T_w)} |C_\ell(p_\ell^*)| \times$$

$$\prod_{i=1}^{w} \int \prod_{\substack{\tilde{k}_j \in S_{v_i} \\ j=1,\cdots,w'}} d\tilde{k}_j |I_{2q_{v_i}}(K_1^*, \cdots, K_{2q_{v_i}}^*)| \tag{8.69}$$

For each $i = 1, \cdots, w$, exactly one of the arguments, $K_1^*, \cdots, K_{2q_{v_i}}^*$, say for convenience the first, is the momentum flowing through the single line that connects $I_{2q_{v_i}}$ to I_*. It is the sum of all external momenta flowing into $I_{2q_{v_i}}$ and some loop momenta. Furthermore, exactly $|S_{v_i}| - 1$ of the arguments $K_1^*, \cdots, K_{q_{v_i}}^*$ appearing in $I_{2q_{v_i}}$ are equal to external momenta in the set $\{\tilde{k}_1, \cdots, \tilde{k}_{w'}\}$. By construction, no other momenta flow through these legs. For convenience, suppose that $K_2^* = \tilde{k}_2, \cdots, K_{|S_{v_i}|}^* = \tilde{k}_{|S_{v_i}|}$. The remaining arguments on the list $K_1^*, \cdots, K_{q_{v_i}}^*$ are sums of loop momenta only. Recall that $K_1^* = -K_2^* - \cdots - K_{2q_{v_i}}^*$. Thus,

$$\int \prod_{\substack{\tilde{k}_j \in S_{v_i} \\ j=1,\cdots,w'}} d\tilde{k}_j |I_{2q_{v_i}}(K_1^*, \cdots, K_{2q_{v_i}}^*)| =$$

$$\int dp \prod_{\substack{\tilde{k}_j \in S_{v_i} \\ j=1,\cdots,w'}} d\tilde{k}_j \, \delta(p + K_2^* + \cdots + K_{2q_{v_i}}^*) |I_{2q_{v_i}}(p, K_2^*, \cdots, K_{2q_{v_i}}^*)|$$

$$= \int dp \prod_{\substack{\tilde{k}_j \in S_{v_i} \\ j=1,\cdots,w'}} d\tilde{k}_j \, \delta\left(p + \tilde{k}_2 + \cdots + \tilde{k}_{|S_{v_i}|} + K_{|S_{v_i}|+1}^* + \cdots + K_{2q_{v_i}}^*\right) \times$$

$$|I_{2q_{v_i}}(p, \tilde{k}_2, \cdots, K_{2q_{v_i}}^*)|$$

$$\leq \sup_{K_{|S_{v_i}|+1}^*, \cdots, K_{2q_{v_i}}^*} \int dp \prod_{\substack{\tilde{k}_j \in S_{v_i} \\ j=1,\cdots,w'}} d\tilde{k}_j \, \delta\left(p + \tilde{k}_2 + \cdots + \tilde{k}_{|S_{v_i}|} + \right.$$

$$\left. + K_{|S_{v_i}|+1}^* + \cdots + K_{2q_{v_i}}^*\right) |I_{2q_{v_i}}(p, \tilde{k}_2, \cdots, K_{2q_{v_i}}^*)|$$

$$= \|I_{2q_{v_i}}\|_{S_{v_i}} \tag{8.70}$$

Combining (8.69) and (8.70), we arrive at

$$\int \prod_{j=1}^{w'} d\tilde{k}_j \int \prod_{\ell \in L_{G^*} \backslash T^*} dp_\ell^* \prod_{\ell \in L \backslash (T_1 \cup \cdots \cup T_w)} |C_\ell(p_\ell^*)| \prod_{i=1}^{w} |I_{2q_{v_i}}(K_1^*, \cdots, K_{2q_{v_i}}^*)|$$

$$\leq \prod_{i=1}^{w} \|I_{2q_{v_i}}\|_{S_{v_i}} \int \prod_{\ell \in L_{G^*} \backslash T^*} dp_\ell^* \prod_{\ell \in L \backslash (T_1 \cup \cdots \cup T_w)} |C_\ell(p_\ell^*)|$$

$$= \prod_{i=1}^{w} \|I_{2q_{v_i}}\|_{S_{v_i}} \prod_{\ell \in L \backslash (T_1 \cup \cdots \cup T_w)} \|C_\ell\|_1 \tag{8.71}$$

Finally,

$$\|G\|_S \leq \prod_{\ell \in T_1 \cup \cdots \cup T_w} \|C_\ell\|_\infty \prod_{\substack{v \neq v_i \\ i=1,\cdots,w}} \|I_{2q_v}\|_\infty \prod_{i=1}^{w} \|I_{2q_{v_i}}\|_{S_{v_i}} \prod_{\ell \in L \backslash (T_1 \cup \cdots \cup T_w)} \|C_\ell\|_1$$

which proves the lemma. ∎

8.3.2 Single Scale Bounds

The Lemmata 8.3.1 and 8.3.2 estimate a diagram in terms of the L^1- and L^∞- norms of its propagators. An L^∞ bound of the type, say, $\frac{1}{1+x^2} \leq 1$ is of course very crude since the information is lost that there is decay for large x. In order to get sharp bounds, one introduces a scale decomposition of the covariance which isolates the singularity and puts it at a certain scale. To this end let M be some constant bigger than one and write

$$
C(k) = \frac{1}{ik_0 - e_{\mathbf{k}}} = \frac{\chi(|ik_0 - e_{\mathbf{k}}| \leq 1)}{ik_0 - e_{\mathbf{k}}} + \frac{\chi(|ik_0 - e_{\mathbf{k}}| > 1)}{ik_0 - e_{\mathbf{k}}}
$$

$$
= \sum_{j=0}^{\infty} \frac{\chi(M^{-j-1} < |ik_0 - e_{\mathbf{k}}| \leq M^{-j})}{ik_0 - e_{\mathbf{k}}} + \frac{\chi(|ik_0 - e_{\mathbf{k}}| > 1)}{ik_0 - e_{\mathbf{k}}}
$$

$$
=: \sum_{j=0}^{\infty} C^j(k) + C^{UV}(k) \tag{8.72}
$$

For each $C^j(k)$ we have the momentum space bounds

$$
\|C^j\|_{\infty} \leq M^j, \quad \|C^j\|_1 \leq c\,M^{-j} \tag{8.73}
$$

since $vol\{(k_0, \mathbf{k}) \mid \sqrt{k_0^2 + e_{\mathbf{k}}^2} \leq M^{-j}\} \leq c\,M^{-2j}$. Here c is some j independent constant. Then the strategy to bound a diagram is the following one. Substitute each covariance C of the diagram in (8.51) by its scale decomposition. Interchange the scale-sums with the momentum space integrals. Then one has to bound a diagram where each propagator has a fixed scale. The L^1- and L^{∞}-bounds on these propagators are now sharp bounds. Thus we apply Lemma 8.3.2. The bounds of Lemma 8.3.2 depend on the choice of a tree for the diagram. Propagators on the tree are bounded by their L^{∞}-norm which is large whereas propagators not on the tree are bounded by their L^1-norm which is small. Thus the tree should be chosen in such a way such that propagators with small scales are on the tree and those with large scales are not on the tree. This is the basic idea of the Gallavotti Nicolo tree expansion [33] which has been applied to the diagrams of the many-electron system in [29],[8]. An application to QED can be found in [17].

Instead of completely multiplying out all scales one can also apply an inductive treatment which is more in the spirit of renormalization group ideas which are discussed in the next section. Let

$$
C^{\leq j}(k) := \sum_{i=0}^{j} C^i(k) \tag{8.74}
$$

such that

$$
C^{\leq j+1}(k) = C^{\leq j}(k) + C^{j+1}(k) \tag{8.75}
$$

We consider a diagram up to a fixed scale j, that is, each covariance is given by $C^{\leq j}$. Then we see how the bounds change if we go from scale j to scale $j+1$ by using (8.75). If the diagram has L lines, application of (8.75) produces 2^L terms. Each term can be considered as a diagram, consisting of subdiagrams which have propagators $C^{\leq j}$, and lines which carry propagators of scale $j+1$. Then we apply Lemma 8.3.2 where the generalized vertices I_{2q} are given by the subdiagrams of scale $\leq j$ which we can bound by a suitable induction hypothesis and all propagators are given by C^{j+1}.

The next lemma specifies Lemmata 8.3.1, 8.3.2 for the case that all propagators have the same scale and satisfy the bounds (8.73) and in Theorem 8.3.4 diagrams are bounded using the induction outlined above. In the following we will consider only the infrared part $\sum_{j=0}^{\infty} C^j$ of the covariance which contains the physically relevant region around the singularity at the Fermi surface $e_{\mathbf{k}} = 0$ and $k_0 = 0$ and neglect the ultraviolet part $C^{UV}(k)$ in (8.72). The proof that all n'th order diagrams with $C^{UV}(k)$ as propagator are bounded by $const^n$ can be found in [29].

Lemma 8.3.3 *Let G be a graph as in Lemma 8.3.1 or 8.3.2 and C_ℓ be some covariance.*

a) **Coordinate Space:** *Suppose that each C_ℓ satisfies the estimates*

$$\|C_\ell(x)\|_\infty \leq c\,M^{-j}, \quad \|C_\ell(x)\|_1 \leq c\,M^{\alpha j} \qquad (8.76)$$

Then there are the bounds

$$\|G\|_\emptyset \leq c^{|L_G|} \prod_{v \in V_G} \left(\|I_{2q_v}\|_\emptyset M^{-(q_v - 1 - \alpha)j} \right) M^{(q-1-\alpha)j} \qquad (8.77)$$

$$\|G\|_S \leq c^{|L_G|} \prod_{v \in V_{G,\text{int.}}} \left(\|I_{2q_v}\|_\emptyset M^{-(q_v - 1 - \alpha)j} \right) \times \qquad (8.78)$$

$$\prod_{v \in V_{G,\text{ext.}}} \left(\|I_{2q_v}\|_{S_v} M^{-(q_v - \frac{|S_v|}{2})j} \right) M^{(q - \frac{|S|}{2})j}$$

and if G has no unpaired legs

$$|G| \leq \beta L^d\, c^{|L_G|} \prod_{v \in V_G} \left(\|I_{2q_v}\|_\emptyset M^{-(q_v - 1 - \alpha)j} \right) M^{-(1+\alpha)j} \qquad (8.79)$$

b) **Momentum Space:** *Suppose that each $C_\ell(k)$ satisfies the estimates*

$$\|C_\ell(k)\|_\infty \leq c\,M^j \quad \|C_\ell(k)\|_1 \leq c\,M^{-j} \qquad (8.80)$$

Then there are the bounds

$$\|G\|_\infty \leq c^{|L_G|} \prod_{v \in V_G} \left(\|I_{2q_v}\|_\infty M^{-(q_v - 2)j} \right) M^{(q-2)j} \qquad (8.81)$$

$$\|G\|_S \leq c^{|L_G|} \prod_{v \in V_{G,\text{int.}}} \left(\|I_{2q_v}\|_\infty M^{-(q_v - 2)j} \right) \times \qquad (8.82)$$

$$\prod_{v \in V_{G,\text{ext.}}} \left(\|I_{2q_v}\|_{S_v} M^{-(q_v - \frac{|S_v|}{2})j} \right) M^{(q - \frac{|S|}{2})j}$$

and if G has no unpaired legs

$$|G| \leq \beta L^d c^{|L_G|} \prod_{v \in V_G} \left(\|I_{2q_v}\|_\infty M^{-(q_v-2)j} \right) M^{-2j} \qquad (8.83)$$

Proof: a) Apply Lemma 8.3.1. Choose a spanning tree T for G. Suppose G is made of $n = \sum_v 1$ vertices. Then

$$|T| = n-1, \quad |L| = \frac{1}{2} \left(\sum_{v \in V} 2q_v - 2q \right) \qquad (8.84)$$

$$|L \setminus T| = \sum_{v \in V} q_v - q - n + 1 = \sum_{v \in V} (q_v - 1) - q + 1 \qquad (8.85)$$

Thus

$$\|G\|_\emptyset \leq \left(c\, M^{\alpha j} \right)^{n-1} \left(c\, M^{-j} \right)^{\sum_{v \in V} q_v - q - n + 1} \prod_{v \in V} \|I_{2q_v}\|_\emptyset$$

$$= c^{|L|} M^{-j \left(\sum_{v \in V} q_v - (1+\alpha)n + 1 + \alpha - q \right)} \prod_{v \in V} \|I_{2q_v}\|_\emptyset$$

$$= c^{|L|} M^{-j \sum_{v \in V} (q_v - 1 - \alpha)} M^{j(q-1-\alpha)} \prod_{v \in V} \|I_{2q_v}\|_\emptyset$$

$$= c^{|L|} \prod_{v \in V} \left(\|I_{2q_v}\|_\emptyset M^{-(q_v-1-\alpha)j} \right) M^{(q-1-\alpha)j} \qquad (8.86)$$

which proves (8.77). To obtain (8.78), we use (8.42). Let w be $|V_{\text{ext}}|$. Choose a union of trees $\bar{T} = \bigcup_{i=1}^w T_i$ such that each T_i contains one external vertex and $\bar{T}$ spans G. Then

$$|\bar{T}| = n - 1 - (w-1),$$

$$|L \setminus \bar{T}| = \sum_{v \in V} q_v - q - (n - 1 - (w-1)) = \sum_{v \in V} q_v - p - n + w$$

Recall that $|S_v|$ is the number of external legs of I_{2q_v} and $|S|$ the number of external legs of G, so $\sum_{v \in V_{\text{ext}}} |S_v| = |S|$. Thus one gets

$$\|G\|_S \leq \left(c\, M^{\alpha j} \right)^{n-w} \left(c\, M^{-j} \right)^{\sum_{v \in V} q_v - q - n + w} \prod_{v \in V} \|I_{2q_v}\|_{S_v}$$

$$= c^{|L|} M^{-j \sum_{v \in V} (q_v - 1 - \alpha)} M^{-j((1+\alpha)w - q)} \prod_{v \in V} \|I_{2q_v}\|_{S_v}$$

$$= c^{|L|} \prod_{v \in V_{\text{int}}} \left(\|I_{2q_v}\|_\emptyset M^{-(q_v-1-\alpha)j} \right) \times$$

$$\prod_{v \in V_{\text{ext}}} \left(\|I_{2q_v}\|_{S_v} M^{-j(q_v-1-\alpha)} M^{-(1+\alpha)j} \right) M^{qj}$$

$$= c^{|L|} \prod_{v \in V_{\text{int}}} \left(\|I_{2q_v}\|_{\emptyset} M^{-(q_v-1-\alpha)j} \right) \times$$

$$\prod_{v \in V_{\text{ext}}} \left(\|I_{2q_v}\|_{S_v} \right) M^{-j\left(\sum_{v \in V_{\text{ext}}} q_v - \frac{|S|}{2}\right)} M^{j\left(q-\frac{|S|}{2}\right)}$$

$$= c^{|L|} \prod_{v \in V_{\text{int}}} \left(\|I_{2q_v}\|_{\emptyset} M^{-(q_v-1-\alpha)j} \right) \times$$

$$\prod_{v \in V_{\text{ext}}} \left(\|I_{2q_v}\|_{S_v} M^{-j\left(q_v-\frac{|S_v|}{2}\right)} \right) M^{j\left(q-\frac{|S|}{2}\right)} \tag{8.87}$$

The bounds of part (b) are obtained in the same way by using Lemma 8.3.2. The L^1- and L^∞-norms are interchanged on the tree and not on the tree but also the covariance bounds for the L^1- and L^∞-norm are interchanged (now $\alpha = 1$) which results in the same graph bounds. ∎

8.3.3 Multiscale Bounds

Theorem 8.3.4 *Let G be a $2q$-legged diagram made from vertices I_{2q_v} as in Lemma 8.3.1. Suppose that each line $\ell \in L_G$, the set of all lines of G, carries the covariance $C_\ell = \sum_{j=0}^{\infty} C^j$ where C^j satisfies the momentum space bounds (8.73),*

$$\|C^j\|_\infty \le c_M M^j, \qquad \|C^j\|_1 \le c_M M^{-j}$$

a) *Suppose that $q \ge 3$, $q_v \ge 3$ for all vertices and that G has no two- and four-legged subgraphs (which is formalized by (8.101) below). Then one has*

$$\|G\|_{\{1,\cdots,2q\}} \le a_M^{|L_G|} \prod_{v \in V_G} \|I_{2q_v}\|_{S_v}, \tag{8.88}$$

where $a_M = c_M \sum_{j=0}^{\infty} M^{-\frac{j}{3}}$.

b) *Suppose that $q \ge 2$, $q_v \ge 2$ for all vertices and that G has no two-legged subgraphs (which is formalized by (8.115) below). Then one has*

$$\|G\|_{\{1,\cdots,2q\}} \le r_{M,|L_G|} c_M^{|L_G|} \prod_{v \in V_G} \|I_{2q_v}\|_{S_v}, \tag{8.89}$$

where

$$r_{M,|L_G|} = 1 + \sum_{j=1}^{\infty} j^{|L_G|} M^{-\frac{j}{2}}\left(1 - M^{-\frac{1}{2}}\right) \le \text{const}_M^{|L_G|} |L_G|! \tag{8.90}$$

152

Proof: Let $C^{\leq j} = \sum_{i=0}^{j} C^i$ and

$$(2\pi)^d \delta(k_1 + \cdots + k_{2q}) \, G^{\leq j}(k_1, \cdots, k_{2q}) =$$

$$\int \prod_{\ell \in L_G} dk_\ell \prod_{\ell \in L_G} C^{\leq j}(k_\ell) \prod_{v \in V_G} (2\pi)^d \delta(k_1^v + \cdots + k_{2q_v}^v) I_{2q_v}(k_1^v, \cdots, k_{2q_v}^v)$$

We write $C^{\leq j+1} = C^{\leq j} + C^{j+1}$ and multiply out:

$$(2\pi)^d \delta(k_1 + \cdots + k_{2q}) \, G^{\leq j+1}(k_1, \cdots, k_{2q})$$

$$= \int \prod_{\ell \in L_G} dk_\ell \prod_{\ell \in L_G} \left(C^{\leq j} + C^{j+1} \right)(k_\ell) \times$$

$$\prod_{v \in V_G} (2\pi)^d \delta(k_1^v + \cdots + k_{2q_v}^v) I_{2q_v}(k_1^v, \cdots, k_{2q_v}^v)$$

$$= \sum_{A \subset L_G} \int \prod_{\ell \in L_G} dk_\ell \prod_{\ell \in L_G \backslash A} C^{\leq j}(k_\ell) \prod_{\ell \in A} C^{j+1}(k_\ell) \times$$

$$\prod_{v \in V_G} (2\pi)^d \delta(\cdots) I_{2q_v}(k_1^v, \cdots, k_{2q_v}^v)$$

$$= \sum_{\substack{A \subset L_G \\ A \neq \emptyset}} \int \prod_{\ell \in L_G} dk_\ell \prod_{\ell \in L_G \backslash A} C^{\leq j}(k_\ell) \prod_{\ell \in A} C^{j+1}(k_\ell) \times$$

$$\prod_{v \in V_G} (2\pi)^d \delta(\cdots) I_{2q_v}(k_1^v, \cdots, k_{2q_v}^v)$$

$$+ (2\pi)^d \delta(k_1 + \cdots + k_{2q}) \, G^{\leq j}(k_1, \cdots, k_{2q}) \tag{8.91}$$

The lines in A have scale $j+1$ propagators while the lines in $L_G \backslash A$ have $C^{\leq j}$ as propagators. Generalized vertices which are connected by $\leq j$ lines we consider as a single subdiagram such that, for given $A \subset L_G$, every term in (8.91) can be considered as a diagram consisting of certain subgraphs H_{2q_w} and only scale $j+1$ lines. See the following figure 8.2 for some examples. Thus, let $W_{G,A}$ be the set of connected components which consist of vertices I_{2q_v} which are connected by lines in $L_G \backslash A$. Each connected component is labelled by $w \in W_{G,A}$ and is itself a connected amputated diagram with, say, $2q_w$ unpaired legs. These legs may be external legs of G or come from scale $j+1$ lines. We write the connected components as $(2\pi)^d \delta(k_1^w + \cdots + k_{2q_w}^w) H_{2q_w}(k_1^w, \cdots, k_{2q_w}^w)$ such that we obtain

$$(2\pi)^d \delta(k_1 + \cdots + k_{2q}) \, G^{\leq j+1}(k_1, \cdots, k_{2q})$$

$$= \sum_{\substack{A \subset L_G \\ A \neq \emptyset}} \int \prod_{\ell \in A} dk_\ell \prod_{\ell \in A} C^{j+1}(k_\ell) \prod_{w \in W_{G,A}} (2\pi)^d \delta(\cdots) H_{2q_w}(k_1^w, \cdots, k_{2q_w}^w)$$

$$+ (2\pi)^d \delta(k_1 + \cdots + k_{2q}) \, G^{\leq j}(k_1, \cdots, k_{2q}) \tag{8.92}$$

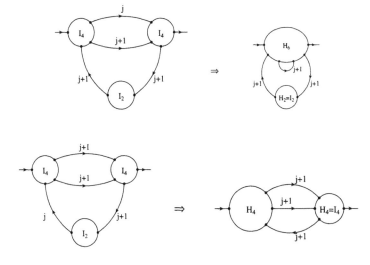

Figure 8.2

where the delta function $\delta(\cdots) = \delta(k_1^w + \cdots + k_{2q_w}^w)$. The above expression can be considered as the value of a diagram with generalized vertices H_{2q_w} and lines $\ell \in A$. Let $T_{G,A} \subset A$ be a tree for that diagram. Then we can eliminate all the momentum-conserving delta functions by a choice of loop momenta:

$$G^{\leq j+1}(k_1, \cdots, k_{2q}) =$$

$$\sum_{\substack{A \subset L_G \\ A \neq \emptyset}} \int \prod_{\ell \in A \setminus T_{G,A}} dk_\ell \prod_{\ell \in A} C^{j+1}(K_\ell) \prod_{w \in W_{G,A}} H_{2q_w}(K_1^w, \cdots, K_{2q_w}^w)$$

$$+ G^{\leq j}(k_1, \cdots, k_{2q}) \tag{8.93}$$

where, as in (8.52), K_ℓ is the sum of momenta flowing through the line ℓ. Then we get with the momentum space bound of Lemma 8.3.3

$$\|G^{\leq j+1} - G^{\leq j}\|_\infty \leq \sum_{\substack{A \subset L_G \\ A \neq \emptyset}} c^{|A|} \prod_{w \in W_{G,A}} \left(M^{-(j+1)(q_w-2)} \|H_{2q_w}\|_\infty \right) M^{(j+1)(q-2)}$$

$$\tag{8.94}$$

$$\|G^{\leq j+1} - G^{\leq j}\|_S \leq \sum_{\substack{A \subset L_G \\ A \neq \emptyset}} c^{|A|} \prod_{\substack{w \in W_{G,A} \\ w \text{ int.}}} \left(M^{-(j+1)(q_w-2)} \|H_{2q_w}\|_\infty \right) \times$$

$$\prod_{\substack{w \in W_{G,A} \\ w \text{ ext.}}} \left(M^{-(j+1)(q_w - \frac{|S_w|}{2})} \|H_{2q_w}\|_{S_w} \right) M^{(j+1)(q - \frac{|S|}{2})}$$

$$\tag{8.95}$$

Part a) We verify the following bounds by induction on j: For each connected amputated $2q$-legged diagram $G_{2q}^{\leq j}$ with $q \geq 3$ one has

$$\|G_{2q}^{\leq j}\|_\infty \leq c^{|L_G|} s(j)^{|L_G|} M^{-j\frac{2}{3}} M^{j(q-2)} \prod_{v \in V_G} \|I_{2q_v}\|_\infty \qquad (8.96)$$

and for all $S \neq \emptyset, \{1, \cdots, 2q\}$:

$$\|G_{2q}^{\leq j}\|_S \leq c^{|L_G|} s(j)^{|L_G|} M^{-\frac{j}{3}(q-\frac{|S|}{2})} M^{j(q-\frac{|S|}{2}-\frac{1}{3})} \prod_{v \in V_G} \|I_{2q_v}\|_{S_v} \quad (8.97)$$

$$\|G_{2q}^{\leq j}\|_{\{1,\cdots,2q\}} \leq c^{|L_G|} s(j)^{|L_G|} \prod_{v \in V_G} \|I_{2q_v}\|_{S_v} \qquad (8.98)$$

where $s(j) = \sum_{i=0}^{j} M^{-\frac{i}{3}}$. Then part (a) is a consequence of (8.98).

For $j = 0$ one has $C = C^0$ for all lines and one obtains with Lemma 8.3.3:

$$\|G\|_\infty \leq c^{|L_G|} \prod_{v \in V_G} \left(\|I_{2q_v}\|_\infty M^{-(q_v-2)0} \right) M^{(q-2)0}$$

$$= c^{|L_G|} s(0)^{|L_G|} \prod_{v \in V_G} \|I_{2q_v}\|_\infty$$

$$\|G\|_S \leq c^{|L_G|} \prod_{v \in V_{G,\text{int.}}} \left(\|I_{2q_v}\|_\infty M^{-(q_v-2)0} \right) \times$$

$$\prod_{v \in V_{G,\text{ext.}}} \left(\|I_{2q_v}\|_{S_v} M^{-(q_v-\frac{|S_v|}{2})0} \right) M^{(q-\frac{|S|}{2})0}$$

$$= c^{|L_G|} s(0)^{|L_G|} \prod_{v \in V_G} \|I_{2q_v}\|_{S_v} \qquad (8.99)$$

Suppose (8.96) through (8.98) are correct for j. Then, to verify (8.96) for $j + 1$, observe that by (8.93) and the induction hypothesis (8.96)

$$\|G_{2q}^{\leq j+1}\|_\infty \leq \|G_{2q}^{\leq j+1} - G_{2q}^{\leq j}\|_\infty + \|G_{2q}^{\leq j}\|_\infty$$

$$\leq \sum_{\substack{A \subset L_G \\ A \neq \emptyset}} c^{|A|} \prod_{w \in W_{G,A}} \left(M^{-(j+1)(q_w-2)} \|H_{2q_w}\|_\infty \right) M^{(j+1)(q-2)} + \|G_{2q}^{\leq j}\|_\infty$$

$$\leq \sum_{\substack{A \subset L_G \\ A \neq \emptyset}} c^{|A|} \prod_{w \in W_{G,A}} \left(M^{-(j+1)(q_w-2)} c^{L_w} s(j)^{L_w} M^{-j\frac{q_w}{3}} M^{j(q_w-2)} \times \right.$$

$$\left. \prod_{v \in V_w} \|I_{2q_v}\|_\infty \right) M^{(j+1)(q-2)} +$$

$$+ c^{|L_G|}s(j)^{|L_G|}M^{-j\frac{q}{3}}M^{j(q-2)}\prod_{v\in V_G}\|I_{2q_v}\|_\infty$$

$$= \sum_{\substack{A\subset L_G \\ A\neq\emptyset}} c^{|L_G|}s(j)^{L_G\setminus A}\prod_{w\in W_{G,A}}\left(M^{-(q_w-2)}M^{-j\frac{q_w}{3}}\right)\prod_{v\in V_G}\|I_{2q_v}\|_\infty M^{(j+1)(q-2)}$$

$$+ c^{|L_G|}s(j)^{|L_G|}M^{-j\frac{q}{3}}M^{j(q-2)}\prod_{v\in V_G}\|I_{2q_v}\|_\infty \tag{8.100}$$

Now the assumption that G has no two- and four-legged subgraphs means

$$q_w \geq 3 \tag{8.101}$$

for all possible connected components H_{2q_w} in (8.93). Thus we have

$$\prod_{w\in W_{G,A}}\left(M^{-(q_w-2)}M^{-j\frac{q_w}{3}}\right) \leq \prod_{w\in W_{G,A}}M^{-(j+1)\frac{q_w}{3}} = \left(M^{-\frac{j+1}{3}}\right)^{|A|+q}$$

and (8.100) is bounded by

$$\sum_{\substack{A\subset L_G \\ A\neq\infty}} c^{|L_G|}s(j)^{|L_G\setminus A|}\left(M^{-\frac{j+1}{3}}\right)^{|A|+q}\prod_{v\in V_G}\|I_{2q_v}\|_\infty M^{(j+1)(q-2)}$$

$$+ c^{|L_G|}s(j)^{|L_G|}M^{-j\frac{q}{3}}M^{(j+1)(q-2)}M^{-\frac{q}{3}}\prod_{v\in V_G}\|I_{2q_v}\|_\infty$$

$$= c^{|L_G|}M^{-(j+1)\frac{q}{3}}\sum_{k=1}^{|L_G|}\binom{|L_G|}{k}s(j)^{|L_G|-k}\left(M^{-\frac{j+1}{3}}\right)^k\prod_{v\in V_G}\|I_{2q_v}\|_\infty M^{(j+1)(q-2)}$$

$$+ c^{|L_G|}s(j)^{|L_G|}M^{-(j+1)\frac{q}{3}}M^{(j+1)(q-2)}\prod_{v\in V_G}\|I_{2q_v}\|_\infty$$

$$= c^{|L_G|}M^{-(j+1)\frac{q}{3}}\sum_{k=0}^{|L_G|}\binom{|L_G|}{k}s(j)^{|L_G|-k}\left(M^{-\frac{j+1}{3}}\right)^k\prod_{v\in V_G}\|I_{2q_v}\|_\infty M^{(j+1)(q-2)}$$

$$= c^{|L_G|}M^{-(j+1)\frac{q}{3}}\left(s(j)+M^{-\frac{j+1}{3}}\right)^{|L_G|}\prod_{v\in V_G}\|I_{2q_v}\|_\infty M^{(j+1)(q-2)}$$

$$= c^{|L_G|}M^{-(j+1)\frac{q}{3}}s(j+1)^{|L_G|}\prod_{v\in V_G}\|I_{2q_v}\|_\infty M^{(j+1)(q-2)} \tag{8.102}$$

which verifies (8.96). To verify (8.97) and (8.98) for scale $j+1$, observe that by (8.95) and the induction hypothesis (8.97,8.98)

$$\|G_{2q}^{\leq j+1}\|_S \leq \|G_{2q}^{\leq j+1} - G_{2q}^{\leq j}\|_S + \|G_{2q}^{\leq j}\|_S$$

$$\leq M^{(j+1)(q-\frac{|S|}{2})} \sum_{\substack{A \subset L_G \\ A \neq \emptyset}} c^{|A|} \prod_{\substack{w \in W_{G,A} \\ w \text{ int.}}} \left(M^{-(j+1)(q_w-2)} \|H_{2q_w}\|_\infty \right) \times$$

$$\prod_{\substack{w \in W_{G,A} \\ w \text{ ext.}}} \left(M^{-(j+1)(q_w-\frac{|S_w|}{2})} \|H_{2q_w}\|_{S_w} \right) + \|G_{2q}^{\leq j}\|_S$$

$$\leq M^{(j+1)(q-\frac{|S|}{2})} \sum_{\substack{A \subset L_G \\ A \neq \emptyset}} c^{|A|} \prod_{\substack{w \in W_{G,A} \\ w \text{ int.}}} \left(M^{-(j+1)(q_w-2)} c^{|L_w|} s(j)^{L_w} M^{-j\frac{q_w}{3}} \times \right.$$

$$M^{j(q_w-2)} \prod_{v \in V_w} \|I_{2q_v}\|_\infty \right) \prod_{\substack{w \in W_{G,A} \\ w \text{ ext.}}} \left(M^{-(j+1)(q_w-\frac{|S_w|}{2})} c^{|L_w|} s(j)^{L_w} \times \right.$$

$$M^{-\frac{j}{3}(q_w-\frac{|S_w|}{2})} M^{j(q_w-\frac{|S_w|}{2}-\frac{1}{3})} \prod_{v \in V_w} \|I_{2q_v}\|_{S_v} \right) + \|G_{2q}^{\leq j}\|_S$$

$$= M^{(j+1)(q-\frac{|S|}{2})} c^{|L_G|} \sum_{\substack{A \subset L_G \\ A \neq \emptyset}} s(j)^{|L_G \backslash A|} \prod_{\substack{w \in W_{G,A} \\ w \text{ int.}}} \left(M^{-(q_w-2)} M^{-j\frac{q_w}{3}} \right) \times$$

$$\prod_{\substack{w \in W_{G,A} \\ w \text{ ext.}}} \left(M^{-(q_w-\frac{|S_w|}{2})} M^{-\frac{j}{3}(q_w-\frac{|S_w|}{2})} M^{-j\frac{1}{3}} \right) \prod_{v \in V_G} \|I_{2q_v}\|_{S_v}$$

$$+ \|G_{2q}^{\leq j}\|_S \tag{8.103}$$

Now, for $S \neq \emptyset$ but $S = \{1, \cdots, 2q\}$ may be allowed, one has

$$\prod_{\substack{w \in W_{G,A} \\ w \text{ int.}}} \left(M^{-(q_w-2)} M^{-j\frac{q_w}{3}} \right) \prod_{\substack{w \in W_{G,A} \\ w \text{ ext.}}} \left(M^{-(q_w-\frac{|S_w|}{2})} M^{-\frac{j}{3}(q_w-\frac{|S_w|}{2})} M^{-j\frac{1}{3}} \right)$$

$$\leq M^{-\frac{j+1}{3} \sum_{\substack{w \in W_{G,A} \\ w \text{ int.}}} q_w} M^{-\frac{j+1}{3} \sum_{\substack{w \in W_{G,A} \\ w \text{ ext.}}} (q_w-\frac{|S_w|}{2})} \times$$

$$M^{-\frac{2}{3} \sum_{\substack{w \in W_{G,A} \\ w \text{ ext.}}} (q_w-\frac{|S_w|}{2})} M^{-\frac{j}{3}|V_{\text{ext.}}|}$$

$$= M^{-\frac{j+1}{3}(|A|+q-\frac{|S|}{2})} M^{-\frac{2}{3} \sum_{\substack{w \in W_{G,A} \\ w \text{ ext.}}} (q_w-\frac{|S_w|}{2})} M^{-\frac{j}{3}|V_{\text{ext.}}|}$$

$$\leq M^{-\frac{j+1}{3}(|A|+q-\frac{|S|}{2})} M^{-\frac{2}{3} \times \frac{1}{2}} M^{-\frac{j}{3}}$$

$$= M^{-\frac{j+1}{3}(|A|+q-\frac{|S|}{2})} M^{-\frac{j+1}{3}} \tag{8.104}$$

Therefore it follows from (8.103) for $S \neq \emptyset$

$$\|G_{2q}^{\leq j+1}\|_S \leq c^{|L_G|} M^{-\frac{j+1}{3}(q-\frac{|S|}{2})} M^{(j+1)(q-\frac{|S|}{2}-\frac{1}{3})} \prod_{v \in V_G} \|I_{2q_v}\|_{S_v} \times$$

$$\sum_{\substack{A \subset L_G \\ A \neq \emptyset}} s(j)^{|L_G \setminus A|} M^{-\frac{j+1}{3}|A|} + \|G_{2q}^{\leq j}\|_S \qquad (8.105)$$

Now suppose first that $S \neq \{1, \cdots, 2q\}$. By the induction hypothesis (8.97) one has

$$\|G_{2q}^{\leq j}\|_S \leq c^{|L_G|} s(j)^{|L_G|} M^{-\frac{j}{3}(q-\frac{|S|}{2})} M^{j(q-\frac{|S|}{2}-\frac{1}{3})} \prod_{v \in V_G} \|I_{2q_v}\|_{S_v}$$

$$= c^{|L_G|} s(j)^{|L_G|} M^{-\frac{2}{3}(q-\frac{|S|}{2})} M^{\frac{1}{3}} M^{-\frac{j+1}{3}(q-\frac{|S|}{2})} M^{(j+1)(q-\frac{|S|}{2}-\frac{1}{3})} \prod_{v \in V_G} \|I_{2q_v}\|_{S_v}$$

$$\leq c^{|L_G|} s(j)^{|L_G|} M^{-\frac{j+1}{3}(q-\frac{|S|}{2})} M^{(j+1)(q-\frac{|S|}{2}-\frac{1}{3})} \prod_{v \in V_G} \|I_{2q_v}\|_{S_v} \qquad (8.106)$$

Substituting this in (8.105), one obtains

$$\|G_{2q}^{\leq j+1}\|_S \leq c^{|L_G|} M^{-\frac{j+1}{3}(q-\frac{|S|}{2})} M^{(j+1)(q-\frac{|S|}{2}-\frac{1}{3})} \times$$

$$\prod_{v \in V_G} \|I_{2q_v}\|_{S_v} \sum_{\substack{A \subset L_G \\ A \neq \emptyset}} s(j)^{|L_G \setminus A|} M^{-\frac{j+1}{3}|A|}$$

$$+ c^{|L_G|} s(j)^{|L_G|} M^{-\frac{j+1}{3}(q-\frac{|S|}{2})} M^{(j+1)(q-\frac{|S|}{2}-\frac{1}{3})} \prod_{v \in V_G} \|I_{2q_v}\|_{S_v}$$

$$= c^{|L_G|} M^{-\frac{j+1}{3}(q-\frac{|S|}{2})} M^{(j+1)(q-\frac{|S|}{2}-\frac{1}{3})} \times$$

$$\prod_{v \in V_G} \|I_{2q_v}\|_{S_v} \sum_{A \subset L_G} s(j)^{|L_G \setminus A|} M^{-\frac{j+1}{3}|A|}$$

$$= c^{|L_G|} M^{-\frac{j+1}{3}(q-\frac{|S|}{2})} M^{(j+1)(q-\frac{|S|}{2}-\frac{1}{3})} \prod_{v \in V_G} \|I_{2q_v}\|_{S_v} s(j+1)^{|L_G|}$$

which verifies (8.97) for $j+1$.

Now let $S = \{1, \cdots, 2q\}$. Then by the induction hypothesis (8.98) one has

$$\|G_{2q}^{\leq j}\|_{\{1,\cdots,2q\}} \leq c^{|L_G|} s(j)^{|L_G|} \prod_{v \in V_G} \|I_{2q_v}\|_{S_v} \qquad (8.107)$$

and (8.105) becomes

$$\|G_{2q}^{\leq j+1}\|_{\{1,\cdots,2q\}} \leq c^{|L_G|} M^{-(j+1)\frac{1}{3}} \prod_{v \in V_G} \|I_{2q_v}\|_{S_v} \sum_{\substack{A \subset L_G \\ A \neq \emptyset}} s(j)^{|L_G \setminus A|} M^{-\frac{j+1}{3}|A|}$$

$$+ c^{|L_G|} s(j)^{|L_G|} \prod_{v \in V_G} \|I_{2q_v}\|_{S_v}$$

$$\leq c^{|L_G|} \prod_{v \in V_G} \|I_{2q_v}\|_{S_v} \sum_{A \subset L_G} s(j)^{|L_G \setminus A|} M^{-\frac{i+1}{3}|A|}$$

$$= c^{|L_G|} s(j+1)^{|L_G|} \prod_{v \in V_G} \|I_{2q_v}\|_{S_v} \qquad (8.108)$$

which verifies the induction hypothesis (8.98) for $j + 1$.

Part b) We verify the following bounds by induction on j: For each connected amputated $2q$-legged diagram $G_{2q}^{\leq j}$ with $q \geq 2$ one has

$$\|G_{2q}^{\leq j}\|_\infty \leq c_M^{|L_G|} j^{|L_G|} M^{j(q-2)} \prod_{v \in V_G} \|I_{2q_v}\|_\infty \qquad (8.109)$$

and for all $S \neq \emptyset, \{1, \cdots, 2q\}$:

$$\|G_{2q}^{\leq j}\|_S \leq c_M^{|L_G|} j^{|L_G|} M^{j(q-\frac{|S|}{2}-\frac{1}{2})} \prod_{v \in V_G} \|I_{2q_v}\|_{S_v} . \qquad (8.110)$$

$$\|G_{2q}^{\leq j}\|_{\{1,\cdots,2q\}} \leq r_M(j) c_M^{|L_G|} \prod_{v \in V_G} \|I_{2q_v}\|_{S_v} \qquad (8.111)$$

where

$$r_M(j) = 1 + \sum_{i=1}^{j-1} i^{|L_G|} (M^{-\frac{i}{2}} - M^{-\frac{i+1}{2}}) + j^{|L_G|} M^{-\frac{j}{2}} \qquad (8.112)$$

Then part (b) is a consequence of (8.111).

For $j = 0$ one has $C = C^0$ for all lines and one obtains with Lemma 8.3.3:

$$\|G\|_\infty \leq c_M^{|L_G|} \prod_{v \in V_G} \left(\|I_{2q_v}\|_\infty M^{-(q_v-2)0} \right) M^{(q-2)0} = c_M^{|L_G|} \prod_{v \in V_G} \|I_{2q_v}\|_\infty$$

$$\|G\|_S \leq c_M^{|L_G|} \prod_{v \in V_{G,\text{int.}}} \left(\|I_{2q_v}\|_\infty M^{-(q_v-2)0} \right) \times$$

$$\prod_{v \in V_{G,\text{ext.}}} \left(\|I_{2q_v}\|_{S_v} M^{-(q_v-\frac{|S_v|}{2})0} \right) M^{(q-\frac{|S|}{2})0}$$

$$= c_M^{|L_G|} \prod_{v \in V_G} \|I_{2q_v}\|_{S_v} \qquad (8.113)$$

Suppose (8.109) through (8.111) are correct for j. Then, to verify (8.109) for

$j + 1$, observe that by (8.93) and the induction hypothesis (8.109)

$$\|G_{2q}^{\leq j+1}\|_\infty \leq \|G_{2q}^{\leq j+1} - G_{2q}^{\leq j}\|_\infty + \|G_{2q}^{\leq j}\|_\infty$$

$$\leq \sum_{\substack{A \subset L_G \\ A \neq \emptyset}} c^{|A|} \prod_{w \in W_{G,A}} \left(M^{-(j+1)(q_w-2)} \|H_{2q_w}\|_\infty \right) M^{(j+1)(q-2)} + \|G_{2q}^{\leq j}\|_\infty$$

$$\leq \sum_{\substack{A \subset L_G \\ A \neq \emptyset}} c_M^{|A|} \prod_{w \in W_{G,A}} \left(M^{-(j+1)(q_w-2)} c_M^{|L_w|} j^{|L_w|} M^{j(q_w-2)} \right. \times$$

$$\prod_{v \in V_w} \|I_{2q_v}\|_\infty \bigg) M^{(j+1)(q-2)}$$

$$+ \; c_M^{|L_G|} j^{|L_G|} M^{j(q-2)} \prod_{v \in V_G} \|I_{2q_v}\|_\infty$$

$$= \sum_{\substack{A \subset L_G \\ A \neq \emptyset}} c_M^{|L_G|} j^{|L_G \setminus A|} \prod_{w \in W_{G,A}} \left(M^{-(q_w-2)} \right) \prod_{v \in V_G} \|I_{2q_v}\|_\infty M^{(j+1)(q-2)}$$

$$+ \; c_M^{|L_G|} j^{|L_G|} M^{j(q-2)} \prod_{v \in V_G} \|I_{2q_v}\|_\infty \tag{8.114}$$

Now the assumption that G has no two-legged subgraphs means

$$q_w \geq 2 \tag{8.115}$$

for all possible connected components H_{2q_w} in (8.93). Thus we have

$$\prod_{w \in W_{G,A}} M^{-(q_w-2)} \leq 1$$

and (8.114) is bounded by

$$\sum_{\substack{A \subset L_G \\ A \neq \infty}} c_M^{|L_G|} j^{|L_G \setminus A|} \prod_{v \in V_G} \|I_{2q_v}\|_\infty M^{(j+1)(q-2)}$$

$$+ \; c_M^{|L_G|} j^{|L_G|} M^{(j+1)(q-2)} \prod_{v \in V_G} \|I_{2q_v}\|_\infty$$

$$= c_M^{|L_G|} \sum_{k=1}^{|L_G|} \binom{|L_G|}{k} j^{|L_G|-k} \prod_{v \in V_G} \|I_{2q_v}\|_\infty M^{(j+1)(q-2)}$$

$$+ \; c_M^{|L_G|} j^{|L_G|} M^{(j+1)(q-2)} \prod_{v \in V_G} \|I_{2q_v}\|_\infty$$

$$= c_M^{|L_G|} \sum_{k=0}^{|L_G|} \binom{|L_G|}{k} j^{|L_G|-k} \prod_{v \in V_G} \|I_{2q_v}\|_\infty M^{(j+1)(q-2)}$$

$$= c_M^{|L_G|} (j+1)^{|L_G|} \prod_{v \in V_G} \|I_{2q_v}\|_\infty M^{(j+1)(q-2)}$$

$$= c_M^{|L_G|} (j+1)^{|L_G|} \prod_{v \in V_G} \|I_{2q_v}\|_\infty M^{(j+1)(q-2)} \tag{8.116}$$

which verifies (8.109). To verify (8.110, 8.111) for scale $j+1$, observe that by (8.95) and the induction hypothesis (8.110, 8.111)

$$\|G_{2q}^{\leq j+1}\|_S \leq \|G_{2q}^{\leq j+1} - G_{2q}^{\leq j}\|_S + \|G_{2q}^{\leq j}\|_S$$

$$\leq M^{(j+1)(q-\frac{|S|}{2})} \sum_{\substack{A \subset L_G \\ A \neq \emptyset}} c_M^{|A|} \prod_{\substack{w \in W_{G,A} \\ w \text{ int.}}} \left(M^{-(j+1)(q_w-2)} \|H_{2q_w}\|_\infty \right) \times$$

$$\prod_{\substack{w \in W_{G,A} \\ w \text{ ext.}}} \left(M^{-(j+1)(q_w - \frac{|S_w|}{2})} \|H_{2q_w}\|_{S_w} \right) + \|G_{2q}^{\leq j}\|_S$$

$$\leq M^{(j+1)(q-\frac{|S|}{2})} \sum_{\substack{A \subset L_G \\ A \neq \emptyset}} c_M^{|A|} \prod_{\substack{w \in W_{G,A} \\ w \text{ int.}}} \left(M^{-(j+1)(q_w-2)} c_M^{|L_w|} j^{|L_w|} \times \right.$$

$$M^{j(q_w-2)} \prod_{v \in V_w} \|I_{2q_v}\|_\infty \right) \prod_{\substack{w \in W_{G,A} \\ w \text{ ext.}}} \left(M^{-(j+1)(q_w - \frac{|S_w|}{2})} c_M^{|L_w|} j^{|L_w|} \times \right.$$

$$M^{j(q_w - \frac{|S_w|}{2} - \frac{1}{2})} \prod_{v \in V_w} \|I_{2q_v}\|_{S_v} \right) + \|G_{2q}^{\leq j}\|_S$$

$$= M^{(j+1)(q-\frac{|S|}{2})} c_M^{|L_G|} \sum_{\substack{A \subset L_G \\ A \neq \emptyset}} j^{|L_G \setminus A|} \prod_{\substack{w \in W_{G,A} \\ w \text{ int.}}} M^{-(q_w-2)} \times$$

$$\prod_{\substack{w \in W_{G,A} \\ w \text{ ext.}}} \left(M^{-(q_w - \frac{|S_w|}{2})} M^{-j\frac{1}{2}} \right) \prod_{v \in V_G} \|I_{2q_v}\|_{S_v} + \|G_{2q}^{\leq j}\|_S \tag{8.117}$$

Now, for $S \neq \emptyset$ but $S = \{1, \cdots, 2q\}$ may be allowed, one has

$$\prod_{\substack{w \in W_{G,A} \\ w \text{ int.}}} M^{-(q_w-2)} \prod_{\substack{w \in W_{G,A} \\ w \text{ ext.}}} \left(M^{-(q_w - \frac{|S_w|}{2})} M^{-j\frac{1}{2}} \right)$$

$$\leq 1 \cdot M^{-\frac{1}{2}} M^{-j\frac{1}{2}} = M^{-\frac{j+1}{2}} \tag{8.118}$$

since there is at least one external vertex and at least one internal line because of $A \neq \emptyset$. Therefore it follows from (8.117) for $S \neq \emptyset$

$$\|G_{2q}^{\le j+1}\|_S \le c_M^{|L_G|} M^{(j+1)(q-\frac{|S|}{2}-\frac{1}{2})} \prod_{v \in V_G} \|I_{2q_v}\|_{S_v} \sum_{\substack{A \subset L_G \\ A \neq \emptyset}} j^{|L_G \backslash A|} + \|G_{2q}^{\le j}\|_S$$

(8.119)

Now suppose first that $S \neq \{1, \cdots, 2q\}$. By the induction hypothesis (8.110) one has

$$\|G_{2q}^{\le j}\|_S \le c_M^{|L_G|} j^{|L_G|} M^{j(q-\frac{|S|}{2}-\frac{1}{2})} \prod_{v \in V_G} \|I_{2q_v}\|_{S_v}$$

$$\le c_M^{|L_G|} j^{|L_G|} M^{(j+1)(q-\frac{|S|}{2}-\frac{1}{2})} \prod_{v \in V_G} \|I_{2q_v}\|_{S_v} \qquad (8.120)$$

Substituting this in (8.119), one obtains

$$\|G_{2q}^{\le j+1}\|_S \le c_M^{|L_G|} M^{(j+1)(q-\frac{|S|}{2}-\frac{1}{2})} \prod_{v \in V_G} \|I_{2q_v}\|_{S_v} \sum_{\substack{A \subset L_G \\ A \neq \emptyset}} j^{|L_G \backslash A|}$$

$$+ c_M^{|L_G|} j^{|L_G|} M^{(j+1)(q-\frac{|S|}{2}-\frac{1}{2})} \prod_{v \in V_G} \|I_{2q_v}\|_{S_v}$$

$$= c_M^{|L_G|} M^{(j+1)(q-\frac{|S|}{2}-\frac{1}{2})} \prod_{v \in V_G} \|I_{2q_v}\|_{S_v} \sum_{A \subset L_G} j^{|L_G \backslash A|}$$

$$= c_M^{|L_G|} M^{(j+1)(q-\frac{|S|}{2}-\frac{1}{2})} \prod_{v \in V_G} \|I_{2q_v}\|_{S_v} (j+1)^{|L_G|} \qquad (8.121)$$

which verifies (8.110) for $j+1$.

Now let $S = \{1, \cdots, 2q\}$. By the induction hypothesis (8.111) one has

$$\|G_{2q}^{\le j}\|_{\{1,\cdots,2q\}} \le r_M(j) c_M^{|L_G|} \prod_{v \in V_G} \|I_{2q_v}\|_{S_v} \qquad (8.122)$$

where

$$r_M(j) = 1 + \sum_{i=1}^{j-1} i^{|L_G|} (M^{-\frac{i}{2}} - M^{-\frac{i+1}{2}}) + j^{|L_G|} M^{-\frac{j}{2}} \qquad (8.123)$$

and (8.119) becomes

$$\|G_{2q}^{\le j+1}\|_{\{1,\cdots,2q\}} \le c_M^{|L_G|} M^{-(j+1)\frac{1}{2}} \prod_{v \in V_G} \|I_{2q_v}\|_{S_v} \sum_{\substack{A \subset L_G \\ A \neq \emptyset}} j^{|L_G \backslash A|}$$

$$+ r_M(j) c_M^{|L_G|} \prod_{v \in V_G} \|I_{2q_v}\|_{S_v}$$

$$= c_M^{|L_G|} M^{-(j+1)\frac{1}{2}} \left((j+1)^{|L_G|} - j^{|L_G|} \right) \prod_{v \in V_G} \| I_{2q_v} \|_{S_v}$$

$$+ r_M(j) \, c_M^{|L_G|} \prod_{v \in V_G} \| I_{2q_v} \|_{S_v}$$

$$= r_M(j+1) \, c_M^{|L_G|} M^{-(j+1)\frac{1}{2}} \prod_{v \in V_G} \| I_{2q_v} \|_{S_v} \qquad (8.124)$$

since

$$r_M(j) + M^{-(j+1)\frac{1}{2}} \left((j+1)^{|L_G|} - j^{|L_G|} \right)$$

$$= 1 + \sum_{i=1}^{j-1} \left(i^{|L_G|} M^{-\frac{i}{2}} - i^{|L_G|} M^{-\frac{i+1}{2}} \right) + j^{|L_G|} M^{-\frac{j}{2}}$$

$$+ \left((j+1)^{|L_G|} - j^{|L_G|} \right) M^{-(j+1)\frac{1}{2}}$$

$$= r_M(j+1) \qquad (8.125)$$

which verifies the induction hypothesis (8.111) for $j + 1$. ∎

8.4 Ladder Diagrams

In this section we explicitly compute n'th order ladder diagrams and show that they produce factorials. Consider the following diagram:

Its value is given by

$$\Lambda_n(s,t,q) := \int \prod_{i=1}^{n} \frac{d^{d+1}k_i}{(2\pi)^{d+1}} \prod_{i=1}^{n} C(q/2 + k_i) C(q/2 - k_i) \times$$

$$V(s - k_1) \prod_{i=1}^{n-1} V(k_i - k_{i+1}) V(k_n - t) \qquad (8.126)$$

Specializing to the case of a delta function interaction in coordinate space or a constant in momentum space, $V(k) = \lambda$, one obtains $\Lambda_n(s, t, q) = \lambda^{n+1}\Lambda_n(q)$ where

$$\Lambda_n(q) = \int \prod_{i=1}^{n} \frac{d^{d+1}k_i}{(2\pi)^{d+1}} \prod_{i=1}^{n} C(q/2 + k_i)C(q/2 - k_i)$$
$$= \left\{ \int \frac{d^{d+1}k}{(2\pi)^{d+1}} C(q/2 + k)C(q/2 - k) \right\}^n = \{\Lambda(q)\}^n \qquad (8.127)$$

and

$$\Lambda(q) = \int \frac{d^{d+1}k}{(2\pi)^{d+1}} C(q/2 + k)C(q/2 - k) \qquad (8.128)$$

is the value of the particle-particle bubble,

The covariance is given by $C(k) = C(k_0, \mathbf{k}) = 1/(ik_0 - e_\mathbf{k})$. For small q the value of $\Lambda(q)$ is computed in the following

Lemma 8.4.1 *Let $C(k) = 1/(ik_0 - \mathbf{k}^2 + 1)$, let $d = 3$ and for $q = (q_0, \mathbf{q})$, $|\mathbf{q}| < 1$, let*

$$\Lambda(q) := \int_{-\infty}^{\infty} \frac{dk_0}{2\pi} \int_{|\mathbf{k}| \leq 2} \frac{d^3 k}{(2\pi)^3} C(q/2 + k)C(q/2 - k) \qquad (8.129)$$

Then for small q

$$\Lambda(q) = -\frac{1}{8\pi^2} \log[q_0^2 + 4|\mathbf{q}|^2] + O(1) \qquad (8.130)$$

Proof: We first compute the k_0-integral. By the residue theorem

$$\int_{-\infty}^{\infty} \frac{dk_0}{2\pi} \frac{1}{i(k_0 + q_0/2) - e_{\mathbf{k}+\mathbf{q}/2}} \frac{1}{i(-k_0 + q_0/2) - e_{-\mathbf{k}+\mathbf{q}/2}}$$

$$= \int_{-\infty}^{\infty} \frac{dk_0}{2\pi} \frac{1}{k_0 + q_0/2 + ie_{\mathbf{k}+\mathbf{q}/2}} \frac{1}{k_0 - q_0/2 - ie_{-\mathbf{k}+\mathbf{q}/2}}$$

$$= \frac{2\pi i}{2\pi} \left\{ \frac{\chi(e_{\mathbf{k}+\mathbf{q}/2} < 0)\chi(e_{-\mathbf{k}+\mathbf{q}/2} < 0)}{-q_0 - ie_{\mathbf{k}+\mathbf{q}/2} - ie_{-\mathbf{k}+\mathbf{q}/2}} + \frac{\chi(e_{\mathbf{k}+\mathbf{q}/2} > 0)\chi(e_{-\mathbf{k}+\mathbf{q}/2} > 0)}{q_0 + ie_{\mathbf{k}+\mathbf{q}/2} + ie_{-\mathbf{k}+\mathbf{q}/2}} \right\}$$

$$= \frac{\chi(e_{\mathbf{k}+\mathbf{q}/2} < 0)\chi(e_{\mathbf{k}-\mathbf{q}/2} < 0)}{iq_0 + |e_{\mathbf{k}+\mathbf{q}/2}| + |e_{\mathbf{k}-\mathbf{q}/2}|} + \frac{\chi(e_{\mathbf{k}+\mathbf{q}/2} > 0)\chi(e_{\mathbf{k}-\mathbf{q}/2} > 0)}{-iq_0 + |e_{\mathbf{k}+\mathbf{q}/2}| + |e_{\mathbf{k}-\mathbf{q}/2}|} \qquad (8.131)$$

To compute the integral over the spatial momenta, we consider the cases $e_{\mathbf{k}+\mathbf{q}/2} < 0$, $e_{\mathbf{k}-\mathbf{q}/2} < 0$ and $e_{\mathbf{k}+\mathbf{q}/2} > 0$, $e_{\mathbf{k}-\mathbf{q}/2} > 0$ separately.

Case (i): $e_{\mathbf{k}+\mathbf{q}/2} < 0$, $e_{\mathbf{k}-\mathbf{q}/2} < 0$. Let $\mathbf{p} = \mathbf{q}/2$ and let $k := |\mathbf{k}|$, $p := |\mathbf{p}|$, and $\cos\theta = \mathbf{kp}/(kp)$. Then

$$k^2 + 2kp\cos\theta + p^2 < 1, \quad k^2 - 2kp\cos\theta + p^2 < 1 \quad \Leftrightarrow$$
$$k^2 < 1 - p^2, \quad 2kp|\cos\theta| < 1 - p^2 - k^2 \tag{8.132}$$

If $2kp < 1 - p^2 - k^2$ or $k < 1 - p$, then the last inequality in (8.132) gives no restriction on θ. For $1 - p \leq k < \sqrt{1 - p^2}$ one gets

$$|\cos\theta| < \frac{1 - p^2 - k^2}{2kp} \tag{8.133}$$

Thus we have

$$\int_{|\mathbf{k}|<2} \frac{d^3k}{(2\pi)^3} \frac{\chi(e_{\mathbf{k}+\mathbf{q}/2} < 0)\chi(e_{\mathbf{k}-\mathbf{q}/2} < 0)}{iq_0 + |e_{\mathbf{k}+\mathbf{q}/2}| + |e_{\mathbf{k}-\mathbf{q}/2}|}$$

$$= \frac{1}{4\pi^2} \int_0^2 dk\, k^2 \int_0^\pi d\theta\, \sin\theta \frac{\chi(e_{\mathbf{k}+\mathbf{q}/2} < 0)\chi(e_{\mathbf{k}-\mathbf{q}/2} < 0)}{iq_0 - 2(k^2 + p^2 - 1)}$$

$$= \frac{1}{4\pi^2} \left\{ \int_0^{1-p} dk\, k^2 \int_0^\pi d\theta\, \sin\theta \frac{1}{iq_0 - 2(k^2 + p^2 - 1)} + \right.$$
$$\left. \int_{1-p}^{\sqrt{1-p^2}} dk\, k^2 \int_0^\pi d\theta\, \sin\theta \frac{\chi\left(|\cos\theta| < \frac{1-p^2-k^2}{2kp}\right)}{iq_0 - 2(k^2 + p^2 - 1)} \right\}$$

$$= \frac{1}{4\pi^2} \left\{ \int_0^{1-p} dk\, k^2 \frac{2}{iq_0 - 2(k^2 + p^2 - 1)} + \right.$$
$$\left. \int_{1-p}^{\sqrt{1-p^2}} dk\, k^2 \frac{1-p^2-k^2}{2kp} \frac{2}{iq_0 - 2(k^2 + p^2 - 1)} \right\} \tag{8.134}$$

Now we substitute $x = 1 - p^2 - k^2$, $k = \sqrt{1 - p^2 - x} = k_x$ to obtain

$$\int_{|\mathbf{k}|<2} \frac{d^3k}{(2\pi)^3} \frac{\chi(e_{\mathbf{k}+\mathbf{q}/2} < 0)\chi(e_{\mathbf{k}-\mathbf{q}/2} < 0)}{iq_0 + |e_{\mathbf{k}+\mathbf{q}/2}| + |e_{\mathbf{k}-\mathbf{q}/2}|}$$

$$= \frac{1}{4\pi^2} \left\{ \int_0^{1-p} dk\, k^2 \frac{2}{iq_0 - 2(k^2 + p^2 - 1)} + \right.$$
$$\left. + \int_{1-p}^{\sqrt{1-p^2}} dk\, k^2 \frac{1 - p^2 - k^2}{2kp} \frac{2}{iq_0 - 2(k^2 + p^2 - 1)} \right\}$$

$$= \frac{1}{4\pi^2} \left\{ \int_{2p(1-p)}^{1-p^2} dx\, k_x \frac{1}{iq_0 + 2x} + \int_0^{2p(1-p)} dx \frac{x}{2p} \frac{1}{iq_0 + 2x} \right\}$$

$$= \frac{1}{4\pi^2} \left\{ -\frac{k_x}{2} \log[iq_0 + 2x] \Big|_{x=2p(1-p)} + O(1) \right.$$

$$\left. + \int_0^{2p(1-p)} dx \, \frac{1}{4p} \left(1 - \frac{iq_0}{iq_0 + 2x} \right) \right\}$$

$$= \frac{1}{4\pi^2} \left\{ -\frac{1-p}{2} \log[iq_0 + 4p(1-p)] + O(1) - \frac{iq_0}{8p} \log[iq_0 + 2x] \Big|_0^{2p(1-p)} \right\}$$

$$= \frac{1}{4\pi^2} \left\{ -\frac{1}{2} \log[iq_0 + 4p(1-p)] + O(1) - \frac{iq_0}{8p} \log[1 + 4p(1-p)/(iq_0)] \right\}$$

$$= -\frac{1}{8\pi^2} \log[iq_0 + 4p(1-p)] + O(1) \tag{8.135}$$

Case (ii): $e_{\mathbf{k}+\mathbf{q}/2} > 0$, $e_{\mathbf{k}-\mathbf{q}/2} > 0$. The computation is similar. As above, let $\mathbf{p} = \mathbf{q}/2$, $k = |\mathbf{k}|$, $p = |\mathbf{p}|$ and $\cos\theta = \mathbf{kp}/(kp)$. Then

$$k^2 + 2kp\cos\theta + p^2 > 1, \quad k^2 - 2kp\cos\theta + p^2 > 1 \quad \Leftrightarrow$$
$$k^2 > 1 - p^2, \quad \pm 2kp\cos\theta > 1 - p^2 - k^2 \quad \Leftrightarrow$$
$$k^2 > 1 - p^2, \quad \mp 2kp\cos\theta < k^2 - (1 - p^2) \quad \Leftrightarrow$$
$$k^2 > 1 - p^2, \quad 2kp|\cos\theta| < k^2 - (1 - p^2) \tag{8.136}$$

If $2kp < k^2 - (1-p^2)$ or $k > 1+p$ (recall that $|\mathbf{q}| < 1$), then the last inequality in (8.136) gives no restriction on θ. For $\sqrt{1-p^2} \le k < 1+p$ one gets

$$|\cos\theta| < \frac{k^2 - (1-p^2)}{2kp} \tag{8.137}$$

Thus we have

$$\int_{|\mathbf{k}|<2} \frac{d^3\mathbf{k}}{(2\pi)^3} \frac{\chi(e_{\mathbf{k}+\mathbf{q}/2} > 0)\chi(e_{\mathbf{k}-\mathbf{q}/2} > 0)}{-iq_0 + |e_{\mathbf{k}+\mathbf{q}/2}| + |e_{\mathbf{k}-\mathbf{q}/2}|}$$

$$= \frac{1}{4\pi^2} \int_0^2 dk\, k^2 \int_0^\pi d\theta \sin\theta \, \frac{\chi(e_{\mathbf{k}+\mathbf{p}} > 0)\chi(e_{\mathbf{k}-\mathbf{p}} > 0)}{-iq_0 + 2(k^2 + p^2 - 1)}$$

$$= \frac{1}{4\pi^2} \left\{ \int_{\sqrt{1-p^2}}^{1+p} dk\, k^2 \int_0^\pi d\theta \sin\theta \, \frac{\chi\left(|\cos\theta| < \frac{k^2-(1-p^2)}{2kp}\right)}{-iq_0 + 2(k^2 + p^2 - 1)} + \right.$$

$$\left. \int_{1+p}^2 dk\, k^2 \int_0^\pi d\theta \sin\theta \, \frac{1}{-iq_0 + 2(k^2 + p^2 - 1)} \right\}$$

$$= \frac{1}{4\pi^2} \left\{ \int_{\sqrt{1-p^2}}^{1+p} dk\, k^2 \frac{k^2-(1-p^2)}{2kp} \frac{2}{-iq_0 + 2(k^2 + p^2 - 1)} + \right.$$

$$\left. \int_{1+p}^2 dk\, k^2 \frac{2}{-iq_0 + 2(k^2 + p^2 - 1)} \right\} \tag{8.138}$$

Now we substitute $x = k^2 - (1 - p^2)$, $k = \sqrt{x + 1 - p^2} = k_x$ to obtain

$$\int_{|\mathbf{k}|<2} \frac{d^3\mathbf{k}}{(2\pi)^3} \frac{\chi(e_{\mathbf{k}+\mathbf{q}/2} > 0)\chi(e_{\mathbf{k}-\mathbf{q}/2} > 0)}{-iq_0 + |e_{\mathbf{k}+\mathbf{q}/2}| + |e_{\mathbf{k}-\mathbf{q}/2}|}$$

$$= \frac{1}{4\pi^2} \left\{ \left[\int_{\sqrt{1-p^2}}^{1+p} dk\, k^2 \frac{k^2 - (1 - p^2)}{2kp} \frac{2}{-iq_0 + 2(k^2 + p^2 - 1)} + \right. \right.$$

$$\left. \int_{1+p}^{2} dk\, k^2 \frac{2}{-iq_0 + 2(k^2 + p^2 - 1)} \right\}$$

$$= \frac{1}{4\pi^2} \left\{ \int_0^{2p(1+p)} dx \frac{x}{2p} \frac{1}{-iq_0 + 2x} + \int_{2p(1+p)}^{3+p^2} dx\, k_x \frac{1}{-iq_0 + 2x} \right\}$$

$$= \frac{1}{4\pi^2} \left\{ \int_0^{2p(1+p)} dx \frac{1}{4p} \left(1 + \frac{iq_0}{-iq_0 + 2x} \right) \right.$$

$$\left. - \left. \frac{k_x}{2} \log[-iq_0 + 2x] \right|_{x=2p(1+p)} + O(1) \right\}$$

$$= \frac{1}{4\pi^2} \left\{ \left. \frac{iq_0}{8p} \log[-iq_0 + 2x] \right|_0^{2p(1+p)} + O(1) \right.$$

$$\left. - \frac{1+p}{2} \log[-iq_0 + 4p(1 + p)] + O(1) \right\}$$

$$= \frac{1}{4\pi^2} \left\{ \frac{iq_0}{8p} \log[1 - 4p(1 + p)/(iq_0)] - \frac{1}{2} \log[-iq_0 + 4p(1 + p)] + O(1) \right\}$$

$$= -\frac{1}{8\pi^2} \log[-iq_0 + 4p(1 + p)] + O(1) \tag{8.139}$$

Combining (8.131), (8.135) and (8.139) we arrive at

$$\Lambda(q) = -\frac{1}{8\pi^2} \left(\log[iq_0 + 4p(1 - p)] + \log[-iq_0 + 4p(1 + p)] \right) + O(1)$$

Since $p = |\mathbf{q}|/2 > 0$ we have $\arg[\pm iq_0 + 4p(1 \mp p)] \in (-\pi/2, \pi/2)$ and therefore

$$\log[iq_0 + 4p(1 - p)] + \log[-iq_0 + 4p(1 + p)]$$
$$= \frac{1}{2} \left\{ \log[q_0^2 + 16p^2(1 - p)^2] + \log[q_0^2 + 16p^2(1 + p)^2] \right\} + iO(1)$$
$$= \log[q_0^2 + 4|\mathbf{q}|^2] + O(1) \tag{8.140}$$

which proves the lemma. $\blacksquare$

Thus the leading order behavior of $\Lambda_n(q)$ for small q is given by

$$\Lambda_n(q) = \{\Lambda(q)\}^n \sim \text{const}^n \left\{ \log[1/(q_0^2 + 4|\mathbf{q}|^2)] \right\}^n \tag{8.141}$$

If $\Lambda_n(q)$ is part of a larger diagram, there may be an integral over q which then leads to an $n!$. Namely, if ω_3 denotes the surface area of S^3,

$$\int_{q_0^2+4|\mathbf{q}|^2\leq 1} dq_0 d^3\mathbf{q} \, \{\log[1/(q_0^2 + 4|\mathbf{q}|^2)]\}^n \quad = \quad \frac{\omega_3}{8} 2^n \int_0^1 d\rho\, \rho^3 \, (\log[1/\rho])^n$$

$$\overset{\rho=e^{-x}}{=\!=} \quad \frac{\omega_3}{8} 2^n \int_0^\infty dx \, e^{-4x} \, x^n$$

$$= \quad const^n \, n! \qquad\qquad (8.142)$$

Chapter 9

Renormalization Group Methods

In the last section we proved that, for the many-electron system with a short range potential (that is, $V(\mathbf{x}) \in L^1$), an n'th order diagram G_n allows the following bounds. If G_n has no two- and four-legged subgraphs, it is bounded by $const^n$ (measured in a suitable norm), if G_n has no two-legged but may have some four-legged subgraphs it is bounded by $n!\, const^n$ (and it was shown that the factorial is really there by computing n'th order ladder diagrams with dispersion relation $e_{\mathbf{k}} = \mathbf{k}^2/2m - \mu$), and if G_n contains two-legged subdiagrams it is in general divergent. The large contributions from two- and four-legged subdiagrams have to be eliminated by some kind of renormalization procedure. After that, one is left with a sum of diagrams where each n'th order diagram allows a $const^n$ bound.

The perturbation series for the logarithm of the partition function and similarly those series for the correlation functions are of the form (8.16)

$$\log Z(\lambda) = \sum_{n=1}^{\infty} \frac{\lambda^n}{n!} \sum_{\substack{\pi \in S_{2n} \\ G_n(\pi) \text{ connected}}} \mathrm{sign}\pi\; G_n(\pi) \tag{9.1}$$

where each permutation $\pi \in S_{2n}$ generates a certain diagram. The sign $\mathrm{sign}\pi$ is present in a fermionic model like the many-electron system but would be absent in a bosonic model. The condition of connectedness and similar conditions like '$G_n(\pi)$ does not contain two-legged subgraphs' do not significantly change the number of diagrams. That is, the number of diagrams which contribute to (9.1) or to a quantity like

$$F(\lambda) = \sum_{n=1}^{\infty} \frac{\lambda^n}{n!} \sum_{\substack{\pi \in S_{2n} \\ G_n(\pi) \text{ connected, without} \\ 2-,4-\text{legged subgraphs}}} \mathrm{sign}\pi\; G_n(\pi) =: \sum_{n=1}^{\infty} g_n\, \lambda^n \tag{9.2}$$

is of the order $(2n)!$ or, ignoring a factor $const^n$, of the order $(n!)^2$, even in the case that diagrams which contain two- and four-legged subgraphs are removed.

If we ignore the sign in (9.2) we would get a bound

$$|F(\lambda)| \le \sum_{n=0}^{\infty} \frac{|\lambda|^n}{n!} const^n\, (n!)^2 = \sum_{n=0}^{\infty} n!\, (const|\lambda|)^n \tag{9.3}$$

and the series on the right hand side of (9.3) has radius of convergence zero. Now there are two possibilities. This also holds for the original series (9.2) or the sign in (9.2) improves the situation. In the first case, which unavoidably would be the case for a bosonic model, there are again two possibilities. Either the series is asymptotic, that is, there is the bound

$$\left| F(\lambda) - \sum_{j=1}^{n} g_j \, \lambda^j \right| \leq n! \, (const|\lambda|)^n \tag{9.4}$$

or it is not. If one is interested only in small coupling, which is the case we restrict to, an asymptotic series would already be sufficient in order to get information on the correlation functions since (9.4) implies that the lowest order terms are a good approximation if the coupling is small.

For the many-electron system with short range interaction the sign in (9.2) indeed improves the situation and one obtains a small positive radius of convergence (at least in two space dimensions) for the series in (9.2). One obtains the bound

$$\left| F(\lambda) - \sum_{j=1}^{n} g_j \, \lambda^j \right| \leq (const|\lambda|)^n \tag{9.5}$$

Concerning the degree of information we can get about the correlation functions or partition function we are interested in (here the quantity $F(\lambda)$), there is practically no difference whether we have (9.4) or (9.5). In both cases we can say that the lowest order terms are a good approximation if the coupling is small and not more. We compute the lowest order terms for, say, $n = 1, 2$ and then it does not make a difference whether the error is $3! \, (const|\lambda|)^3$ or just $(const|\lambda|)^3$. However, from a technical point of view, it is much easier to rigorously prove a bound for a series with a small positive radius of convergence, as in (9.5), which eventually may hold for the sum of convergent diagrams for a fermionic model, than to prove a bound for an expansion which is only asymptotic, as in (9.4), which typically holds for the sum of convergent diagrams for a bosonic model. The goal of this section is to rigorously prove a bound of the form (9.5) on the sum of convergent diagrams for a typical fermionic model.

By 'convergent diagrams' we mean diagrams which allow a $const^n$-bound. In particular, for the many-electron system, this excludes for example ladder diagrams which, although being finite, behave like $n! \, const^n$. The reason is the following. There are two sources of factorials which may influence the convergence properties of the perturbation series. The number of diagrams which is of order $(n!)^2$, which, including the prefactor of $1/n!$ in (9.2), may produce (if the sign is absent) an $n!$ and there may be an $n!$ due to the values of certain n'th order diagrams, like the ladder diagrams. Now, roughly one can expect the following:

- A sum of convergent diagrams, that is, a sum of diagrams where each diagram allows a $const^n$ bound, is at least asymptotic. That is, its lowest order terms are a good approximation if the coupling is small, regardless whether the model is fermionic or bosonic.

- A sum of finite diagrams which contains diagrams which behave like $n!\, const^n$ is usually not asymptotic. That is, its lowest order contributions are not a good approximation. Those diagrams which behave like $n!\, const^n$ have to be resummed, for example by the use of integral equations as described in the next chapter.

For the many-electron system in two space dimensions a bound of the form (9.5) on the sum of convergent diagrams has been rigorously proven in [18, 16]. The restriction to two space dimensions comes in because in the proof one has to use conservation of momentum for momenta which are close to the Fermi surface $\mathbf{k}^2/2m - \mu = 0$ which is the place of the singularity of the free propagator. Then in two dimensions one gets more restrictive conditions than in three dimensions [24]. If the singularity of the covariance would be at a single point, say at $k = 0$, these technicalities due to the implementation of conservation of momentum are absent and the proof of (9.5) becomes more transparent. Therefore in this section we choose a model with covariance

$$C(k) = \frac{1}{|k|^{\frac{d}{2}}}, \quad k \in \mathbb{R}^d \tag{9.6}$$

and, as before, a short range interaction $V \in L^1$. This C has the same power counting as the propagator of the many-electron system which means that the bounds on Feynman diagrams in Lemma 8.3.3 and Theorem 8.3.4 for both propagators are the same (with M substituted by $M^{\frac{d}{2}}$ in case of (9.6)). However, as mentioned above, the proof of (9.5), that is, the proof of the Theorems 9.2.2 and 9.3.1 below will become more transparent.

9.1 Integrating Out Scales

In this subsection we set up the scale by scale formalism which allows one to compute the sum of all diagrams at scale j from the sum at scale $j - 1$. We start with the following

Lemma 9.1.1 *Let* $C, C^1, C^2 \in \mathbb{C}^{N \times N}$ *be invertible complex matrices and* $C = C^1 + C^2$. *Let* $d\mu_C, d\mu_{C^1}, d\mu_{C^2}$ *be the corresponding Grassmann Gaussian measures,* $d\mu_C = \det C \, e^{-\sum_{i,j=1}^{N} \bar{\psi}_i C_{ij}^{-1} \psi_j} \prod_{i=1}^{N} (d\psi_i d\bar{\psi}_i)$. *Let* $P(\psi, \bar{\psi})$ *be some* *polynomial of Grassmann variables. Then*

$$\int P(\psi, \bar{\psi}) \, d\mu_C(\psi, \bar{\psi}) = \iint P(\psi^1 + \psi^2, \bar{\psi}^1 + \bar{\psi}^2) \, d\mu_{C^1}(\psi^1, \bar{\psi}^1) \, d\mu_{C^2}(\psi^2, \bar{\psi}^2)$$
(9.7)

Proof: Since the integral is linear and every monomial can be obtained from the function $e^{-i\sum_{j=1}^{N}(\bar{\eta}_j \psi_j + \eta_j \bar{\psi}_j)}$ by differentiation with respect to the Grassmann variables η and $\bar{\eta}$, it suffices to prove (9.7) for $P(\psi, \bar{\psi}) = e^{-i\langle \bar{\eta}, \psi \rangle - i\langle \eta, \bar{\psi} \rangle}$, $\langle \bar{\eta}, \psi \rangle := \sum_{j=1}^{N} \bar{\eta}_j \psi_j$. We have

$$\int e^{-i\langle \bar{\eta}, \psi \rangle - i\langle \eta, \bar{\psi} \rangle} \, d\mu_C = e^{-i\langle \bar{\eta}, C\eta \rangle}$$

$$= e^{-i\langle \bar{\eta}, C^1\eta \rangle} \, e^{-i\langle \bar{\eta}, C^2\eta \rangle}$$

$$= \int e^{-i\langle \bar{\eta}, \psi^1 \rangle - i\langle \eta, \bar{\psi}^1 \rangle} \, d\mu_{C^1} \int e^{-i\langle \bar{\eta}, \psi^2 \rangle - i\langle \eta, \bar{\psi}^2 \rangle} \, d\mu_{C^2}$$

$$= \iint e^{-i\langle \bar{\eta}, \psi^1 + \psi^2 \rangle - i\langle \eta, \bar{\psi}^1 + \bar{\psi}^2 \rangle} \, d\mu_{C^1} \, d\mu_{C^2}$$
(9.8)

which proves the lemma. ∎

Now let

$$C^{\leq j} = \sum_{i=0}^{j} C^i$$
(9.9)

be a scale decomposition of some covariance C and let

$$d\mu_{\leq j} := d\mu_{C^{\leq j}} = \det[C^{\leq j}] e^{-\left\langle \bar{\psi}, (C^{\leq j})^{-1} \psi \right\rangle} \Pi(d\psi d\bar{\psi})$$
(9.10)

be the Grassmann Gaussian measure with covariance $C^{\leq j}$. We consider a general interaction of the form

$$\lambda \mathcal{V}(\psi, \bar{\psi}) = \lambda \sum_{q=1}^{m} \int \prod_{i=1}^{2q} d\xi_i \, V_{2q}(\xi_1, \cdots \xi_{2q}) \, \psi(\xi_1) \bar{\psi}(\xi_2) \cdots \psi(\xi_{2q-1}) \bar{\psi}(\xi_{2q})$$
(9.11)

which we assume to be short range. That is,

$$\sum_{q=1}^{m} \|V_{2q}\|_{\emptyset} < \infty$$
(9.12)

The connected amputated correlation functions up to scale j are generated by the functional (from now on we suppress the $\bar{\psi}$-dependence, that is, for brevity we write $F(\psi)$ instead of $F(\psi, \bar{\psi})$)

$$\mathcal{W}^{\leq j}(\psi) := \log \frac{1}{Z^{\leq j}} \int e^{\lambda \mathcal{V}(\psi + \psi^{\leq j})} d\mu_{\leq j}(\psi^{\leq j}) \tag{9.13}$$

$$Z^{\leq j} := \int e^{\lambda \mathcal{V}(\psi^{\leq j})} d\mu_{\leq j}(\psi^{\leq j}) \tag{9.14}$$

and the sum of all $2q$-legged, connected amputated diagrams up to scale j is given by the coefficient $W_{2q}^{\leq j}$ in the expansion

$$\mathcal{W}^{\leq j}(\psi) = \sum_{q=1}^{\infty} \int \prod_{i=1}^{2q} d\xi_i \, W_{2q}^{\leq j}(\xi_1, \cdots \xi_{2q}) \, \psi(\xi_1) \bar{\psi}(\xi_2) \cdots \psi(\xi_{2q-1}) \bar{\psi}(\xi_{2q}) \tag{9.15}$$

$\mathcal{W}^{\leq j}$ can be computed inductively from $\mathcal{W}^{\leq j-1}$ in the following way.

Lemma 9.1.2 *For some function F let*

$$F(\psi; \eta) := F(\psi) - F(\eta) \tag{9.16}$$

Let $d\mu_j := d\mu_{C^j}$ be the Gaussian measure with covariance C^j. Define the quantities $\mathcal{V}^j$ inductively by

$$\mathcal{V}^j(\psi) := \log \frac{1}{Y_j} \int e^{(\sum_{i=0}^{j-1} \mathcal{V}^i + \lambda \mathcal{V})(\psi + \psi^j; \psi)} d\mu_j(\psi^j) \tag{9.17}$$

$$Y_j = \int e^{(\sum_{i=0}^{j-1} \mathcal{V}^i + \lambda \mathcal{V})(\psi^j)} d\mu_j(\psi^j) \tag{9.18}$$

where, for $j = 0$, $\sum_{i=0}^{-1} \mathcal{V}^i := 0$ and $\lambda \mathcal{V}$ is given by (9.11). Then

$$\mathcal{W}^{\leq j} = \sum_{i=0}^{j} \mathcal{V}^i + \lambda \mathcal{V} = \mathcal{W}^{\leq j-1} + \mathcal{V}^j \tag{9.19}$$

Proof: Induction on j. For $j = 0$

$$\mathcal{W}^{\leq 0}(\psi) = \log \frac{1}{Z^{\leq 0}} \int e^{\lambda \mathcal{V}(\psi + \psi^{\leq 0})} d\mu_{\leq 0}$$

$$= \log \frac{1}{Y_0} \int e^{\lambda \mathcal{V}(\psi + \psi^0; \psi)} d\mu_0 + \lambda \mathcal{V}(\psi)$$

$$= \mathcal{V}^0(\psi) + \lambda \mathcal{V}(\psi) \tag{9.20}$$

since

$$Z^{\leq 0} = \int e^{\lambda \mathcal{V}(\psi^0)} d\mu_0 = Y_0 \tag{9.21}$$

Furthermore $\mathcal{W}^{\leq 0}(0) = \mathcal{V}^0(0) = 0$. Suppose (9.19) holds for j and $\mathcal{V}^i(0) = 0$ for all $0 \leq i \leq j$. Then, using Lemma 9.1.1 in the second line,

$$
\begin{aligned}
\mathcal{W}^{\leq j+1}(\psi) &= \log \int e^{\lambda \mathcal{V}(\psi + \psi^{\leq j+1})} d\mu_{\leq j+1} - \log Z^{\leq j+1} \\
&= \log \int \exp\left\{\log \int e^{\lambda \mathcal{V}(\psi + \psi^{j+1} + \psi^{\leq j})} d\mu_{\leq j}\right\} d\mu_{j+1} - \log Z^{\leq j+1} \\
&= \log \int e^{\mathcal{W}^{\leq j}(\psi + \psi^{j+1})} d\mu_{j+1} + \log Z^{\leq j} - \log Z^{\leq j+1} \\
&= \log \int e^{\mathcal{W}^{\leq j}(\psi + \psi^{j+1};\psi)} d\mu_{j+1} + \mathcal{W}^{\leq j}(\psi) + \log \frac{Z^{\leq j}}{Z^{\leq j+1}} \\
&= \log \frac{1}{Y_{j+1}} \int e^{\sum_{i=0}^{j} \mathcal{V}^i(\psi + \psi^{j+1};\psi)} d\mu_{j+1} + \mathcal{W}^{\leq j}(\psi) + \log \left[Y_{j+1} \frac{Z^{\leq j}}{Z^{\leq j+1}}\right] \\
&= \mathcal{V}^{j+1}(\psi) + \mathcal{W}^{\leq j}(\psi) + \log \left[Y_{j+1} \frac{Z^{\leq j}}{Z^{\leq j+1}}\right]
\end{aligned}
\tag{9.22}
$$

and, since $\mathcal{V}^i(0)$ for $0 \leq i \leq j$,

$$
\mathcal{V}^{j+1}(0) = \log \frac{1}{Y_{j+1}} \int e^{\sum_{i=0}^{j} \mathcal{V}^i(\psi^{j+1})} d\mu_{j+1} = \log 1 = 0
\tag{9.23}
$$

Since also by definition $\mathcal{W}^{\leq j+1}(0) = \mathcal{W}^{\leq j}(0) = 0$ the constant $\log\left[Y_{j+1} \frac{Z^{\leq j}}{Z^{\leq j+1}}\right]$ in (9.22) must vanish and the lemma is proven ∎

9.2 A Single Scale Bound on the Sum of All Diagrams

We consider the model with generating functional

$$
\mathcal{W}(\eta) = \log \frac{1}{Z} \int e^{\lambda \mathcal{V}(\eta + \psi)} d\mu_C(\psi)
\tag{9.24}
$$

$$
Z = \int e^{\lambda \mathcal{V}(\psi)} d\mu_C(\psi)
$$

and covariance $C(\xi, \xi') = \delta_{\sigma,\sigma'} C(x - x')$ where

$$
C(x) = \int \frac{d^d k}{(2\pi)^d} e^{ikx} \frac{\chi(|k| \leq 1)}{|k|^{\frac{d}{2}}}
\tag{9.25}
$$

The interaction is given by (9.11) which we assume to be short range, that is, $\|V_{2q}\|_0 < \infty$ for all $1 \leq q \leq m$. The norm $\| \cdot \|_0$ is defined in (8.38). The strategy to controll $W_{2q,n}$, the sum of all n'th order connected amputated $2q$-legged diagrams, is basically the same as those of the last section. We introduce a scale decomposition of the covariance and then we see how the bounds change if we go from scale j to scale $j+1$. Thus, if $\rho \in C_0^\infty$, $0 \le \rho \le 1$, $\rho(x) = 1$ for $x \le 1$ and $\rho(x) = 0$ for $x \ge 2$ is some ultraviolet cutoff, let

$$C(k) := \frac{\rho(|k|)}{|k|^{\frac{d}{2}}} = \sum_{j=0}^\infty \frac{\rho(M^j|k|) - \rho(M^{j+1}|k|)}{|k|^{\frac{d}{2}}}$$

$$= \sum_{j=0}^\infty \frac{f(M^j|k|)}{|k|^{\frac{d}{2}}} =: \sum_{j=0}^\infty C^j(k) \qquad (9.26)$$

where $f(x) := \rho(x) - \rho(Mx)$ has support in $\frac{1}{M} \le x \le 2$ which implies

$$supp\, C^j \subset \left\{ k \in \mathbb{R}^d \mid \frac{1}{M} M^{-j} \le |k| \le 2M^{-j} \right\} \qquad (9.27)$$

Lemma 9.2.1 *Let* $C^j(k) = \frac{f(M^j|k|)}{|k|^{\frac{d}{2}}}$ *be given by (9.26) and let* $C^j(x) = \int \frac{d^dk}{(2\pi)^d} e^{ikx} C^j(k)$. *Then there are the following bounds:*

a) *Momentum space:*

$$\|C^j(k)\|_\infty \le c_M\, M^{\frac{d}{2}j}, \qquad \|C^j(k)\|_1 \le c_M\, M^{-\frac{d}{2}j} \qquad (9.28)$$

b) *Coordinate space:*

$$\|C^j(x)\|_\infty \le c_M\, M^{-\frac{d}{2}j}, \qquad \|C^j(x)\|_1 \le c_M\, M^{\frac{d}{2}j} \qquad (9.29)$$

where the constant $c_M = c(M, d, \|f\|_\infty, \cdots, \|f^{(2d)}\|_\infty)$ *is independent of* j.

Proof: The momentum space bounds are an immediate consequence of (9.27). Furthermore $\|C^j(x)\|_\infty \le (2\pi)^{-\frac{d}{2}} \|C^j(k)\|_1$. To obtain the L^1 bound on $C^j(x)$, observe that

$$|(M^{-2j}x^2)^N C^j(x)| = \left| \int \frac{d^dk}{(2\pi)^d} \left\{ (-M^{-2j}\Delta)^N e^{ikx} \right\} \frac{f(M^j|k|)}{|k|^{\frac{d}{2}}} \right|$$

$$= \left| \int \frac{d^dk}{(2\pi)^d} e^{ikx} (-M^{-2j}\Delta)^N \frac{f(M^j|k|)}{|k|^{\frac{d}{2}}} \right|$$

$$\le (M^{-2j})^N \int \frac{d^dk}{(2\pi)^d} \left| \left(\frac{\partial^2}{\partial|k|^2} + \frac{d-1}{|k|} \frac{\partial}{\partial|k|} \right)^N \frac{f(M^j|k|)}{|k|^{\frac{d}{2}}} \right|$$

$$\le const^N \sup\{\|f\|_\infty, \cdots, \|f^{(2N)}\|_\infty\} \int_{|k| \le 2M^{-j}} d^dk \frac{1}{|k|^{\frac{d}{2}}}$$

$$\le const^N \sup\{\|f\|_\infty, \cdots, \|f^{(2N)}\|_\infty\} M^{-\frac{d}{2}j} \qquad (9.30)$$

since each derivative either acts on $f(M^j|k|)$, producing an M^j by the chain rule, or acts on $|k|^{-\frac{d}{2}}$, producing an additional factor of $1/|k|$ which can be estimated against an M^j on the support of the integrand. Thus, including the supremum over the derivatives of f into the constant,

$$|[1 + (M^{-2j}x^2)^d]C^j(x)| \leq const\, M^{-\frac{d}{2}j}$$

or

$$|C^j(x)| \leq const\, \frac{M^{-\frac{d}{2}j}}{1 + (M^{-j}|x|)^{2d}} \tag{9.31}$$

from which the L^1 bound on $C^j(x)$ follows by integration. ∎

Let

$$\mathcal{W}^{\leq j}(\psi) := \log \frac{1}{Z^{\leq j}} \int e^{\lambda V(\psi + \psi^{\leq j})} d\mu_{\leq j}(\psi^{\leq j}) \tag{9.32}$$

$$= \sum_{q=1}^{\infty} \int \prod_{i=1}^{2q} d\xi_i\, W_{2q}^{\leq j}(\xi_1, \cdots \xi_{2q})\, \psi(\xi_1)\bar{\psi}(\xi_2) \cdots \psi(\xi_{2q-1})\bar{\psi}(\xi_{2q})$$

be as in (9.13), then $W_{2q}^{\leq j}$ is given by the sum of all connected amputated $2q$-legged diagrams up to scale j, that is, with covariance $C^{\leq j} = \sum_{i=0}^{j}$. We want to control all n'th order contributions $W_{2q,n}^{\leq j}$ which are given by $W_{2q}^{\leq j} = \sum_{n=1}^{\infty} \lambda^n W_{2q,n}^{\leq j}$.

According to Lemma 9.1.2 we have, for all $n \geq 2$,

$$W_{2q,n}^{\leq j} = \sum_{i=0}^{j} V_{2q,n}^i \tag{9.33}$$

where the $\mathcal{V}^i$'s are given by (9.17). The goal of this subsection is to prove the following bounds on the $V_{2q,n}^i$.

Theorem 9.2.2 *Let $V_{2q,n}^j$ be given by*

$$\mathcal{V}^j(\psi) = \log \frac{1}{Y_j} \int e^{\left(\sum_{i=0}^{j-1} \mathcal{V}^i + \lambda V\right)(\psi + \psi^j; \psi)} d\mu_j(\psi^j) \tag{9.34}$$

$$= \sum_{q=1}^{\infty} \sum_{n=1}^{\infty} \lambda^n \int \prod_{l=1}^{2q} d\xi_l\, V_{2q,n}^j(\xi_1, \cdots, \xi_{2q})\, \psi(\xi_1) \cdots \bar{\psi}(\xi_{2q})$$

where C^j is given by (9.26) and let the interaction λV be given by (9.11). Define $\|V_{2q}^j\|_{S, \leq n} := \sum_{\ell=1}^{n} |\lambda|^{\ell} \|V_{2q,\ell}^j\|_S$ where the norms $\| \cdot \|_{\emptyset}$ and $\| \cdot \|_S$ are *defined by (8.38) and (8.40). Then there are the following bounds:*

$$\|V_{2q}^j\|_{\emptyset,\,\leq n} \leq M^{\frac{d}{2}j(q-2)} \sup_{\substack{1\leq r\leq n \\ q_v\geq 1,\ \Sigma\, q_v\leq nm}} \left\{ 2^{11\Sigma_v q_v}\, c_M^{\Sigma_v q_v - q} \times \right. \tag{9.35}$$

$$\left. \prod_{v=1}^{r}\left(M^{-\frac{d}{2}j(q_v-2)}\|W_{2q_v}^{\leq j-1}\|_{\emptyset,\,\leq n}\right)\right\}$$

and for $S \subset \{1,\cdots,2q\}$, $S \neq \emptyset$

$$\|V_{2q}^j\|_{S,\,\leq n} \leq M^{\frac{d}{2}j(q-\frac{|S|}{2})} \sup_{\substack{1\leq r+s\leq n,\ s\geq 1 \\ q_v\geq 1,\ \Sigma\, q_v\leq nm \\ |S_v|<2q_v}} \left\{ 2^{11\Sigma_v q_v}\, c_M^{\Sigma_v q_v - q} \times \right. \tag{9.36}$$

$$\prod_{v=1}^{r}\left(M^{-\frac{d}{2}j(q_v-2)}\|W_{2q_v}^{\leq j-1}\|_{\emptyset,\,\leq n}\right) \times$$

$$\left. \prod_{v=r+1}^{r+s}\left(M^{-\frac{d}{2}j(q_v-\frac{|S_v|}{2})}\|W_{2q_v}^{\leq j-1}\|_{S_v,\,\leq n}\right)\right\}$$

where c_M is the constant of Lemma 9.2.1.

Proof: To isolate all contributions up to n'th order, we apply n times the R-operation of Theorem 4.4.8. This operation has been introduced in [22] and is an improvement of the integration by parts formula used in [18]. Since the scale j is kept fixed in this proof, we use, only in this proof, the following notation to shorten the formulae a bit.

$$\mathcal{W} := \mathcal{W}^{\leq j-1} = \sum_{i=0}^{j-1}\mathcal{V}^i + \lambda\mathcal{V} \tag{9.37}$$
$$\mathcal{V} := \mathcal{V}^j \tag{9.38}$$
$$d\mu(\psi) := d\mu_j(\psi_j) \tag{9.39}$$

The fields which are integrated over are now called ψ and the external fields, in the statement of the theorem denoted as ψ, will now be denoted as η. In this new notation, the functional integral to be controlled reads

$$\mathcal{V}(\eta) = \log\frac{1}{Y}\int e^{\mathcal{W}(\eta+\psi;\eta)}d\mu(\psi) \tag{9.40}$$

and Y such that $\mathcal{V}(0) = 0$. Furthermore we let $\xi = (x,\sigma,b)$ where $b \in \{0,1\}$ indicates whether the field is a barred one, $\bar\psi$, for $b = 1$, or an unbarred one, ψ, for $b = 0$. Recall that $\mathcal{W}(\eta + \psi;\eta) = \mathcal{W}(\eta + \psi) - \mathcal{W}(\eta)$. We start the expansion as follows:

$$\mathcal{V}(\eta) = \log \tfrac{1}{Y} \int e^{\mathcal{W}(\eta+\psi;\eta)} d\mu(\psi) \tag{9.41}$$

$$= \int_0^1 dt \, \frac{d}{dt} \log \tfrac{1}{Y} \int e^{\mathcal{W}(t\eta+\psi;t\eta)} d\mu(\psi)$$

$$= \int_0^1 \frac{\int \left(\frac{d}{dt}\mathcal{W}(t\eta+\psi;t\eta)\right) e^{\mathcal{W}(t\eta+\psi;t\eta)} d\mu(\psi)}{\int e^{\mathcal{W}(t\eta+\psi;t\eta)} d\mu(\psi)}$$

$$= \sum_{q=1}^{\infty} \int \prod_{l=1}^{2q} d\xi_l \, W_{2q}(\underline{\xi}) \sum_{\substack{I\subset\{1,\cdots,2q\} \\ I\neq\emptyset}} \int_0^1 dt \, (\tfrac{d}{dt} t^{|I^c|}) \prod_{i\in I^c} \eta(\xi_i) \times$$

$$\frac{\int \prod_{i\in I} \psi(\xi_i) \, e^{\mathcal{W}(t\eta+\psi;t\eta)} d\mu(\psi)}{\int e^{\mathcal{W}(t\eta+\psi;t\eta)} d\mu(\psi)}$$

where $I^c := \{1,\cdots,2q\} \setminus I$ denotes the complement of I. For a fixed choice of q and I, we may represent the term in (9.41) graphically as in figure 9.1.

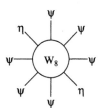

Figure 9.1

Since the bare interaction (9.11) is by assumption at most $2m$-legged, $m \geq 2$ some fixed chosen number, a $2q$-legged contribution is at least of order $\max\{1, q/m\}$. Since we are interested only in contributions up to n'th order, we can cut off the q-sum at nm.

To the functional integral in the last line of (9.41) we apply the **R**-operation of Theorem 4.4.8. In order to do so we write $\mathcal{W}_{t\eta}(\psi) := \mathcal{W}(\psi + t\eta) - \mathcal{W}(t\eta)$ such that with Theorem 4.4.8 we obtain

$$\frac{\int \prod_{i\in I} \psi(\xi_i) \, e^{\mathcal{W}_{t\eta}(\psi)} d\mu(\psi)}{\int e^{\mathcal{W}_{t\eta}(\psi)} d\mu(\psi)} = \tag{9.42}$$

$$\int \prod_{i\in I} \psi(\xi_i) \, d\mu(\psi) \; + \; \frac{\int \mathbf{R}\left(\prod_{i\in I} \psi(\xi_i)\right)(\psi) \, e^{\mathcal{W}_{t\eta}(\psi)} d\mu(\psi)}{\int e^{\mathcal{W}_{t\eta}(\psi)} d\mu(\psi)}$$

where

$$\mathbf{R}\Big(\prod_{i\in I}\psi(\xi_i)\Big)(\tilde{\psi}) = \sum_{k=1}^{\infty}\frac{1}{k!}\int \prod_{i\in I}\psi(\xi_i) \; :[\mathcal{W}_{t\eta}(\psi+\tilde{\psi})-\mathcal{W}_{t\eta}(\tilde{\psi})]^k: d\mu(\psi) \quad (9.43)$$

$$= \sum_{k=1}^{\infty}\frac{1}{k!}\int \prod_{i\in I}\psi(\xi_i) \; :[\mathcal{W}(\psi+\tilde{\psi}+t\eta)-\mathcal{W}(\tilde{\psi}+t\eta)]^k: d\mu(\psi)$$

Since we are interested only in contributions up to n'th order, we could cut off the k-sum at n. Letting $\tilde{\eta} := \tilde{\psi}+t\eta$, we get

$$\mathbf{R}\Big(\prod_{i\in I}\psi(\xi_i)\Big)(\tilde{\psi}) = \quad\quad (9.44)$$

$$\sum_{k=1}^{\infty}\frac{1}{k!}\sum_{q_1,\cdots,q_k=1}^{\infty}\int d\underline{\xi}_1\cdots d\underline{\xi}_k W_{2q_1}(\underline{\xi}_1)\cdots W_{2q_k}(\underline{\xi}_k)\times$$

$$\sum_{\substack{I_1\subset\{1,\cdots,2q_1\}\\I_1\neq\emptyset}}\cdots\sum_{\substack{I_k\subset\{1,\cdots,2q_k\}\\I_k\neq\emptyset}}\prod_{r=1}^{k}\prod_{i\in I_r^c}\tilde{\eta}(\xi_i)\int\prod_{i\in I}\psi(\xi_i)\;:\prod_{r=1}^{k}\prod_{i\in I_r}\psi(\xi_i): d\mu(\psi)$$

Substituting this in (9.42), we obtain

$$\frac{\int \prod_{i\in I}\psi(\xi_i)\,e^{\mathcal{W}_{t\eta}(\psi)}d\mu(\psi)}{\int e^{\mathcal{W}_{t\eta}(\psi)}d\mu(\psi)} = \int\prod_{i\in I}\psi(\xi_i)\,d\mu(\psi)\; + \quad\quad (9.45)$$

$$\sum_{k=1}^{\infty}\frac{1}{k!}\sum_{q_1,\cdots,q_k=1}^{\infty}\int d\underline{\xi}_1\cdots d\underline{\xi}_k W_{2q_1}(\underline{\xi}_1)\cdots W_{2q_k}(\underline{\xi}_k)\times$$

$$\sum_{\substack{I_1\subset\{1,\cdots,2q_1\}\\I_1\neq\emptyset}}\cdots\sum_{\substack{I_k\subset\{1,\cdots,2q_k\}\\I_k\neq\emptyset}}\int\prod_{i\in I}\psi(\xi_i)\;:\prod_{r=1}^{k}\prod_{i\in I_r}\psi(\xi_i): d\mu(\psi)\times$$

$$\frac{\int \prod_{r=1}^{k}\prod_{i\in I_r^c}(\psi+t\eta)(\xi_i)\,e^{\mathcal{W}(\psi+t\eta;t\eta)}d\mu(\psi)}{\int e^{\mathcal{W}(\psi+t\eta;t\eta)}d\mu(\psi)}$$

The first term in (9.45) we can define as the $k=0$ term of the k-sum in (9.45). By multiplying out the $(\psi+t\eta)$ brackets in the last line of (9.45) we obtain

$$\frac{\int \prod_{i\in I}\psi(\xi_i)\,e^{\mathcal{W}_{t\eta}(\psi)}d\mu(\psi)}{\int e^{\mathcal{W}_{t\eta}(\psi)}d\mu(\psi)} = \quad\quad (9.46)$$

$$\sum_{k=0}^{\infty}\frac{1}{k!}\sum_{q_1,\cdots,q_k=1}^{\infty}\int d\underline{\xi}_1\cdots d\underline{\xi}_k W_{2q_1}(\underline{\xi}_1)\cdots W_{2q_k}(\underline{\xi}_k)\times$$

$$\sum_{\substack{I_1\subset\{1,\cdots,2q_1\}\\I_1\neq\emptyset}}\cdots\sum_{\substack{I_k\subset\{1,\cdots,2q_k\}\\I_k\neq\emptyset}}\int\prod_{i\in I}\psi(\xi_i)\;:\prod_{r=1}^{k}\prod_{i\in I_r}\psi(\xi_i): d\mu(\psi)\times$$

$$\sum_{J_1\subset I_1^c}\cdots\sum_{J_k\subset I_k^c}\prod_{r=1}^{k}\prod_{i\in J_r^c}t\eta(\xi_i)\frac{\int \prod_{r=1}^{k}\prod_{i\in J_r}\psi(\xi_i)\,e^{\mathcal{W}(\psi+t\eta;t\eta)}d\mu(\psi)}{\int e^{\mathcal{W}(\psi+t\eta;t\eta)}d\mu(\psi)}$$

Graphically we may represent this as follows. The functional integral on the left hand side of (9.46) comes with a generalized vertex W_{2q}, see (9.41) and figure 9.1. $|I^c|$ legs of W_{2q} are external (η-fields) and are no longer integrated over. $|I|$ legs of W_{2q} are internal (ψ-fields) and are integrated over, that is, they have to be contracted. The value of k in the first sum of (9.46) is the number of new vertices produced by one step of the **R**-operation to which the $|I|$ legs of W_{2q} may contract. In figure 9.2, $k = 3$. If $k = 0$, then all these fields have to contract among themselves. $q_1, \cdots, q_k$ in the second sum of (9.46) fix the number of legs of the new vertices. $I_1, \cdots, I_k$ specify which of those legs have to contract to the $|I|$ legs of W_{2q}. These legs are not allowed to contract among themselves. Observe that, because of the restriction $I_r \neq \emptyset$, there is at least one contraction between a new vertex and W_{2q}. This contraction will be made explicit in Lemma 9.2.3, part (a), below and also has been made explicit in figure 9.2.

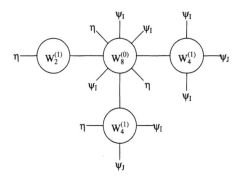

Figure 9.2

Finally, $J_1, \cdots, J_k$ specify which legs of the new vertices which have not been contracted to the W_{2q} fields will be external (those in J_r^c) or will produce new vertices in a second step of the **R**-operation.

The improvement of this expansion step compared to the integration by part formula used in [18], Lemma II.4, is the following. The only place where 'dangerous' factorials due to the number of diagrams (or number of contractions) may arise is the integral

$$\int \prod_{i \in I} \psi(\xi_i) : \prod_{r=1}^{k} \prod_{i \in I_r} \psi(\xi_i) : d\mu(\psi) \tag{9.47}$$

In [18], this corresponds to the 'primed integral' in formula (II.4). Now the point is that to the integral in (9.47) we can apply Gram's inequality of Theorem 4.4.9 which immediately eliminates potential factorials and gives a bound $const^{\text{number of fields}}$ times the right power counting factor whereas in [18] we could not use Gram's inequality because the determinant which is given by the 'primed integral' has a less ordered structure.

To produce all contributions which are made of at most n generalized vertices W_{2q_v}, we repeat the **R**-operation n-times. This produces a big sum of the form

$$\sum_{q=1}^{nm}\sum_{I}\Big[\sum_{k^1=1}^{n}\sum_{q_1^1\cdots q_{k1}^1=1}\sum_{I_1^1\cdots I_{k1}^1}\sum_{J_1^1\cdots J_{k1}^1}\Big]\Big[\sum_{k^2=0}^{n}\sum_{q_1^2\cdots q_{k2}^2=1}\sum_{I_1^2\cdots I_{k2}^2}\sum_{J_1^2\cdots J_{k2}^2}\Big]\cdots$$

$$\cdots\Big[\sum_{k^n=0}^{n}\sum_{q_1^n\cdots q_{kn}^n=1}\sum_{I_1^n\cdots I_{kn}^n}\Big]\mathcal{K}\big(q,I,\{k^\ell,\underline{q}^\ell,\underline{I}^\ell,\underline{J}^\ell\}\big)(\eta) \quad (9.48)$$

The purpose of the square brackets above is only to group sums together which belong to the same step of **R**-operation. In the n'th application of **R** we have $J_1^n=\cdots=J_{kn}^n=\emptyset$ such that the corresponding functional integral in the last line of (9.46) is equal to 1 since otherwise this would be a contribution with at least $n+1$ generalized vertices W_{2q_v}. Figure 9.3 shows the term in figure 9.2 after a second step of **R**-operation.

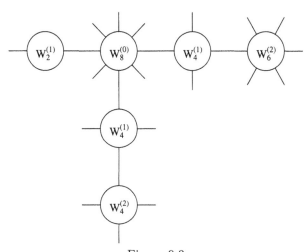

Figure 9.3

The contributions $\mathcal{K}$ look as follows:

$$\mathcal{K}\big(q,I,\{k^\ell,\underline{q}^\ell,\underline{I}^\ell,\underline{J}^\ell\}\big)(\eta) \quad (9.49)$$

$$=\prod_{\ell=1}^{n}\frac{1}{k^\ell!}\int d\underline{\xi}\prod_{\ell=1}^{n}d\underline{\xi}^\ell W_{2q}(\underline{\xi})\prod_{\ell=1}^{n}\prod_{r=1}^{k^\ell}W_{2q_r^\ell}(\underline{\xi}_r^\ell)\times$$

$$\mathrm{Int}\big(I,\underline{I}^1\big)(\underline{\xi})\prod_{\ell=2}^{n}\mathrm{Int}\big(\underline{J}^{\ell-1},\underline{I}^\ell\big)(\underline{\xi})\int_0^1 dt\Big(\frac{d}{dt}\prod_{i\in I^c}t\eta(\xi_i)\Big)\prod_{\ell=1}^{n}\prod_{r=1}^{k^\ell}\prod_{i\in J_r^{\ell c}}t\eta(\xi_{r,i}^\ell)$$

$$=: \int \prod_{i \in I^c} d\xi_i \, \eta(\xi_i) \prod_{r=1}^{k^\ell} \prod_{i \in J_r^{\ell c}} d\xi_{r,i}^\ell \, \eta(\xi_{r,i}^\ell) \, K\big(q, I, \{k^\ell, \underline{q}^\ell, \underline{I}^\ell, \underline{J}^\ell\}\big)(\xi_i, \xi_{r,i}^\ell)$$

where the last equation defines the kernels K. The integrals Int in the second line of (9.49) are given by

$$\mathrm{Int}\big(\underline{J}^{\ell-1}, \underline{I}^\ell\big)(\underline{\xi}) := \int \prod_{r=1}^{k^\ell} \prod_{i \in J_r^\ell} \psi(\xi_{r,i}^\ell) : \prod_{r=1}^{k^\ell} \prod_{i \in I_r^\ell} \psi(\xi_{r,i}^\ell) : d\mu_C(\psi) \quad (9.50)$$

Let $K = \sum_{l=1}^{\infty} \lambda^l K_l$ be the power series expansion of K with respect to the coupling λ (which enters (9.49) only through $W_{2q} = \sum_{l=1}^{\infty} \lambda^l W_{2q,l}$) and let

$$|K(\underline{\xi})|_{\le n} := \sum_{l=1}^{n} |\lambda|^l |K_l(\underline{\xi})| \quad (9.51)$$

Then we have, suppressing the argument $q, I, \{k^\ell, \underline{q}^\ell, \underline{I}^\ell, \underline{J}^\ell\}$ of K at the moment,

$$|K(\tilde{\underline{\xi}})|_{\le n} \le \quad (9.52)$$

$$\prod_{\ell=1}^{n} \frac{1}{k^\ell!} \int d\underline{\xi} \prod_{\ell=1}^{n} d\underline{\xi}^\ell |W_{2q}(\underline{\xi})|_{\le n} \prod_{\ell=1}^{n} \prod_{r=1}^{k^\ell} |W_{2q_r^\ell}(\underline{\xi}_r^\ell)|_{\le n} \left|\mathrm{Int}\big(I, \underline{I}^1\big)(\underline{\xi})\right| \times$$

$$\prod_{\ell=2}^{n} \left|\mathrm{Int}\big(\underline{J}^{\ell-1}, \underline{I}^\ell\big)(\underline{\xi})\right| \int_0^1 dt \Big(\frac{d}{dt} \prod_{i \in I^c} t\delta(\xi_i - \tilde{\xi}_i)\Big) \prod_{\ell=1}^{n} \prod_{r=1}^{k^\ell} \prod_{i \in J_r^{\ell c}} t\delta(\xi_{r,i}^\ell - \tilde{\xi}_{r,i}^\ell)$$

If we abbreviate the sum in (9.48) by $\sum_{q,I,\{k^\ell,\underline{q}^\ell,\underline{I}^\ell,\underline{J}^\ell\}}$, then a first bound on all $2p$-legged contributions up to n'th order, $\sum_{l=1}^{n} \lambda^l V_{2p,l}$, reads as follows:

$$|V_{2p}(\underline{\xi})|_{\le n} \le \quad (9.53)$$

$$\sum_{q,I,\{k^\ell,\underline{q}^\ell,\underline{I}^\ell,\underline{J}^\ell\}} \chi\Big(\sum_{\ell=1}^{n} k^\ell \le n-1, \, |I^c| + \sum_{\ell=1}^{n} \sum_{r=1}^{k^\ell} |J_r^{\ell c}| = 2p\Big) |K(\underline{\xi})|_{\le n}$$

with the bound (9.52) on $|K(\underline{\xi})|_{\le n}$. Here we introduced

$$\chi(A) = \begin{cases} 1 \text{ if } A \text{ is true} \\ 0 \text{ if } A \text{ is false} \end{cases} \quad (9.54)$$

to enforce a $2p$-legged contribution and to restrict the number of generalized vertices to at most n. By integrating (9.53) over the ξ-variables, we obtain the analog bounds for the $\|\cdot\|_{S,\le n}$-norms, $S = \emptyset$ or $S \subset \{1, \cdots, 2p\}$.

$$\|V_{2p}(\underline{\xi})\|_{S,\le n} \le \sum_{q,I,\{k^\ell,\underline{q}^\ell,\underline{I}^\ell,\underline{J}^\ell\}} \chi\big(\cdots\big) \|K(\underline{\xi})\|_{S,\le n} \quad (9.55)$$

Although K is not a single diagram, but, if we would expand the integrals Int in (9.52), is given by a sum of diagrams, we can obtain a bound on $\|K\|_S$ in the same way as we obtained the bounds in Lemma 8.3.1 for single diagrams. In that lemma first we had to choose a tree for the diagram and then, in (8.41, 8.42) one has to take the L^∞-norm for propagators not on the tree and the L^1-norm for propagators on the tree. A natural choice of a tree for K is the following one. We choose the vertex W_{2q} which was produced at the very beginning in (9.41) as the root. Every vertex W_{2q^ℓ} which is produced at the ℓ'th step of **R**-operation is connected by at least one line to some vertex $W_{2q^{\ell-1}_s}$ which is the meaning of $I_r^\ell \neq \emptyset$. For each $W_{2q^\ell_r}$ we choose exactly one of these lines to be on the tree for K. The point is that this tree would be a tree for all the diagrams which can be obtained from (9.52) by expanding the integrals Int, but now we can use the sign cancellations by applying a determinant bound, that is, Gram's inequality, to the integrals $\text{Int}\big(\underline{J}^{\ell-1}, \underline{I}^\ell\big)(\underline{\xi})$ in (9.52). This is done in part (b) of Lemma 9.2.3 below. First however, we must make explicit the propagators for lines on the tree which are also contained in $\text{Int}\big(\underline{J}^{\ell-1}, \underline{I}^\ell\big)(\underline{\xi})$ (the factors in the wavy brackets in (9.56) below) since they are needed to absorb some of the coordinate space ξ-integrals (the remaining ξ-integrals are absorbed by the generalized vertices $W_{2q^\ell_r}$). This is done in part (a) of the following

Lemma 9.2.3 a) *Let* $J, I_1, \cdots, I_k \subset \mathbb{N}$ *be some finite sets, let* $I_r \neq \emptyset$ *for* $1 \leq r \leq k$ *and let* $i_1^r := \min I_r$. *Recall the definition (9.50) for the Integrals* Int. *Then*

$$\big|\text{Int}\big(J, \cup_{r=1}^k I_r\big)(\underline{\xi})\big| \leq \tag{9.56}$$

$$\sum_{\substack{j_1,\cdots,j_k \in J \\ j_r \neq j_s}} \Big\{ \prod_{r=1}^k |C(\xi_{i_1^r}, \xi_{j_r})| \Big\} \Big| \text{Int}\Big(J \setminus \{j_1, \cdots, j_k\}, \cup_{r=1}^k (I_r \setminus \{i_1^r\}) \Big) \Big|$$

b) *Let* $J, I \subset \mathbb{N}$ *be some finite sets and let*

$$C(x, x') = C^j(x - x') = \int \frac{d^d k}{(2\pi)^d} e^{ik(x-x')} \frac{f(M^j |k|)}{|k|^{\frac{d}{2}}}$$

be the covariance of Lemma 9.2.1. Then

$$\sup_\xi \big|\text{Int}(J, I)(\underline{\xi})\big| \leq \Big(\sqrt{2c}\, M^{-\frac{d}{4}j} \Big)^{|I|+|J|} \tag{9.57}$$

Proof: a) We have, using the notation of Definition 4.4.6,

$$\text{Int}\big(J, \cup_{r=1}^k I_r\big)(\underline{\xi}) = \int \prod_{j \in J} \psi(\xi_j) : \prod_{r=1}^k \prod_{i^r \in I_r} \psi(\xi_{i^r}) : d\mu_C \tag{9.58}$$

$$= \int \prod_{j \in J} \psi(\xi_j) \prod_{r=1}^k \prod_{i^r \in I_r} \psi^\sharp(\xi_{i^r}) : d\mu_{C^\sharp}$$

$$= \pm \int \psi^\sharp(\xi_{i_1^1}) \cdots \psi^\sharp(\xi_{i_1^r}) \prod_{j \in J} \psi(\xi_j) \prod_{r=1}^k \prod_{i^r \in I_r \setminus \{i_1^r\}} \psi^\sharp(\xi_{i^r}) : d\mu_{C^\sharp}$$

$$= \sum_{j_1 \in J} (\pm) C(\xi_{i_1^1}, \xi_{j_1}) \int \psi^\sharp(\xi_{i_1^2}) \cdots \psi^\sharp(\xi_{i_1^r}) \times$$

$$\prod_{j \in J \setminus \{j_1\}} \psi(\xi_j) \prod_{r=1}^{k} \prod_{i^r \in I_r \setminus \{i_1^r\}} \psi^\sharp(\xi_{i^r}) \!: d\mu_{C^\sharp}$$

$$= \sum_{\substack{j_1, \cdots, j_k \in J \\ j_r \neq j_s}} (\pm) C(\xi_{i_1^1}, \xi_{j_1}) \cdots C(\xi_{i_k}, \xi_{j_k}) \times$$

$$\int \prod_{j \in J \setminus \{j_1, \cdots, j_k\}} \psi(\xi_j) \prod_{r=1}^{k} \prod_{i^r \in I_r \setminus \{i_1^r\}} \psi^\sharp(\xi_{i^r}) \!: d\mu_{C^\sharp}$$

which proves part (a) by taking absolute value. Part (b) is an immediate consequence of (4.112) of Theorem 4.4.9: We have

$$C(x, x') = \frac{1}{(2\pi)^d} \int d^d k \, f_x(k) \, g_{x'}(k) \tag{9.59}$$

where $f_x(k) = e^{ikx} \sqrt{f(M^j|k|)}/|k|^{\frac{d}{4}}$, $g_{x'}(k) = e^{-ikx'} \sqrt{f(M^j|k|)}/|k|^{\frac{d}{4}}$ and $\mathcal{H} = L^2(\mathbb{R}^d)$ with norms

$$\|f_x\|_{L^2}, \|g_{x'}\|_{L^2} \leq \sqrt{\int d^d k \, \frac{f(M^j|k|)}{|k|^{\frac{d}{2}}}} = \sqrt{\|C^j(k)\|_{L^1}} \leq \sqrt{c} \, M^{-\frac{d}{4}j} \tag{9.60}$$

Application of (4.112) gives (9.57). ∎

We now continue with the proof of Theorem 9.2.2. The combination of (9.56) and (9.57) gives the following bound on the integrals $\mathrm{Int}(\underline{J}^{\ell-1}, \underline{I}^\ell)$. Recall that $\underline{I}^\ell = (I_1^\ell, \cdots, I_{k^\ell}^\ell)$. Let $i_{r,1}^\ell := \min I_r^\ell$. Then

$$\left| \mathrm{Int}(\underline{J}^{\ell-1}, \underline{I}^\ell)(\underline{\xi}) \right| \leq \tag{9.61}$$

$$\sum_{j_1, \cdots, j_{k^\ell} \in \underline{J}^{\ell-1}} \left\{ \prod_{r=1}^{k^\ell} |C^j(\xi_{i_{r,1}^\ell}, \xi_{j_r})| \right\} \left(\sqrt{2c} \, M^{-\frac{d}{4}j} \right)^{|\underline{J}^{\ell-1}| + |\underline{I}^\ell| - 2k^\ell}$$

Now that we have made explicit the propagators on the spanning tree for K we can estimate as in Lemma 8.3.1 to obtain the following bound. We substitute (9.61) in (9.52) and get ($k^0 := 1$, $q_1^0 := q$, $\underline{J}^0 = \underline{I}^0 := I$)

$$\|K\|_{\emptyset, \leq n} \leq \prod_{\ell=0}^{n} \left(\frac{1}{k^\ell!} \prod_{r=1}^{k^\ell} \|W_{2q_r^\ell}\|_{\emptyset, \leq n} \right) \int_0^1 dt \left(\frac{d}{dt} t^{|I^c|} \right) t^{\sum_{\ell=1}^{n} \sum_{r=1}^{k^\ell} |J_r^{\ell c}|} \times$$

$$\prod_{\ell=1}^{n} \left\{ \sum_{j_1^{\ell-1}, \cdots, j_{k^\ell}^{\ell-1} \in \underline{J}^{\ell-1}} \prod_{r=1}^{k^\ell} \|C^j(x)\|_{L^1} \left(\sqrt{2c} \, M^{-\frac{d}{4}j} \right)^{|\underline{J}^{\ell-1}| + |\underline{I}^\ell| - 2k^\ell} \right\}$$

$$\leq |I^c| \prod_{\ell=1}^{n} \binom{|\underline{J}^{\ell-1}|}{k^\ell} \prod_{\ell=0}^{n} \|W_{2q_r^\ell}\|_{\emptyset, \leq n} \left(c \, M^{\frac{d}{2}j} \right)^{\sum_{\ell=1}^{n} k^\ell} \times$$

$$\left(2c \, M^{-\frac{d}{2}j} \right)^{\sum_{\ell=1}^{n} \left(\frac{|\underline{J}^{\ell-1}| + |\underline{I}^\ell|}{2} - k^\ell \right)} \tag{9.62}$$

The sum in (9.53) or more explicitly, the sum in (9.48) we bound as follows:

$$\sum_{J_r^\ell \subset I_r^\ell} \cdots \leq 2^{|I_r^\ell|} \sup_{J_r^\ell \subset I_r^\ell} \cdots \leq 2^{2q_r^\ell} \sup_{J_r^\ell \subset I_r^\ell} \cdots \tag{9.63}$$

$$\sum_{I_r^\ell \subset \{1,\cdots,2q_r^\ell\}} \cdots \leq 2^{2q_r^\ell} \sup_{I_r^\ell \subset \{1,\cdots,2q_r^\ell\}} \cdots \tag{9.64}$$

$$\sum_{q_r^\ell=1}^{\infty} \cdots \leq \sum_{q_r^\ell=1}^{\infty} \frac{1}{2^{q_r^\ell}} \sup_{q_r^\ell} 2^{q_r^\ell} \cdots = \sup_{q_r^\ell} 2^{q_r^\ell} \cdots \tag{9.65}$$

$$\sum_{k^\ell} \cdots \leq \sum_{k^\ell} \binom{|J^{\ell-1}|}{k^\ell} \sup_{k^\ell} \frac{1}{\binom{|J^{\ell-1}|}{k^\ell}} \cdots$$

$$\leq 2^{|J^{\ell-1}|} \sup_{k^\ell} \frac{1}{\binom{|J^{\ell-1}|}{k^\ell}} \cdots$$

$$\leq 2^{\sum_{r=1}^{k^{\ell-1}} 2q_r^{\ell-1}} \sup_{k^\ell} \frac{1}{\binom{|J^{\ell-1}|}{k^\ell}} \cdots \tag{9.66}$$

Thus, combining (9.62), (9.55) and the above estimates we arrive at

$$\|V_{2p}\|_{\emptyset,\,\leq n} \leq \sup_{q,I,\{k^\ell,\underline{q}^\ell,\underline{I}^\ell,\underline{J}^\ell\}} \left\{ \chi(\cdots) \prod_{\ell=0}^{n} \prod_{r=1}^{k^\ell} 2^{7q_r^\ell} |I^c| \prod_{\ell=0}^{n} \prod_{r=1}^{k^\ell} \|W_{2q_r^\ell}\|_{\emptyset,\,\leq n} \times \right.$$
$$\left. \left(c\,M^{\frac{d}{2}j}\right)^{\sum_{\ell=1}^{n} k^\ell} \left(2c\,M^{-\frac{d}{2}j}\right)^{\sum_{\ell=1}^{n} \left(\frac{|J^{\ell-1}|+|I^\ell|}{2} - k^\ell\right)} \right\}$$

$$\leq \sup_{q,I,\{k^\ell,\underline{q}^\ell,\underline{I}^\ell,\underline{J}^\ell\}} \left\{ \chi(\cdots) \prod_{\ell=0}^{n} \prod_{r=1}^{k^\ell} 2^{9q_r^\ell} \prod_{\ell=0}^{n} \prod_{r=1}^{k^\ell} \|W_{2q_r^\ell}\|_{\emptyset,\,\leq n} \times \right.$$
$$\left. \left(2cM^{\frac{d}{2}j}\right)^{\sum_{\ell=1}^{n} k^\ell} \left(2cM^{-\frac{d}{2}j}\right)^{\sum_{\ell=1}^{n} \left(\frac{|J^{\ell-1}|+|I^\ell|}{2} - k^\ell\right)} \right\} \tag{9.67}$$

The factors of $M^{\pm\frac{d}{2}j}$ are treated exactly in the same way as in the proof of Lemma 8.3.3. To this end observe that a fixed choice of $q, I, \{k^\ell, \underline{q}^\ell, \underline{I}^\ell, \underline{J}^\ell\}$ fixes the number of generalized vertices and the number of ψ's which have to be paired. In other words, all diagrams which contribute to a given $K(q, I, \{k^\ell, \underline{q}^\ell, \underline{I}^\ell, \underline{J}^\ell\})$ have the same number of generalized vertices $W_{2q_r^\ell} \equiv W_{2q_v}$, $1 \leq v \leq \rho$, $\rho = \rho(q, I, \{k^\ell, \underline{q}^\ell, \underline{I}^\ell, \underline{J}^\ell\})$ the number of generalized vertices, and the same number of lines,

$$L = \sum_{\ell=1}^{n} \frac{|J^{\ell-1}| + |I^\ell|}{2} = \frac{1}{2}\left(\sum_{\ell=0}^{n}\sum_{r=1}^{k^\ell} 2q_r^\ell - 2p\right) = \sum_{v=1}^{\rho} q_v - p \tag{9.68}$$

although the individual positions of the lines are not fixed, only the lines on the tree are fixed. The number of lines on the tree is given by

$$T = \sum_{\ell=1}^{n} k^\ell = \rho - 1 \tag{9.69}$$

Thus (9.67) can be written as

$$\|V_{2p}\|_{\emptyset, \le n} \le \sup_{\substack{1 \le \rho \le n, \, q_v \ge 1 \\ \sum_v q_v \le nm}} \left\{ 2^{11 \sum_v q_v} \, c^{\sum_v q_v - p} \prod_{v=1}^{\rho} \|W_{2q_v}\|_{\emptyset, \le n} \times \right. \tag{9.70}$$

$$\left. \left(M^{\frac{d}{2}j}\right)^{\rho-1} \left(M^{-\frac{d}{2}j}\right)^{\sum_v q_v - p - \rho + 1} \right\}$$

In the same way we obtain for $S \ne \emptyset$, if w counts the number of external vertices,

$$\|V_{2p}\|_{S, \le n} \le \sup_{\substack{1 \le \rho + w \le n, \, w \ge 1 \\ q_v \ge 1, \, \sum_v q_v \le nm \\ |S_v| < 2q_v}} \left\{ 2^{11 \sum_v q_v} \, c^{\sum_v q_v - p} \prod_{v=1}^{\rho} \|W_{2q_v}\|_{\emptyset, \le n} \times \right. \tag{9.71}$$

$$\left. \prod_{v=\rho+1}^{\rho+w} \|W_{2q_v}\|_{S_v, \le n} \left(M^{\frac{d}{2}j}\right)^{\rho-w} \left(M^{-\frac{d}{2}j}\right)^{\sum_v q_v - p - \rho + w} \right\}$$

The rearrangement of the factors of $M^{\pm \frac{d}{2}j}$ in the above inequalities into the form of (9.35) and (9.36) is then exactly done as in the proof of Lemma 8.3.3. Thus we have reached the **end of proof of Theorem 9.2.2.** ∎

9.3 A Multiscale Bound on the Sum of Convergent Diagrams

We saw already in section 8.3 that all diagrams without two- and four-legged subdiagrams were finite. In this subsection we show that the sum of all those diagrams is also finite and has in fact a small positive radius of convergence. This is a nontrivial result since the number of diagrams is of order $(2n)! \sim const^n (n!)^2$ but there is only a factor of $1/n!$ which comes from the expansion of the exponential. Thus, if we would expand down to all diagrams, or more precisely, if we would expand the integrals Int in (9.52), we would be left with a series of the form $\sum_n n! \, \lambda^n$ which has zero radius of convergence. However, in the last subsection we generated all n'th order diagrams in the following way. First we chose a tree. For an n'th order diagram, there are about $n!$ ways of doing this. Then we treated the sum of all diagrams obtained by pairing all legs which are not on the tree as a single term K. Thus K corresponds to a sum of $\sim n!$ diagrams. However, by using the sign cancellations between these diagrams, that is, by applying Lemma 9.2.3 which results from Gram's inequality, we were able to eliminate the factorial coming from pairing all fields (or legs or halflines) which were not on the tree.

In this subsection we use Theorem 9.2.2 to give an inductive proof that the sum of all diagrams without two- and four-legged subdiagrams has positive radius of convergence. As in the last section, we consider the simplified model with propagator $1/|k|^{\frac{d}{2}}$. The proof for the many-electron system is given in [18] (for $d = 2$).

First we give a precise mathematical definition of 'sum of all diagrams without two and four legged subdiagrams'. The sum of all connected $2q$-legged diagrams W_{2q} is given by

$$
\begin{aligned}
W(\eta) &= \log \tfrac{1}{Z} \int e^{\lambda V(\psi + \eta)} d\mu_C(\psi) \\
&= \sum_{q=1}^{\infty} \int \prod_{i=1}^{2q} d\xi_i \, W_{2q}(\xi_1, \cdots, \xi_{2q}) \, \eta(\xi_1) \bar\eta(\xi_2) \cdots \eta(\xi_{2q-1}) \bar\eta(\xi_{2q})
\end{aligned} \tag{9.72}
$$

According to Lemma 9.1.2, W_{2q} can be inductively computed:

$$
W = \sum_{j=0}^{\infty} \mathcal{V}^j + \lambda \mathcal{V} \tag{9.73}
$$

where

$$
\mathcal{V}^j(\psi) = \log \tfrac{1}{Y_j} \int e^{\left(\sum_{i=0}^{j-1} \mathcal{V}^i + \lambda \mathcal{V}\right)(\psi + \psi^j ; \psi)} d\mu_{C^j}(\psi^j) \tag{9.74}
$$

where, for $j = 0$, $\sum_{i=0}^{-1} \mathcal{V}^i := 0$ and $\lambda \mathcal{V}$ is given by (9.11). Let $Q_{2,4}$ be the projection operator acting on Grassmann monomials which takes out two- and four-legged contributions. That is,

$$
\begin{aligned}
&Q_{2,4} \sum_{q=1}^{\infty} \int \prod_{i=1}^{2q} d\xi_i \, W_{2q}(\xi_1, \cdots, \xi_{2q}) \, \psi(\xi_1) \bar\psi(\xi_2) \cdots \psi(\xi_{2q-1}) \bar\psi(\xi_{2q}) \\
&:= \sum_{q=3}^{\infty} \int \prod_{i=1}^{2q} d\xi_i \, W_{2q}(\xi_1, \cdots, \xi_{2q}) \, \psi(\xi_1) \bar\psi(\xi_2) \cdots \psi(\xi_{2q-1}) \bar\psi(\xi_{2q})
\end{aligned} \tag{9.75}
$$

The sum of diagrams $W^{2,4}$ which is estimated in Theorem 9.3.1 below is inductively defined as

$$
W^{2,4} = \sum_{j=0}^{\infty} \mathcal{V}^{2,4,j} + \lambda \mathcal{V} \tag{9.76}
$$

where

$$\mathcal{V}^{2,4,j}(\psi) := Q_{2,4} \log \frac{1}{Y_j^{2,4}} \int e^{\left(\sum_{i=0}^{j-1} \mathcal{V}^{2,4,i}+\lambda\mathcal{V}\right)(\psi+\psi^j;\psi)} d\mu_{C^j}(\psi^j) \quad (9.77)$$

the covariance given by (9.27), $C^j(k) = f(M^j|k|)/|k|^{\frac{d}{2}}$, and the interaction $\lambda\mathcal{V}$ given by (9.11).

This is actually a little bit more than the sum of diagrams without two- and four-legged subgraphs, namely: Consider a diagram. For each line, substitute the covariance by its scale decomposition $C = \sum_j C^j$, interchange the scale sum with the coordinate or momentum space integrals which come with the diagram. This gives a scale sum of labelled diagrams. Then $\mathcal{W}^{2,4}$ is the sum of all labelled diagrams, the sum going over all diagrams and over all scales, which do not contain any two- and four-legged subdiagrams where the highest scale of the subdiagram is smaller than the lowest scale of its (two or four) external legs.

Theorem 9.3.1 *Let $W_{2q}^{2,4}(\xi)$ be the sum of all $2q$-legged connected diagrams without two- and four-legged subgraphs defined inductively in (9.76) and (9.77) above. The interaction given by (9.11) is assumed to be short range and at most $2m$-legged, $\|V\|_\emptyset := \sum_{p=1}^m \|V_{2p}\|_\emptyset < \infty$. Let $f_1, \cdots, f_{2q}$ be some test functions with $\|f_i\|_{1,\infty} := \|f_i\|_1 + \|f_i\|_\infty < \infty$. Let $M := \left(3 \cdot 2^{11}\right)^{\frac{4}{d}(2q+1)}$, let c_M be the constant of Lemma 9.2.1 and let*

$$|\lambda| \le \min\left\{\frac{1}{4\cdot 2^{22m}\|V\|_\emptyset^2}, 1\right\} \frac{1}{c_M^{2m}} \quad (9.78)$$

Then there is the bound

$$\int \prod_{i=1}^{2q} d\xi_i \, f_i(\xi_i) \, |W_{2q}^{2,4}(\xi_1, \cdots, \xi_{2q})| \le 3|\lambda|^{\frac{q}{2m}} \prod_{i=1}^{2q} \|f_i\|_{1,\infty} \quad (9.79)$$

Proof: According to (9.76) and (9.77) we have $\mathcal{W}^{2,4} = \sum_{j=0}^\infty \mathcal{V}^{2,4,j} + \lambda\mathcal{V}$ where

$$\mathcal{V}^{2,4,j}(\psi) := Q_{2,4} \log \frac{1}{Y_j^{2,4}} \int e^{\left(\sum_{i=0}^{j-1} \mathcal{V}^{2,4,i}+\lambda\mathcal{V}\right)(\psi+\psi^j;\psi)} d\mu_{C^j}(\psi^j) \quad (9.80)$$

We now prove the following bounds by induction on j. For λ satisfying (9.78), for all $q \ge 3$ and for $S \subset \{1, \cdots, 2q\}$, $S \ne \emptyset$,

$$\|V_{2q}^{2,4,j}\|_\emptyset \le |\lambda|^{\frac{q}{2m}} M^{\frac{d}{2}j(q-2)} \quad (9.81)$$

$$\|V_{2q}^{2,4,j}\|_S \le |\lambda|^{\frac{q}{2m}} M^{\frac{d}{2}j\left(q-\frac{|S|}{2}-\frac{1}{4}\right)} \prod_{i\in S} \|f_i\|_{1,\infty} \quad (9.82)$$

We start with (9.81). For $j = 0$, every vertex is given by some λV_{2q_v}, $1 \le q_v \le m$. Let $V_{2q,n}^{j=0}$ be the n'th order contribution to $V_{2q}^{j=0} = \sum_{n=\max\{1,q/m\}}^\infty \lambda^n V_{2q,n}^{j=0}$.

Since the interaction is at most $2m$-legged, a $2q$-legged contribution must be at least of order $(2q)/(2m) = q/m$. From the proof of Theorem 9.2.2 we get

$$\|V_{2q,n}^{j=0}\|_\emptyset \le M^{\frac{d}{2}0(q-2)} \sup_{q_v \ge 1,\, \Sigma\, q_v \le nm} \left\{ 2^{11\Sigma_v\, q_v}\, c_M^{\Sigma_v\, q_v - q} \prod_{v=1}^n \left(M^{-\frac{d}{2}0(q_v-2)}\|V_{2q_v}\|_\emptyset \right) \right\}$$

$$\le \left(2^{11m} c_M^m \|V\|_\emptyset \right)^n \tag{9.83}$$

which gives, for, say,

$$\sqrt{|\lambda|} \le \tfrac{1}{2}\, \frac{1}{2^{11m} c_M^m \|V\|_\emptyset}, \tag{9.84}$$

$$\|V_{2q}^{j=0}\|_\emptyset \le \sum_{n=\max\{1,q/m\}} \left(2^{11m} c_M^m \|V\|_\emptyset \right)^n |\lambda|^n \le |\lambda|^{\frac{q}{2m}} \tag{9.85}$$

and proves (9.81) for $j = 0$. Suppose (9.81) holds for $0 \le i \le j$. From (9.35) and (9.77) we have

$$\|V_{2q}^{2,4,j+1}\|_{\emptyset,\, \le n} \le M^{\frac{d}{2}(j+1)(q-2)} \times \tag{9.86}$$

$$\sup_{\substack{1 \le r \le n \\ q_v \ge 3,\, \Sigma\, q_v \le nm}} \left\{ 2^{11\Sigma_v\, q_v}\, c_M^{\Sigma_v\, q_v - q} \prod_{v=1}^r \left(M^{-\frac{d}{2}(j+1)(q_v-2)}\|W_{2q_v}^{2,4,\le j}\|_{\emptyset,\, \le n} \right) \right\}$$

By the induction hypothesis (9.81),

$$\|W_{2q_v}^{2,4,\le j}\|_{\emptyset,\, \le n} \;\le\; \|W_{2q_v}^{2,4,\le j}\|_\emptyset$$

$$\le \sum_{i=0}^j \|V_{2q_v}^{2,4,i}\|_\emptyset + \chi(q_v \le m)\,|\lambda|\|V_{2q_v}\|_\emptyset$$

$$\le \sum_{i=0}^j |\lambda|^{\frac{q_v}{2m}} M^{\frac{d}{2}i(q_v-2)} + \chi(q_v \le m)\,|\lambda|\,\|V\|_\emptyset$$

$$\stackrel{q_v \ge 3}{\le} M^{\frac{d}{2}j(q_v-2)}\, \frac{|\lambda|^{\frac{q_v}{2m}}}{1 - M^{-\frac{d}{2}(q_v-2)}} + \chi(q_v \le m)\,|\lambda|^{\frac{1}{2}}$$

$$\stackrel{M^{\frac{d}{2}} \ge 2}{\le} 2\, M^{\frac{d}{2}j(q_v-2)}|\lambda|^{\frac{q_v}{2m}} + |\lambda|^{\frac{q_v}{2m}}$$

$$\le 3\, M^{\frac{d}{2}j(q_v-2)}|\lambda|^{\frac{q_v}{2m}} \tag{9.87}$$

Substituting this in (9.86), the supremum becomes

$$\sup_{\substack{1 \le r \le n \\ q_v \ge 3, \, \sum q_v \le nm}} \left\{ 2^{11 \sum_v q_v} \, c_M^{\sum_v q_v - q} \prod_{v=1}^{r} \left(M^{-\frac{d}{2}(j+1)(q_v-2)} 3 \, M^{\frac{d}{2}j(q_v-2)} |\lambda|^{\frac{q_v}{2m}} \right) \right\}$$

$$= |\lambda|^{\frac{q}{2m}} \sup_{\substack{1 \le r \le n \\ q_v \ge 3, \, \sum q_v \le nm}} \left\{ 2^{11 \sum_v q_v} \, c_M^{\sum_v q_v - q} |\lambda|^{\sum_v \frac{q_v}{2m} - \frac{q}{2m}} \prod_{v=1}^{r} \left(3 \, M^{-\frac{d}{2}(q_v-2)} \right) \right\}$$

$$\le |\lambda|^{\frac{q}{2m}} \sup_{\substack{1 \le r \le n \\ q_v \ge 3, \, \sum q_v \le nm}} \left\{ 2^{11 \sum_v q_v} \, c_M^{\sum_v q_v - q} |\lambda|^{\sum_v \frac{q_v}{2m} - \frac{q}{2m}} \prod_{v=1}^{r} \left(3^{q_v} \, M^{-\frac{d}{2}\frac{q_v}{3}} \right) \right\}$$

$$= |\lambda|^{\frac{q}{2m}} \sup_{\substack{1 \le r \le n \\ q_v \ge 3, \, \sum q_v \le nm}} \left\{ \left(3 \cdot 2^{11} M^{-\frac{d}{6}} \right)^{\sum_v q_v} \left(|\lambda|^{\frac{1}{2m}} c_M \right)^{\sum_v q_v - q} \right\}$$

$$\le |\lambda|^{\frac{q}{2m}} \tag{9.88}$$

if we first choose

$$M^{\frac{d}{6}} \ge 3 \cdot 2^{11} \tag{9.89}$$

and then let

$$|\lambda| \le \frac{1}{c_M^{2m}} \tag{9.90}$$

This proves (9.81).

We now turn to (9.82). For $j = 0$, we obtain as in (9.83)

$$\|V_{2q,n}^{j=0}\|_S \le \left(2^{11m} c_M^m \right)^n \prod_{v=1}^{n} \|V\|_{S_v}$$

$$\le \left(2^{11m} c_M^m \|V\|_\emptyset \right)^n \prod_{i \in S} \|f_i\|_{1,\infty} \tag{9.91}$$

which proves (9.82) for $j = 0$ if we choose λ as in (9.84). Suppose now that (9.82) holds for $0 \le i \le j$. From (9.36) and (9.77) we get

$$\|V_{2q}^{2,4,j+1}\|_{S, \le n} \le M^{\frac{d}{2}(j+1)(q - \frac{|S|}{2})} \sup_{\substack{1 \le r+s \le n, \, s \ge 1 \\ q_v \ge 3, \, \sum q_v \le nm \\ |S_v| < 2q_v}} \left\{ 2^{11 \sum_v q_v} \, c_M^{\sum_v q_v - q} \times \right.$$

$$\prod_{v=1}^{r} \left(M^{-\frac{d}{2}(j+1)(q_v-2)} \|W_{2q_v}^{2,4,\le j}\|_{\emptyset, \le n} \right) \times$$

$$\left. \prod_{v=r+1}^{r+s} \left(M^{-\frac{d}{2}(j+1)(q_v - \frac{|S_v|}{2})} \|W_{2q_v}^{2,4,\le j}\|_{S_v, \le n} \right) \right\} \tag{9.92}$$

By the induction hypothesis (9.82) we obtain, for $|S_v| < 2q_v$,

$$\|W_{2q_v}^{2,4,\,\leq j}\|_{S_v,\,\leq n} \leq \|W_{2q_v}^{2,4,\,\leq j}\|_{S_v}$$

$$\leq \sum_{i=0}^{j} \|V_{2q_v}^{2,4,i}\|_{S_v} + \chi(q_v \leq m)\,|\lambda|\,\|V_{2q_v}\|_{S_v}$$

$$\leq \left\{ \sum_{i=0}^{j} |\lambda|^{\frac{q_v}{2m}} M^{\frac{d}{2}i(q_v - \frac{|S_v|}{2} - \frac{1}{4})} + \chi(q_v \leq m)\,|\lambda|^{\frac{1}{2}} \right\} \prod_{i \in S_v} \|f_i\|_{1,\infty}$$

$$\leq \left\{ M^{\frac{d}{2}j(q_v - \frac{|S_v|}{2} - \frac{1}{4})} \frac{|\lambda|^{\frac{q_v}{2m}}}{1 - M^{-\frac{d}{2}(q_v - \frac{|S_v|}{2} - \frac{1}{4})}} + |\lambda|^{\frac{q_v}{2m}} \right\} \prod_{i \in S_v} \|f_i\|_{1,\infty}$$

$$\leq \left\{ M^{\frac{d}{2}j(q_v - \frac{|S_v|}{2} - \frac{1}{4})} \frac{|\lambda|^{\frac{q_v}{2m}}}{1 - M^{-\frac{d}{8}}} + |\lambda|^{\frac{q_v}{2m}} \right\} \prod_{i \in S_v} \|f_i\|_{1,\infty}$$

$$\leq 3|\lambda|^{\frac{q_v}{2m}} M^{\frac{d}{2}j(q_v - \frac{|S_v|}{2} - \frac{1}{4})} \prod_{i \in S_v} \|f_i\|_{1,\infty} \tag{9.93}$$

if we choose $M^{\frac{d}{8}} \geq 2$. Substituting (9.93) and (9.87) into (9.92), we obtain

$$\|V_{2q}^{2,4,j+1}\|_{S,\,\leq n} \leq M^{\frac{d}{2}(j+1)(q - \frac{|S|}{2})} \sup_{\substack{1 \leq r+s \leq n,\, s \geq 1 \\ q_v \geq 3,\, \sum q_v \leq nm \\ |S_v| < 2q_v}} \left\{ 2^{11 \sum_v q_v}\, c_M^{\sum_v q_v - q} \times \right.$$

$$\prod_{v=1}^{r} \left(3M^{-\frac{d}{2}(q_v - 2)}|\lambda|^{\frac{q_v}{2m}} \right) \times$$

$$\left. \prod_{v=r+1}^{r+s} \left(3M^{-\frac{d}{2}(q_v - \frac{|S_v|}{2})} M^{-\frac{d}{8}j}|\lambda|^{\frac{q_v}{2m}} \prod_{i \in S_v} \|f_i\|_{1,\infty} \right) \right\}$$

$$\leq M^{\frac{d}{2}(j+1)(q - \frac{|S|}{2})} M^{-\frac{d}{8}(j+1)}|\lambda|^{\frac{q}{2m}} \times$$

$$\sup_{\substack{1 \leq r+s \leq n,\, s \geq 1 \\ q_v \geq 3,\, \sum q_v \leq nm \\ |S_v| < 2q_v}} \left\{ 2^{11 \sum_v q_v} \left(|\lambda|^{\frac{1}{2m}} c_M \right)^{\sum_v q_v - q} \prod_{v=1}^{r} \left(3M^{-\frac{d}{6}q_v} \right) \times \right.$$

$$\left. \prod_{v=r+1}^{r+s} \left(3M^{-\frac{d}{2}(q_v - \frac{|S_v|}{2})} M^{\frac{d}{8}} \right) \right\} \prod_{i \in S} \|f_i\|_{1,\infty} \tag{9.94}$$

where we used the fact that there is at least one external vertex, $s \geq 1$, to make the factor $M^{-\frac{d}{8}(j+1)}$ explicit. Since $|S_v| < 2q_v$, we have

$$\prod_{v=r+1}^{r+s} \left(M^{-\frac{d}{2}(q_v - \frac{|S_v|}{2})} M^{\frac{d}{8}} \right) \leq \prod_{v=r+1}^{r+s} M^{-\frac{d}{4}(q_v - \frac{|S_v|}{2})} \overset{!}{\leq} M^{-\varepsilon \sum_{v=r+1}^{r+s} q_v} \tag{9.95}$$

if

$$\frac{d}{4} \sum_{v=r+1}^{r+s} \left(q_v - \frac{|S_v|}{2} \right) \geq \varepsilon \sum_{v=r+1}^{r+s} q_v$$

$$\Leftrightarrow \quad \sum_{v=r+1}^{r+s} q_v - \frac{|S|}{2} \geq \frac{4\varepsilon}{d} \sum_{v=r+1}^{r+s} q_v$$

$$\Leftrightarrow \quad \sum_{v=r+1}^{r+s} q_v \geq \frac{1}{1-\frac{4\varepsilon}{d}} \frac{|S|}{2} \qquad (9.96)$$

Again, since $|S_v| < 2q_v$, we have

$$\sum_{v=r+1}^{r+s} 2q_v \geq |S| + 1 \qquad (9.97)$$

Thus, (9.96) is satisfied if we can choose ε such that

$$\frac{|S| + 1}{2} \geq \frac{1}{1-\frac{4\varepsilon}{d}} \frac{|S|}{2}$$

$$\Leftrightarrow \quad \varepsilon \leq \frac{d}{4} \frac{1}{|S|+1} \qquad (9.98)$$

Thus, choosing $\varepsilon = \frac{d}{4} \frac{1}{|S|+1} \leq \frac{d}{8}$ we obtain from (9.94)

$$\|V_{2q}^{2,4,j+1}\|_{S,\,\leq n} \leq M^{\frac{d}{2}(j+1)(q-\frac{|S|}{2}-\frac{1}{4})} |\lambda|^{\frac{q}{2m}} \prod_{i\in S} \|f_i\|_{1,\infty} \times$$

$$\sup_{\substack{1\leq r+s\leq n,\, s\geq 1 \\ q_v\geq 3,\, \sum q_v\leq nm \\ |S_v|<2q_v}} \left\{ (3\cdot 2^{11} M^{-\varepsilon})^{\sum_v q_v} \left(|\lambda|^{\frac{1}{2m}} c_M \right)^{\sum_v q_v - q} \right\}$$

$$\leq M^{\frac{d}{2}(j+1)(q-\frac{|S|}{2}-\frac{1}{4})} |\lambda|^{\frac{q}{2m}} \prod_{i\in S} \|f_i\|_{1,\infty} \qquad (9.99)$$

if we first choose M sufficiently large such that

$$M \geq \left(3\cdot 2^{11} \right)^{\frac{1}{\varepsilon}} = \left(3\cdot 2^{11} \right)^{\frac{4}{d}(|S|+1)} \qquad (9.100)$$

and then choose λ sufficiently small such that (9.90) holds. This proves (9.82) provided that

$$|\lambda| \leq \min \left\{ \frac{1}{4\cdot 2^{22m} c_M^{2m} \|V\|_\emptyset^2}, \frac{1}{c_M^{2m}} \right\} \qquad (9.101)$$

In particular, from (9.82) we infer

$$\|W_{2q}^{2,4}\|_{\{1,\cdots,2q\}} \leq \sum_{j=0}^{\infty} \|V^{2,4,j}\|_{\{1,\cdots,2q\}} + \chi(q \leq m)|\lambda|\|V_{2q}\|_{\{1,\cdots,2q\}}$$

$$\leq \left\{ |\lambda|^{\frac{q}{2m}} \sum_{j=0}^{\infty} M^{-\frac{d}{8}j} + |\lambda|^{\frac{q}{2m}} \right\} \prod_{i=1}^{2q} \|f_i\|_{1,\infty}$$

$$\leq 3|\lambda|^{\frac{q}{2m}} \prod_{i=1}^{2q} \|f_i\|_{1,\infty}$$

which proves the theorem. ∎

9.4 Elimination of Divergent Diagrams

In this section we show that one possibility to get rid of two-legged subdiagrams is the addition of a suitable chosen counterterm. The corresponding lemma reads as follows.

Lemma 9.4.1 *Let $\mathcal{V}$ be a quartic interaction,*

$$\mathcal{V}(\psi, \bar{\psi}) = \int d\xi_1 d\xi_2 d\xi_3 d\xi_4 \, V(\xi_1, ..., \xi_4) \bar{\psi}(\xi_1) \bar{\psi}(\xi_2) \psi(\xi_3) \psi(\xi_4) \quad (9.102)$$

and let $\mathcal{Q}$ be a quadratic counterterm,

$$\mathcal{Q}(\psi, \bar{\psi}) = \int d\xi_1 d\xi_2 \, Q(\xi_1, \xi_2) \bar{\psi}(\xi_1) \psi(\xi_2) \quad (9.103)$$

where $Q(\xi_1, \xi_2) = \sum_{n=1}^{\infty} Q_n(\xi_1, \xi_2) \lambda^n$. Consider the generating functional

$$\mathcal{G}(\eta) = \log \int e^{\lambda \mathcal{V}(\psi+\eta) + \mathcal{Q}(\psi+\eta)} d\mu_C(\psi) \quad (9.104)$$
$$= \mathcal{W}(\eta) + \lambda \mathcal{V}(\eta) + \mathcal{Q}(\eta)$$

where, using the notation $\mathcal{V}(\psi; \eta) := \mathcal{V}(\psi) - \mathcal{V}(\eta)$,

$$\mathcal{W}(\eta) = \log \int e^{\lambda \mathcal{V}(\psi+\eta;\eta) + \mathcal{Q}(\psi+\eta;\eta)} d\mu_C(\psi) \quad (9.105)$$

$$= \sum_{q=0}^{\infty} \int d\xi_1 \cdots d\xi_{2q} \, W_{2q}(\xi_1, ..., \xi_{2q}) \bar{\eta}(\xi_1) \cdots \eta(\xi_{2q})$$

and $d\mu_C$ is the Grassmann Gaussian measure for some covariance C. Then:

a) *Let $W_{2,n} = W_{2,n}(\xi_1, \xi_2)$ be the n'th order contribution to the expansion $W_2 = \sum_{n=1}^{\infty} W_{2,n} \lambda^n$. Then, as a function of the Q_j's, $W_{2,n}$ depends only on $Q_1, ..., Q_{n-1}$,*

$$W_{2,n} = W_{2,n}(Q_1, ..., Q_{n-1}) \quad (9.106)$$

b) *Let ℓ be some operation on the $W_{2,n}$'s, for example $\ell = id$ or $\ell W_2 = \int d\xi \, W_2(\xi, 0)$. Let $\mathbf{r} = id - \ell$ be the renormalization operator. Determine the Q_n's inductively by the equation*

$$Q_n = -\ell W_{2,n}(Q_1, ..., Q_{n-1}) \qquad (9.107)$$

Then $G_2 = \mathbf{r}W_2$ and W_{2q} is given by the sum of all $2q$-legged connected amputated diagrams where each two-legged subdiagram has to be evaluated with the $\mathbf{r}$ operator. In particular, for $\ell = id$ one has $G_2 = 0$, the full two-point function of the changed model (by the addition of the counterterm Q) coincides with the free propagator, and G_{2q} is the sum of all $2q$-legged, connected amputated diagrams without two-legged subdiagrams.

Proof: a) W_2 is given by the sum of all two-legged connected amputated diagrams made from four-legged vertices $\mathcal{V}$ coming from the interaction and two-legged vertices $\mathcal{Q}$ coming from the counterterm. To obtain an n'th order diagram, there may be k four-legged vertices and one Q_{n-k} or $n - k$ Q_1's for example. If there is a Q_n, there cannot be anything else. Since all diagrams contributing to W have at least one contraction, this diagram would be a loop with no external legs, thus it does not contribute to $W_{2,n}$. This proves part (a).

b) Consider a diagram $\Gamma_{2q,r}$ of some order r contributing to $W_{2q,r}$. It is made from four-legged vertices $\mathcal{V}$ and two-legged vertices $\mathcal{Q}$. Suppose this diagram has a two-legged subdiagram $T = T_n$ which is of order n. We write $\Gamma_{2q,r} = S_{2q,r-n} \cup T_n$ to distinguish between the two-legged subdiagram and the rest of the diagram. Now consider the sum of all diagrams which are obtained by combining $S_{2q,r-n}$ with an arbitrary n'th order two-legged diagram. Apparently, this sum is given by $S_{2q,r-n} \cup W_{2,n} + S_{2q,r-n} \cup Q_n = S_{2q,r-n} \cup (W_{2,n} + Q_n) = S_{2q,r-n} \cup \mathbf{r}W_{2,n}$. Thus, part (b) follows. ■

Thus, we can eliminate two-legged subdiagrams by the addition of a suitable chosen counterterm. Typically, for infrared problems the counterterms are finite. Using the techniques of this chapter (plus a sectorization of the Fermi surface which is needed to implement conservation of momentum) one can show [18, 20, 16] that for the many-electron system with short-range interaction the renormalized sum of all diagrams without four-legged subgraphs is still well behaved, that is, its lowest order terms are a good approximation. This holds for $e_{\mathbf{k}} = \mathbf{k}^2/(2m) - \mu$ as well as for an $e_{\mathbf{k}}$ with an anisotropic Fermi surface. The theorem reads as follows:

Theorem 9.4.2 *Let $d = 2$, $V(x) \in L^1$ and $C(k) = 1/(ik_0 - e_{\mathbf{k}})$. Then there is a finite counterterm $\delta e(\lambda, \mathbf{k}) = \sum_{n=1}^{\infty} \delta e_n(\mathbf{k}) \lambda^n$, $\sum_{n=1}^{\infty} \|\delta e_n\|_{\infty} |\lambda|^n < \infty$, such that the perturbation theory of the model inductively defined by*

$$\mathcal{W}^0(\psi) = -\frac{\lambda}{\beta L^d} \sum_{k,p,q} v_{\mathbf{k}-\mathbf{p}} \bar{\psi}_{k\uparrow} \bar{\psi}_{q-k\downarrow} \psi_{p\uparrow} \psi_{q-p\downarrow} + \sum_{k,\sigma} \delta e(\mathbf{k}, \lambda) \bar{\psi}_{k\sigma} \psi_{k\sigma} \quad (9.108)$$

$$\mathcal{W}^{j+1}(\psi^{\geq j+1}) = (Id - P_4) \log \int e^{\mathcal{W}^j(\psi^{\geq j+1} + \psi^j)} d\mu_{C^j}(\psi^j) \quad (9.109)$$

has positive radius of convergence independent of volume and temperature. The projection P_4 is defined by

$$P_4 \sum_{q=2}^{\infty} \sum_{k_1,\dots,k_{2q}} W_{2q}(k_1, \dots, k_{2q}) \bar{\psi}_{k_1} \psi_{k_2} \cdots \bar{\psi}_{k_{2q-1}} \psi_{k_{2q}} := \quad (9.110)$$

$$\sum_{k_1,\dots,k_4} W_4(k_1, \dots, k_4) \bar{\psi}_{k_1} \psi_{k_2} \bar{\psi}_{k_3} \psi_{k_4}$$

Proof: The proof can be found in the research literature [18, 20, 16]. ∎

9.5 The Feldman-Knörrer-Trubowitz Fermi Liquid Construction

In the preceding sections we saw that the sum of all diagrams without two- and four-legged subgraphs is analytic. Two-legged subgraphs have to be renormalized which can be done by the addition of a counterterm. By doing this one obtains the theorem above, the renormalized sum of all diagrams without four-legged subgraphs is also analytic for sufficiently small coupling, in two dimensions and for short-range interaction. This is true for the model with kinetic energy $e_{\mathbf{k}} = \mathbf{k}^2/(2m) - \mu$ which has a round Fermi surface $F = \{\mathbf{k} \,|\, e_{\mathbf{k}} = 0\}$ but also holds for models with a more general $e_{\mathbf{k}}$ which may have an anisotropic Fermi surface. Thus, the last and the most complicated step in the perturbative analysis consists of adding in the four-legged diagrams. These diagrams determine the physics of the model.

In section 8.4 we saw that ladder diagrams, for $e_{\mathbf{k}} = \mathbf{k}^2/(2m) - \mu$, have a logarithmic singularity at zero transfer momentum which may lead to an $n!$. This singularity is a consequence of the possibility of forming Cooper pairs: two electrons, with opposite momenta $\mathbf{k}$ and $-\mathbf{k}$, with an effective interaction which has an attractive part, may form a bound state. Since at low temperatures only those momenta close to the Fermi surface are relevant, the formation of Cooper pairs can be suppressed, if one substitutes (by hand) the energy momentum relation $e_{\mathbf{k}} = \mathbf{k}^2/(2m) - \mu$ by a more general expression with an anisotropic Fermi surface. That is, if momentum $\mathbf{k}$ is on the Fermi surface, then momentum $-\mathbf{k}$ is not on F for almost all $\mathbf{k}$, see the following figure:

196

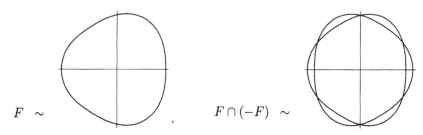

$F \sim$, $F \cap (-F) \sim$

Figure 9.4: An allowed Fermi surface for the FKT-Fermi liquid theorem

For such an $e_\mathbf{k}$ one can prove that four-legged subdiagrams no longer produce any factorials, an n'th order diagram without two-legged but not necessarily without four-legged subgraphs is bounded by $const^n$. As a result, Feldman, Knörrer and Trubowitz could prove that, in two space dimensions, the renormalized perturbation series for such a model has in fact a small positive radius of convergence and that the momentum distribution $\langle a^+_{\mathbf{k}\sigma} a_{\mathbf{k}\sigma} \rangle$ has a jump discontinuity across the Fermi surface of size $1 - \delta_\lambda$ where $\delta_\lambda > 0$ can be chosen arbitrarily small if the coupling λ is made small.

The complete rigorous proof of this fact is a larger technical enterprise [20]. It is distributed over a series of ten papers with a total volume of 680 pages. Joel Feldman has set up a webpage www.math.ubc.ca/~feldman/fl.html where all the relevant material can be found. The introductory paper 'A Two Dimensional Fermi Liquid, Part 1: Overview' gives the precise statement of results and illustrates, in several model computations, the main ingredients and difficulties of the proof.

As FKT remark in that paper, this theorem is still not the complete story. Since two-legged subdiagrams have been renormalized by the addition of a counterterm, the model has been changed. Because $e_\mathbf{k}$ has been chosen anisotropic, also the counterterm $\delta e(\mathbf{k})$ is a nontrivial function of $\mathbf{k}$, not just a constant. Thus, one is led to an invertability problem: for given $e(\mathbf{k})$, is there a $\tilde{e}(\mathbf{k})$ such that $\tilde{e}(\mathbf{k}) + \delta\tilde{e}(\mathbf{k}) = e(\mathbf{k})$? This is a very difficult problem if it is treated on a rigorous level [28, 55]. In fact, if we would choose $\ell = id$ in (9.107) (which we only could do in the Grassmann integral formulation of the model, since in that case the counterterm would depend on the $d + 1$-dimensional momenta $(k_0, \mathbf{k})$, as the quadratic action in the Grassmann integral, but $e_\mathbf{k}$ in the Hamiltonian only depends on the d-dimensional momentum $\mathbf{k}$), then $\mathbf{r} = 0$ and two-legged subgraphs were eliminated completely. In that case the interacting two-point function would be exactly given by $\left(ik_0 - e_\mathbf{k}\right)^{-1}$ if the quadratic part of the action is chosen to be $ik_0 - e_\mathbf{k} +$ counterterm$(k_0, \mathbf{k})$, the counterterm being the sum of all two-legged diagrams without two-legged subdiagrams. Thus in that case the computation of the two-point function would be completely equivalent to the solution of an invertibility problem.

Chapter 10

Resummation of Perturbation Series

10.1   Starting Point and Typical Examples

In the preceding chapters we wrote down the perturbation series for the partition function and for various correlation functions of the many-electron system, and we identified the large contributions of the perturbation series. These contributions have the effect that, as usual, the lowest order terms of the naive perturbation expansion are not a good approximation.

We found that, for a short range interaction, n'th order diagrams without two- and four-legged subgraphs allow a $const^n$ bound, diagrams with four- but without two-legged subgraphs are also finite but they may produce a factorial, that is, there are diagrams with four-legged subgraphs which behave like $n!\, const^n$. And finally diagrams with two-legged subdiagrams are in general infinite if the cutoffs are removed.

Typically one can expect the following: If one has a sum of diagrams where each diagram allows a $const^n$ bound, then this sum is asymptotic. That is, its lowest order contributions are a good approximation if the coupling is small. If the model is fermionic, the situation may even be better and, due to sign cancellations, the series may have a positive radius of convergence which is independent of the cutoffs. If one has a sum of diagrams where all diagrams are finite but certain diagrams behave like $n!\, const^n$, one can no longer expect that the lowest order terms are a good approximation. And, of course, the same conclusion holds if the series contains divergent contributions.

Basically there are two strategies to eliminate the anomalously large contributions of the perturbation series, the use of counterterms and the use of integral equations. In the first case one changes the model and the computation of the correlation functions of the original model requires the solution of an invertability problem which, depending on the model, may turn out to be very difficult [28, 55]. In the second case one resums in particular the divergent contributions of the perturbation series, if this is possible somehow, which leads to integral equations for the correlation functions. Conceptually, this is nothing else than a rearrangement of the perturbation series, like in the example of section 8.1 (which, of course, is an oversimplification since in that case one does not have to solve any integral equation).

The good thing in having integral equations is that the renormalization is done more or less automatically. The correlation functions are obtained from a system of integral equations whose solution can have all kinds of nonanalytic terms (which are responsible for the divergence of the coefficients in the naive perturbation expansion). If one works with counterterms one more or less has to know the answer in advance in order to choose the right counterterms. However, the bad thing with integral equations is that usually it is impossible to get a closed system of equations without making an uncontrolled approximation. If one tries to get an integral equation for a two-point function, one gets an expression with two- and four-point functions. Then, dealing with the four-point function, one obtains an expression with two-, four- and six-point functions and so on. Thus, in order to get a closed system of equations, at some place one is forced to approximate a, say, six-point function by a sum of products of two- and four-point functions.

In this chapter, which is basically the content of [46], we present a somewhat novel approach which also results in integral equations for the correlation functions. It applies to models where a two point function can be written as

$$S(x,y) = \int [P + Q]_{x,y}^{-1} \, d\mu(Q) \, . \tag{10.1}$$

Here P is some operator diagonal in momentum space, typically determined by the unperturbed Hamiltonian, and Q is diagonal in coordinate space. The functional integral is taken with respect to some probability measure $d\mu(Q)$ and goes over the matrix elements of Q. $[\cdot]_{x,y}^{-1}$ denotes the x, y-entry of the matrix $[P + Q]^{-1}$. The starting point is always a model in finite volume and with positive lattice spacing in which case the operator $P + Q$ and the functional integral in (10.1) becomes huge- but finite-dimensional. In the end we take the infinite volume limit and, if wanted, the continuum limit.

Models of this type are for example the many-electron system, the φ^4-model and QED. In Theorem 5.1.2 we showed, using a Hubbard-Stratonovich transformation, that the two-point function of the many-electron system with delta interaction can be written in momentum space as

$$\frac{1}{\beta L^d} \langle \bar\psi_{k\sigma} \psi_{k\sigma} \rangle = \int \left[\begin{matrix} a_p \, \delta_{p,p'} & \frac{ig}{\sqrt{\beta L^d}} \, \bar\phi_{p'-p} \\ \frac{ig}{\sqrt{\beta L^d}} \, \phi_{p-p'} & a_{-p} \, \delta_{p,p'} \end{matrix} \right]^{-1}_{k\sigma,k\sigma} dP(\{\phi_q\}) \tag{10.2}$$

where $a_p = ip_0 - (\mathbf{p}^2/2m - \mu)$ and $dP(\{\phi_q\})$ is the normalized measure

$$dP(\{\phi_q\}) = \frac{1}{Z} \det \left[\begin{matrix} a_p \, \delta_{p,p'} & \frac{ig}{\sqrt{\beta L^d}} \, \bar\phi_{p'-p} \\ \frac{ig}{\sqrt{\beta L^d}} \, \phi_{p-p'} & a_{-p} \, \delta_{p,p'} \end{matrix} \right] e^{-\sum_q |\phi_q|^2} \prod_q d\phi_q d\bar\phi_q \tag{10.3}$$

For the φ^4-model, one can also make a Hubbard-Stratonovich transformation to make the action quadratic in the φ's and then integrate out the φ's. One ends up with (see section 10.5 below)

$$\langle \varphi_k \varphi_{-k} \rangle = \int \left[a_p \, \delta_{p,p'} - \frac{ig}{\sqrt{L^d}} \gamma_{p-p'} \right]_{k,k}^{-1} dP(\{\gamma_q\}) \qquad (10.4)$$

where now $a_p = 4M^2 \sum_{i=1}^{d} \sin^2(p_i/2M) + m^2$, $1/M$ being the lattice spacing in coordinate space, and $dP(\{\gamma_q\})$ is the normalized measure $(\gamma_{-q} = \bar{\gamma}_q)$

$$dP(\{\gamma_q\}) = \frac{1}{Z} \det \left[a_p \, \delta_{p,p'} - \frac{ig}{\sqrt{L^d}} \gamma_{p-p'} \right]^{-\frac{1}{2}} e^{-\sum_q |\gamma_q|^2} \prod_q d\gamma_q d\bar{\gamma}_q \quad (10.5)$$

Thus for both models one has to integrate an inverse matrix element. The only difference is that in the bosonic model, the functional determinant in the integration measure has a negative exponent, here -1/2 since we consider the real scalar φ^4-model, and for the fermionic model the exponent is positive, +1. An example where the integrand is still an inverse matrix element but where the integration measure is simply Gaussian without a functional determinant is given by the averaged Green's function of the Anderson model. In momentum space, it reads (see section 10.3 below)

$$\langle G \rangle (k) = \langle G \rangle (k; E, \lambda, \varepsilon) = \int_{\mathbb{R}^{L^d}} \left[a_p \, \delta_{p,p'} + \frac{\lambda}{\sqrt{L^d}} v_{p-p'} \right]_{k,k}^{-1} \prod_q e^{-|v_q|^2} \frac{dv_q d\bar{v}_q}{\pi}$$

where, for unit lattice spacing in coordinate space, $a_p = 4 \sum_{i=1}^{d} \sin^2(p_i/2) - E - i\varepsilon$. The rigorous control of $\langle G \rangle (k)$ for small disorder λ and energies inside the spectrum of the unperturbed Laplacian, $E \in [0, 4d]$, in which case a_p has a root if $\varepsilon \to 0$, is still an open problem [2, 41, 48, 52, 60]. It is expected that $\lim_{\varepsilon \searrow 0} \lim_{L \to \infty} \langle G \rangle (k) = 1/(a_k - \sigma_k)$ where $\text{Im}\sigma = O(\lambda^2)$.

In all three cases mentioned above one has to invert a matrix and perform a functional integral. Now the main idea of our method is to invert the above matrices with the following formula:

Let $B = (B_{kp})_{k,p \in \mathcal{M}} \in \mathbb{C}^{N \times N}$, $\mathcal{M}$ some index set, $|\mathcal{M}| = N$ and let

$$G(k) := \left[B^{-1} \right]_{kk} \qquad (10.6)$$

Then one has

$$G(k) = \qquad (10.7)$$

$$\frac{1}{B_{kk} + \sum_{r=2}^{N} (-1)^{r+1} \sum_{\substack{p_2 \cdots p_r \in \mathcal{M} \setminus \{k\} \\ p_i \neq p_j}} B_{kp_2} G_k(p_2) B_{p_2 p_3} \cdots B_{p_{r-1} p_r} G_{kp_2 \cdots p_{r-1}}(p_r) B_{p_r k}}$$

where $G_{k_1 \cdots k_j}(p) = \left[(B_{st})_{s,t \in \mathcal{M} \setminus \{k_1 \cdots k_j\}} \right]_{pp}^{-1}$ is the p, p entry of the inverse of the matrix which is obtained from B by deleting the rows and columns labelled by $k_1, \cdots, k_j$. For the Anderson model, the matrix B is of the form $B = $ self adjoint $+ i\varepsilon \, Id$, which, for $\varepsilon \neq 0$, has the property that all submatrices $(B_{st})_{s,t \in \mathcal{M} \setminus \{k_1 \cdots k_j\}}$ are invertible.

There is also a formula for the off-diagonal inverse matrix elements. It reads

$$\left[B^{-1}\right]_{kp} = -G(k)B_{kp}G_k(p) + \qquad\qquad\qquad (10.8)$$

$$\sum_{r=3}^{N} (-1)^{r+1} \sum_{\substack{t_3 \cdots t_r \in \mathcal{M} \setminus \{k,p\} \\ t_i \neq t_j}} G(k)B_{kt_3}G_k(t_3)B_{t_3 t_4} \cdots B_{t_r p}G_{kt_3 \cdots t_r}(p)$$

These formulae also hold in the case where the matrix B has a block structure $B_{kp} = (B_{k\sigma, p\tau})$ where, say, $\sigma, \tau \in \{\uparrow, \downarrow\}$ are some spin variables. In that case the B_{kp} are small matrices, the $G_{k_1 \cdots k_j}(p)$ are matrices of the same size and the $1/\cdot$ in (10.7) means inversion of matrices.

In the next section we prove the inversion formulae (10.7), (10.8) and in the subsequent sections we apply them to the above mentioned models.

10.2 Computing Inverse Matrix Elements

10.2.1 An Inversion Formula

In this section we prove the inversion formulae (10.7), (10.8). We start with the following

Lemma 10.2.1 *Let $B \in \mathbb{C}^{k \times n}$, $C \in \mathbb{C}^{n \times k}$ and let Id_k denote the identity in $\mathbb{C}^{k \times k}$. Then:*

(i) $Id_k - BC$ *invertible* $\Leftrightarrow$ $Id_n - CB$ *invertible.*

(ii) *If the left or the right hand side of (i) fullfilled, then $C \frac{1}{Id_k - BC} = \frac{1}{Id_n - CB} C$.*

Proof: Let

$$B = \begin{pmatrix} - \, \vec{b}_1 \, - \\ \vdots \\ - \, \vec{b}_k \, - \end{pmatrix}, \qquad C = \begin{pmatrix} | & & | \\ \vec{c}_1 & \cdots & \vec{c}_k \\ | & & | \end{pmatrix}$$

where the $\vec{b}_j$ are n-component row vectors and the $\vec{c}_j$ are n-component column vectors. Let $\vec{x} \in \mathrm{Kern}(Id - CB)$. Then $\vec{x} = CB\vec{x} = \sum_j \lambda_j \vec{c}_j$ if we define $\lambda_j := (\vec{b}_j, \vec{x})$. Let $\vec{\lambda} = (\lambda_j)_{1 \le j \le k}$. Then $[(Id - BC)\vec{\lambda}]_i = \lambda_i - \sum_j (\vec{b}_i, \vec{c}_j)\lambda_j = $

$(\vec{b}_i, \vec{x}) - \sum_j (\vec{b}_i, \vec{c}_j)\lambda_j = 0$ since $\vec{x} = \sum_j \lambda_j \vec{c}_j$, thus $\vec{\lambda} \in \mathrm{Kern}(Id - BC)$. On the other hand, if some $\vec{\lambda} \in \mathrm{Kern}(Id - BC)$, then $\vec{x} := \sum_j \lambda_j \vec{c}_j \in \mathrm{Kern}(Id - CB)$ which proves (i). Part (ii) then follows from $C = \frac{1}{Id_n - CB}(Id_n - CB)C = \frac{1}{Id_n - CB}C(Id_k - BC)$ ∎

The inversion formula (10.7) is obtained by iterative application of the next lemma, which states the Feshbach formula for finite dimensional matrices. For a more general version one may look in [4], Theorem 2.1.

Lemma 10.2.2 *Let* $h = \begin{pmatrix} A & B \\ C & D \end{pmatrix} \in \mathbb{C}^{n\times n}$ *where* $A \in \mathbb{C}^{k\times k}$, $D \in \mathbb{C}^{(n-k)\times(n-k)}$ *are invertible and* $B \in \mathbb{C}^{k\times(n-k)}$, $C \in \mathbb{C}^{(n-k)\times k}$. *Then*

$$h \text{ invertible} \quad \Leftrightarrow \quad A - BD^{-1}C \text{ invertible} \tag{10.9}$$
$$\Leftrightarrow \quad D - CA^{-1}B \text{ invertible}$$

and if one of the conditions in (10.9) is fullfilled, one has $h^{-1} = \begin{pmatrix} E & F \\ G & H \end{pmatrix}$ *where*

$$E = \frac{1}{A - BD^{-1}C}, \qquad H = \frac{1}{D - CA^{-1}B}, \tag{10.10}$$
$$F = -EBD^{-1} = -A^{-1}BH, \qquad G = -HCA^{-1} = -D^{-1}CE. \tag{10.11}$$

Proof: We have, using Lemma 10.2.1 in the second line,

$$A - BD^{-1}C \text{ inv.} \Leftrightarrow Id_k - A^{-1}BD^{-1}C \text{ inv.}$$
$$\Leftrightarrow Id_{n-k} - D^{-1}CA^{-1}B \text{ inv.}$$
$$\Leftrightarrow D - CA^{-1}B \text{ inv.}$$

Furthermore, again by Lemma 10.2.1,

$$D^{-1}C\frac{1}{Id - A^{-1}BD^{-1}C} = \frac{1}{Id - D^{-1}CA^{-1}B}D^{-1}C = \frac{1}{D - CA^{-1}B}C = HC$$

and

$$A^{-1}B\frac{1}{Id - D^{-1}CA^{-1}B} = \frac{1}{Id - A^{-1}BD^{-1}C}A^{-1}B = \frac{1}{A - BD^{-1}C}B = EB$$

which proves the last equalities in (10.11), $HCA^{-1} = D^{-1}CE$ and $EBD^{-1} = A^{-1}BH$. Using these equations and the definition of E, F, G and H one computes

$$\begin{pmatrix} A & B \\ C & D \end{pmatrix}\begin{pmatrix} E & F \\ G & H \end{pmatrix} = \begin{pmatrix} E & F \\ G & H \end{pmatrix}\begin{pmatrix} A & B \\ C & D \end{pmatrix} = \begin{pmatrix} Id & 0 \\ 0 & Id \end{pmatrix}$$

It remains to show that the invertibility of h implies the invertibility of $A - BD^{-1}C$. To this end let $P = \begin{pmatrix} Id & 0 \\ & 0 \end{pmatrix}$, $\bar{P} = \begin{pmatrix} 0 & \\ & Id \end{pmatrix}$ such that $A - BD^{-1}C = PhP - Ph\bar{P}(\bar{P}h\bar{P})^{-1}\bar{P}hP$. Then

$$
\begin{aligned}
(A - BD^{-1}C)Ph^{-1}P &= PhPh^{-1}P - Ph\bar{P}(\bar{P}h\bar{P})^{-1}\bar{P}hPh^{-1}P \\
&= Ph(1 - \bar{P})h^{-1}P - Ph\bar{P}(\bar{P}h\bar{P})^{-1}\bar{P}h(1 - \bar{P})h^{-1}P \\
&= P - Ph\bar{P}h^{-1}P + Ph\bar{P}h^{-1}P = P
\end{aligned}
$$

and similarly $Ph^{-1}P(A - BD^{-1}C) = P$ which proves the invertibility of $A - BD^{-1}C$ ∎

Theorem 10.2.3 *Let $B \in \mathbb{C}^{nN \times nN}$ be given by $B = (B_{kp})_{k,p \in \mathcal{M}}$, $\mathcal{M}$ some index set, $|\mathcal{M}| = N$, and $B_{kp} = (B_{k\sigma,p\tau})_{\sigma,\tau \in I} \in \mathbb{C}^{n \times n}$ where I is another index set, $|I| = n$. Suppose that B and, for any $\mathcal{N} \subset \mathcal{M}$, the submatrix $(B_{kp})_{k,p \in \mathcal{N}}$ is invertible. For $k \in \mathcal{M}$ let*

$$
G(k) := [B^{-1}]_{kk} \in \mathbb{C}^{n \times n} \tag{10.12}
$$

and, if $\mathcal{N} \subset \mathcal{M}$, $k \notin \mathcal{N}$,

$$
G_{\mathcal{N}}(k) := \left[\{(B_{st})_{s,t \in \mathcal{M} \setminus \mathcal{N}}\}^{-1} \right]_{kk} \in \mathbb{C}^{n \times n} \tag{10.13}
$$

Then one has the following:

(i) *The on-diagonal block matrices of B^{-1} are given by*

$$
G(k) = \tag{10.14}
$$

$$
\frac{1}{B_{kk} - \sum_{r=2}^{N}(-1)^r \sum_{\substack{p_2 \cdots p_r \in \mathcal{M} \setminus \{k\} \\ p_i \neq p_j}} B_{kp_2} G_k(p_2) B_{p_2 p_3} \cdots B_{p_{r-1} p_r} G_{kp_2 \cdots p_{r-1}}(p_r) B_{p_r k}}
$$

where $1/\cdot$ is inversion of $n \times n$ matrices.

(ii) *Let $k, p \in \mathcal{M}$, $k \neq p$. Then the off-diagonal block matrices of B^{-1} can be expressed in terms of the $G_{\mathcal{N}}(s)$ and the B_{st},*

$$
[B^{-1}]_{kp} = -G(k)B_{kp}G_k(p) \tag{10.15}
$$

$$
- \sum_{r=3}^{N}(-1)^r \sum_{\substack{t_3 \cdots t_r \in \mathcal{M} \setminus \{k,p\} \\ t_i \neq t_j}} G(k)B_{kt_3}G_k(t_3)B_{t_3 t_4} \cdots B_{t_r p}G_{kt_3 \cdots t_r}(p)
$$

Proof: Let k be fixed and let $p, p' \in \mathcal{M} \setminus \{k\}$ below label columns and rows. By Lemma 10.2.2 we have

$$
\begin{pmatrix} G(k) & \\ & * \end{pmatrix} = \begin{pmatrix} B_{kk} & - & B_{kp} & - \\ & | & \\ B_{p'k} & & B_{p'p} & \\ & | & \end{pmatrix}^{-1} = \begin{pmatrix} E & - & F & - \\ & | & \\ G & & H & \\ & | & \end{pmatrix}
$$

where

$$
G(k) = E = \cfrac{1}{B_{kk} - \sum_{p,p' \neq k} B_{kp} \left[\{(B_{p'p})_{p',p \in \mathcal{M}\setminus\{k\}}\}^{-1} \right]_{pp'} B_{p'k}} \tag{10.16}
$$

$$
= \cfrac{1}{B_{kk} - \sum_{p \neq k} B_{kp} G_k(p) B_{pk} - \sum_{\substack{p,p' \neq k \\ p \neq p'}} B_{kp} \left[\{(B_{p'p})_{p',p \in \mathcal{M}\setminus\{k\}}\}^{-1} \right]_{pp'} B_{p'k}}
$$

and

$$
F_{kp} = \left[B^{-1} \right]_{kp} = -G(k) \sum_{t \neq k} B_{kt} \left[\{(B_{p'p})_{p',p \in \mathcal{M}\setminus\{k\}}\}^{-1} \right]_{tp} \tag{10.17}
$$

$$
= -G(k) B_{kp} G_k(p) - G(k) \sum_{t \neq k,p} B_{kt} \left[\{(B_{p'p})_{p',p \in \mathcal{M}\setminus\{k\}}\}^{-1} \right]_{tp}
$$

Apply Lemma 10.2.2 now to the matrix $\{(B_{p'p})_{p',p \in \mathcal{M}\setminus\{k\}}\}^{-1}$ and proceed by induction to obtain after ℓ steps

$$
G(k) = \tag{10.18}
$$

$$
\cfrac{1}{B_{kk} - \sum_{r=2}^{\ell} (-1)^r \sum_{\substack{p_2 \cdots p_r \in \mathcal{M}\setminus\{k\} \\ p_i \neq p_j}} B_{kp_2} G_k(p_2) B_{p_2 p_3} \cdots B_{p_{r-1} p_r} G_{kp_2 \cdots p_{r-1}}(p_r) B_{p_r k} - R_{\ell+1}}
$$

$$
F_{kp} = -G(k) B_{kp} G_k(p) \tag{10.19}
$$

$$
- \sum_{r=3}^{\ell} (-1)^r \sum_{\substack{t_3 \cdots t_r \in \mathcal{M}\setminus\{k,p\} \\ t_i \neq t_j}} G(k) B_{kt_3} G_k(t_3) B_{t_3 t_4} \cdots B_{t_r p} G_{kt_3 \cdots t_r}(p) - \tilde{R}_{\ell+1}
$$

where

$$R_{\ell+1} = (-1)^\ell \sum_{\substack{p_2 \cdots p_{\ell+1} \in \mathcal{M} \backslash \{k\} \\ p_i \neq p_j}} B_{kp_2} G_k(p_2) \cdots B_{p_{\ell-1}p_\ell} \times \tag{10.20}$$

$$\left[\{(B_{p'p})_{p',p \in \mathcal{M} \backslash \{kp_2 \cdots p_\ell\}} \}^{-1} \right]_{p_\ell p_{\ell+1}} B_{p_{\ell+1}k}$$

$$\tilde{R}_{\ell+1} = (-1)^\ell \sum_{\substack{t_3 \cdots t_{\ell+1} \in \mathcal{M} \backslash \{k,p\} \\ t_i \neq t_j}} G(k) B_{kt_3} \cdots G_{kt_3 \cdots t_{\ell-1}}(t_\ell) B_{t_\ell t_{\ell+1}} \times \tag{10.21}$$

$$\left[\{(B_{p'p})_{p',p \in \mathcal{M} \backslash \{kt_3 \cdots t_\ell\}} \}^{-1} \right]_{t_{\ell+1}p}$$

Since $R_{N+1} = \tilde{R}_{N+1} = 0$ the theorem follows. ∎

10.2.2 Field Theoretical Motivation of the Inversion Formula

In many field theoretical models one has to deal with determinants of the form $\det[Id + B]$. The standard approximation formula is

$$\det[Id + B] = e^{Tr \log[Id+B]} \approx e^{\sum_{r=1}^{n} \frac{(-1)^{r+1}}{r} Tr B^r} \tag{10.22}$$

with $n = 1$ or $n = 2$. For large B, this approximation is pretty bad. The reason is that the approximation, say, $\det[Id + B] \approx e^{TrB - \frac{1}{2}TrB^2}$ does not keep track of the fact that in the expansion of $\det[Id + B]$ with respect to B every matrix element of B can only come with a power of 1, but in the expansion of $e^{TrB - \frac{1}{2}TrB^2}$ matrix elements of B with arbitrary high powers are produced. This can be seen already in the one-dimensional situation. In $\det[Id + x] = 1 + x$ the x comes with power 1, but in $e^{\log[1+x]} \approx e^{x - \frac{1}{2}x^2}$, the x comes with arbitrary high powers. However, one can eliminate these higher powers by introducing anticommuting variables, since they have the property that $(\bar{\psi}\psi)^n = 0$ for all $n \geq 2$.

Consider first the one-dimensional case. Introduce the Grassmann algebra with generators ψ and $\bar{\psi}$. We have $e^{\bar{\psi}\psi} = 1 + \bar{\psi}\psi$ and $\int e^{\bar{\psi}\psi} d\bar{\psi} d\psi = 1$ as well as $\int \bar{\psi}\psi \, e^{\bar{\psi}\psi} d\bar{\psi} d\psi = 1$. Thus, abbreviating $d\mu := e^{\bar{\psi}\psi} d\bar{\psi} d\psi$ we may write

$$1 + x = \int (1 + \bar{\psi}\psi \, x) d\mu = \int e^{\log[1 + \bar{\psi}\psi \, x]} d\mu \tag{10.23}$$

Now, if we apply (10.22) for, say, $n = 2$ to the right hand side of (10.23), we end up with

$$\int e^{\log[1 + \bar{\psi}\psi \, x]} d\mu \approx \int e^{\bar{\psi}\psi \, x - \frac{1}{2}(\bar{\psi}\psi \, x)^2} d\mu = \int e^{\bar{\psi}\psi \, x} d\mu = 1 + x \tag{10.24}$$

which is the exact result. The following lemma is the generalization of (10.23) to N-dimensional matrices.

Lemma 10.2.4 *Let* $B = (B_{kp})_{k,p \in \mathcal{M}} \in \mathbb{C}^{N \times N}$, $N = |\mathcal{M}|$. *Then*

a) $$\det[Id + \lambda B] = \sum_{n=0}^{N} \lambda^n \sum_{\{k_1, \cdots, k_n\} \subset M} \det_{k_1 \cdots k_n} B \qquad (10.25)$$

where $\det_{k_1 \cdots k_n} B := \det[(B_{kp})_{k,p \in \{k_1, \cdots, k_n\}}]$.

b) $$\det[Id + B] = \int \det\left[Id + (\bar\psi_k \psi_k B_{kp})_{k,p \in \mathcal{M}}\right] d\mu \qquad (10.26)$$

where $d\mu := e^{\sum_{k \in \mathcal{M}} \bar\psi_k \psi_k} \prod_{k \in \mathcal{M}} d\bar\psi_k d\psi_k$ *and the integral is a Grassmann inte-gral.*

Proof: a) Let $b_p = (B_{kp})_{k \in \mathcal{M}}$ denote the column vectors of B and let $\{e_1, \cdots, e_N\}$ be the standard basis of $\mathbb{C}^N$. Then one has

$$\left(\tfrac{d}{d\lambda}\right)^n_{|\lambda=0} \det[Id + \lambda B] = \sum_{\substack{p_1 \cdots p_n \\ p_i \neq p_j}} \det\begin{bmatrix} | & | & & | & & | \\ e_1 & b_{p_1} & \cdots & b_{p_n} & \cdots & e_N \\ | & | & & | & & | \end{bmatrix}$$

$$= \sum_{\substack{p_1 \cdots p_n \\ p_i \neq p_j}} \det_{p_1 \cdots p_n} B$$

$$= n! \sum_{\{p_1 \cdots p_n\} \subset M} \det_{p_1 \cdots p_n} B$$

b) By part (a)

$$\det[Id + B] = \sum_{n=0}^{N} \sum_{\{k_1, \cdots, k_n\} \subset M} \det_{k_1 \cdots k_n} B$$

In the last sum over $k_1, \cdots, k_n$ one only gets a contribution if all the k's are distinct, because of the determinant $\det_{k_1 \cdots k_n} B$. Since

$$\int \bar\psi_{k_1} \psi_{k_1} \cdots \bar\psi_{k_n} \psi_{k_n} d\mu = \begin{cases} 1 \text{ if all k are distinct} \\ 0 \text{ otherwise} \end{cases}$$

one may write

$$\det[Id + B] = \int \sum_{n=0}^{N} \sum_{\{k_1, \cdots, k_n\}} \bar{\psi}_{k_1} \psi_{k_1} \cdots \bar{\psi}_{k_n} \psi_{k_n} \det{}_{k_1 \cdots k_n} B \, d\mu$$

$$= \int \sum_{n=0}^{N} \sum_{\{k_1, \cdots, k_n\}} \det{}_{k_1 \cdots k_n} \left[(\bar{\psi}_k \psi_k B_{kp}) \right] d\mu$$

$$= \int \det \left[Id + (\bar{\psi}_k \psi_k B_{kp}) \right] d\mu$$

which proves the lemma. ∎

The utility of the above lemma lies in the fact that application of (10.22) to (10.26) gives a better result than a direct application to $\det[Id + B]$. The reason is that in the expansion

$$\det[Id + \lambda B] = e^{Tr \log[Id + \lambda B]} = e^{\sum_{r=1}^{\infty} \frac{(-1)^{r+1}}{r} Tr B^r}$$

$$= \sum_{b_1, b_2, \cdots = 0}^{\infty} \frac{(-1)^{2b_1 + 3b_2 + \cdots}}{b_1! \, 1^{b_1} b_2! \, 2^{b_2} \cdots} [\lambda Tr B]^{b_1} \left[\lambda^2 Tr B^2 \right]^{b_2} \cdots \qquad (10.27)$$

$$= \sum_{n=0}^{\infty} (-\lambda)^n \sum_{\substack{b_1, \cdots, b_n = 0 \\ 1 b_1 + \cdots + n b_n = n}} \frac{(-1)^{b_1 + \cdots + b_n}}{b_1! \, 1^{b_1} \cdots b_n! \, n^{b_n}} \left[\sum_k B_{kk} \right]^{b_1} \times \cdots$$

$$\cdots \times \left[\sum_{k_1 \cdots k_n} B_{k_1 k_2} B_{k_2 k_3} \cdots B_{k_n k_1} \right]^{b_n}$$

all terms in the sums

$$\left[\sum_k \cdot \right]^{b_1} \cdots \left[\sum_{k_1 \cdots k_n} \cdot \right]^{b_n} \equiv \sum_{k_1^1 \cdots k_{b_1}^1 \, k_1^2 \cdots k_{2 b_2}^2 \, k_1^3 \cdots \, k_{n b_n}^n}$$

for which at least two k_j^i's are equal cancel out. The approximation, say,

$$\det[Id + B] \approx e^{Tr B - \frac{1}{2} Tr B^2} \qquad (10.28)$$

$$= \sum_{n=0}^{\infty} (-\lambda)^n \sum_{\substack{b_1, b_2 = 0 \\ 1 b_1 + 2 b_2 = n}} \frac{(-1)^{b_1 + b_2}}{b_1! \, 1^{b_1} b_2! \, 2^{b_2}} \left[\sum_k B_{kk} \right]^{b_1} \left[\sum_{k_1, k_2} B_{k_1 k_2} B_{k_2 k_1} \right]^{b_2}$$

does not keep track of that information, but the approximation

$$\det[Id + B] \approx \int e^{Tr(\bar{\psi}_k \psi_k B_{kp}) - \frac{1}{2}Tr[(\bar{\psi}_k \psi_k B_{kp})^2]} d\mu \qquad (10.29)$$

$$= \int \sum_{n=0}^{\infty} (-\lambda)^n \sum_{\substack{b_1, b_2 = 0 \\ 1b_1 + 2b_2 = n}} \frac{(-1)^{b_1 + b_2}}{b_1! 1^{b_1} b_2! 2^{b_2}} \left[\sum_k \bar{\psi}_k \psi_k B_{kk} \right]^{b_1} \times$$

$$\left[\sum_{k_1, k_2} \bar{\psi}_{k_1} \psi_{k_1} \bar{\psi}_{k_2} \psi_{k_2} B_{k_1 k_2} B_{k_2 k_1} \right]^{b_2} d\mu$$

does.

We consider the lowest order approximations. For $n = 1$, application of (10.22) to (10.26) gives

$$\det[Id + B] \approx \int e^{\sum_k B_{kk} \bar{\psi}_k \psi_k} d\mu = \prod_k (1 + B_{kk}) \qquad (10.30)$$

For $n = 2$ we have (for, say, even N)

$$\det[Id + B] \approx \int e^{\sum_k B_{kk} \bar{\psi}_k \psi_k - \frac{1}{2} \sum_{k,p} B_{kp} B_{pk} \bar{\psi}_k \psi_k \bar{\psi}_p \psi_p} d\mu \qquad (10.31)$$

$$= \int e^{-\frac{1}{2} \sum_{k,p} B_{kp} B_{pk} \bar{\psi}_k \psi_k \bar{\psi}_p \psi_p} e^{\sum_k (1 + B_{kk}) \bar{\psi}_k \psi_k} \prod_k d\bar{\psi}_k d\psi_k$$

$$= \sum_{n=0}^{N/2} \frac{(-1)^n}{n!} \frac{1}{2^n} \sum_{\substack{k_1, \cdots, k_n \\ p_1, \cdots, p_n}} \prod_{i=1}^{n} B_{k_i p_i} B_{p_i k_i} \times$$

$$\int \prod_{i=1}^{n} \bar{\psi}_{k_i} \psi_{k_i} \bar{\psi}_{p_i} \psi_{p_i} e^{\sum_k (1 + B_{kk}) \bar{\psi}_k \psi_k} \prod_k d\bar{\psi}_k d\psi_k$$

$$= \sum_{n=0}^{N/2} (-1)^n \sum_{\substack{\{(k_1, p_1), \cdots, (k_n, p_n)\} \\ \text{disjoint ordered pairs}}} \prod_{i=1}^{n} B_{k_i p_i} B_{p_i k_i} \times$$

$$\int \prod_{i=1}^{n} \bar{\psi}_{k_i} \psi_{k_i} \bar{\psi}_{p_i} \psi_{p_i} e^{\sum_k (1 + B_{kk}) \bar{\psi}_k \psi_k} \prod_k d\bar{\psi}_k d\psi_k$$

$$= \sum_{n=0}^{N/2} (-1)^n \sum_{\substack{\{(k_1, p_1), \cdots, (k_n, p_n)\} \\ \text{disjoint ordered pairs}}} \prod_{i=1}^{n} B_{k_i p_i} B_{p_i k_i} \prod_{q \in \mathcal{M} \setminus \{k_1, \cdots, p_n\}} (1 + B_{qq})$$

$$= \sum_{n=0}^{N/2} \sum_{\substack{\pi \in S_N, \ \pi \text{ consists of } n \\ 2-\text{cycles and } N-2n \ 1-\text{cycles}}} \text{sign}\pi \ (Id + B)_{1\pi 1} (Id + B)_{2\pi 2} \cdots (Id + B)_{N\pi N}$$

$$= \sum_{\substack{\pi \in S_N \\ \pi \text{ consists only of} \\ 1- \text{ and } 2-\text{cycles}}} \text{sign}\pi \ (Id + B)_{1\pi 1} (Id + B)_{2\pi 2} \cdots (Id + B)_{N\pi N}$$

to which we refer in the following as two-cycle or two-loop approximation,

$$\det[Id + B] \approx \sum_{\substack{\pi \in S_N \\ \pi \text{ consists only of} \\ 1- \text{ and } 2-\text{cycles}}} \text{sign}\pi \, (Id + B)_{1\pi 1}(Id + B)_{2\pi 2}\cdots(Id + B)_{N\pi N}$$

(10.32)

Now consider inverse matrix elements. By Cramer's rule one has

$$[Id + B]_{tt}^{-1} = \frac{\det_t[Id + B]}{\det[Id + B]}$$

(10.33)

if we define (opposite to the definition in (10.25))

$$\det_{I,J} B := \det[(B_{kp})_{k \in I^c, p \in J^c}]$$

for $I, J \subset \mathcal{M}$, $|I| = |J|$, $I^c = \mathcal{M} \setminus I$ and

$$\det_I B := \det_{I,I} B.$$

In the following we give an alternative proof of (10.7), (10.8) by using the representation (10.26) and the tool of Grassmann integration.

One has

$$\frac{\det_t[Id + B]}{\det[Id + B]} = \frac{\int e^{\sum_{r=1}^{N-1} \frac{(-1)^{r+1}}{r} \sum_{k_1 \cdots k_r \in \mathcal{M}_t} B_{k_1 k_2} \cdots B_{k_r k_1} (\bar\psi\psi)_{k_1} \cdots (\bar\psi\psi)_{k_r}} d\mu_t}{\int e^{\sum_{r=1}^{N} \frac{(-1)^{r+1}}{r} \sum_{k_1 \cdots k_r \in \mathcal{M}} B_{k_1 k_2} \cdots B_{k_r k_1} (\bar\psi\psi)_{k_1} \cdots (\bar\psi\psi)_{k_r}} d\mu}$$

$$\equiv \frac{Z_t}{Z}$$

(10.34)

where $\mathcal{M}_I := \mathcal{M} \setminus I$ and $d\mu_I := e^{\sum_{k \in \mathcal{M}_I} \bar\psi_k \psi_k} \prod_{k \in \mathcal{M}_I} d\bar\psi_k d\psi_k$. Separating the $(\bar\psi\psi)_t$ variables in the denominator, we obtain

$$Z = \int \left\{ e^{\sum_{r=1}^{N-1} \frac{(-1)^{r+1}}{r} \sum_{k_1 \cdots k_r \in \mathcal{M}_t} B_{k_1 k_2} \cdots B_{k_r k_1} (\bar\psi\psi)_{k_1} \cdots (\bar\psi\psi)_{k_r}} e^{B_{tt}(\bar\psi\psi)_t} \times \right.$$

$$\left. e^{\sum_{r=2}^{N} (-1)^{r+1} \sum_{k_2 \cdots k_r \in \mathcal{M}_t} B_{tk_2} \cdots B_{k_r t}(\bar\psi\psi)_t (\bar\psi\psi)_{k_2} \cdots (\bar\psi\psi)_{k_r}} \right\} e^{\bar\psi_t \psi_t} d\bar\psi_t d\psi_t \, d\mu_t$$

$$= \left(1 + B_{tt} + \sum_{r=2}^{N} (-1)^{r+1} \sum_{k_2 \cdots k_r \in \mathcal{M}_t} B_{tk_2} \cdots B_{k_r t} \langle(\bar\psi\psi)_{k_2} \cdots (\bar\psi\psi)_{k_r}\rangle_t\right) Z_t$$

(10.35)

where

$$\langle(\bar\psi\psi)_{k_2} \cdots (\bar\psi\psi)_{k_r}\rangle_t =$$

$$\frac{1}{Z_t} \int (\bar\psi\psi)_{k_2} \cdots (\bar\psi\psi)_{k_r} e^{\sum_{s=1}^{N-1} \frac{(-1)^{s+1}}{s} \sum_{p_1 \cdots p_s \in \mathcal{M}_t} B_{p_1 p_2} \cdots B_{p_s p_1} (\bar\psi\psi)_{p_1} \cdots (\bar\psi\psi)_{p_s}} d\mu_t$$

Since

$$\langle (\bar\psi\psi)_{k_2} \cdots (\bar\psi\psi)_{k_r} \rangle_t$$

$$= \frac{1}{Z_t} \int e^{\sum_{s=1}^{N-r} \frac{(-1)^{s+1}}{s}} \sum_{p_1 \cdots p_s \in \mathcal{M}_{tk_2 \cdots k_s}} B_{p_1 p_2} \cdots B_{p_s p_1} (\bar\psi\psi)_{p_1} \cdots (\bar\psi\psi)_{p_s} \, d\mu_{tk_2 \cdots k_r}$$

$$\equiv \frac{Z_{tk_2 \cdots k_r}}{Z_t} = \frac{Z_{tk_2}}{Z_t} \frac{Z_{tk_2 k_3}}{Z_{tk_2}} \cdots \frac{Z_{tk_2 \cdots k_r}}{Z_{tk_2 \cdots k_{r-1}}}$$

$$= \langle \bar\psi_{k_2} \psi_{k_2} \rangle_t \langle \bar\psi_{k_3} \psi_{k_3} \rangle_{tk_2} \cdots \langle \bar\psi_{k_r} \psi_{k_r} \rangle_{tk_2 \cdots k_{r-1}} \qquad (10.36)$$

one arrives at

$$\langle \bar\psi_t \psi_t \rangle = \qquad\qquad\qquad\qquad\qquad\qquad\qquad\qquad (10.37)$$

$$\frac{1}{1 + B_{tt} + \displaystyle\sum_{r=2}^{N} (-1)^{r+1} \sum_{\substack{k_2 \cdots k_r \in \mathcal{M}_t \\ k_i \neq k_j}} B_{tk_2} \cdots B_{k_r t} \langle \bar\psi_{k_2} \psi_{k_2} \rangle_t \cdots \langle \bar\psi_{k_r} \psi_{k_r} \rangle_{tk_2 \cdots k_{r-1}}}$$

which coincides with (10.7).

The Off-Diagonal Elements: Let $s \neq t$. One has

$$[Id + B]_{st}^{-1} = \frac{\det_{s,t}[Id + B]}{\det[Id + B]}$$

and one may write

$$\det_{s,t}[Id + B] = \det[(Id + B)^{(s,t)}]$$

where

$$(Id + B)_{kp}^{(s,t)} = \begin{cases} (Id + B)_{kp} = \delta_{kp} + B_{kp} & \text{if } k \neq s \wedge p \neq t \\ 0 & \text{if } (k = s \wedge p \neq t) \vee \\ & \qquad (k \neq s \wedge p = t) \\ 1 & \text{if } k = s \wedge p = t \end{cases}$$

To apply (10.26), we define $\tilde B^{(s,t)} \in C^{N \times N}$ according to

$$(Id + B)^{(s,t)} = Id + \tilde B^{(s,t)}$$

That is, since $s \neq t$ by assumption,

$$\tilde B_{k,p}^{(s,t)} = \begin{cases} B_{kp} & \text{if } k \neq s \wedge p \neq t \\ -\delta_{kp} & \text{if } (k = s \wedge p \neq t) \vee (k \neq s \wedge p = t) \\ 1 & \text{if } k = s \wedge p = t \end{cases}$$

Then

$$[Id + B]_{st}^{-1} = \frac{\det[Id + \tilde B^{(s,t)}]}{\det[Id + B]} = \frac{\int \det[Id + \bar\psi\psi \tilde B^{(s,t)}] d\mu}{\int \det[Id + \bar\psi\psi B] d\mu}$$

and one has

$$[Id + B]_{st}^{-1} = \frac{\int e^{\sum_{r=1}^{N} \frac{(-1)^{r+1}}{r} Tr[(\bar{\psi}\psi \tilde{B}^{(s,t)})^r]} d\mu}{\int e^{\sum_{r=1}^{N} \frac{(-1)^{r+1}}{r} Tr[(\bar{\psi}\psi B)^r]} d\mu} \equiv \frac{Z^{(s,t)}}{Z}$$

The numerator becomes

$$Z^{(s,t)} = \tag{10.38}$$

$$\int e^{\sum_{k \in \mathcal{M}} \tilde{B}_{kk}^{(s,t)} (\bar{\psi}\psi)_k + \sum_{r=2}^{N} \frac{(-1)^{r+1}}{r} \sum_{k_1 \cdots k_r \in \mathcal{M}} \tilde{B}_{k_1 k_2}^{(s,t)} \cdots \tilde{B}_{k_r k_1}^{(s,t)} (\bar{\psi}\psi)_{k_1} \cdots (\bar{\psi}\psi)_{k_r}} \times$$

$$e^{\sum_{k \in \mathcal{M}} (\bar{\psi}\psi)_k} \prod_{k} d\bar{\psi}_k d\psi_k$$

$$= \int e^{\sum_{k \in \mathcal{M}_{st}} B_{kk} (\bar{\psi}\psi)_k + \sum_{r=2}^{N} \frac{(-1)^{r+1}}{r} \sum_{k_1 \cdots k_r \in \mathcal{M}} \tilde{B}_{k_1 k_2}^{(s,t)} \cdots \tilde{B}_{k_r k_1}^{(s,t)} (\bar{\psi}\psi)_{k_1} \cdots (\bar{\psi}\psi)_{k_r}} \times$$

$$e^{\sum_{k \in \mathcal{M}_{st}} (\bar{\psi}\psi)_k} \prod_{k} d\bar{\psi}_k d\psi_k$$

Consider the sum

$$\sum_{k_1 \cdots k_r \in \mathcal{M}} \tilde{B}_{k_1 k_2}^{(s,t)} \cdots \tilde{B}_{k_r k_1}^{(s,t)} (\bar{\psi}\psi)_{k_1} \cdots (\bar{\psi}\psi)_{k_r}$$

which is part of the exponent in (10.38). To get a nonzero contribution, all k_j have to be different. Furthermore, if, say, $k_1 = s$ then k_2 has to be equal to t and one gets a contribution

$$\sum_{k_3 \cdots k_r \in \mathcal{M}_{st}} 1 \cdot \tilde{B}_{tk_3}^{(s,t)} \cdots \tilde{B}_{k_r s}^{(s,t)} (\bar{\psi}\psi)_s (\bar{\psi}\psi)_t (\bar{\psi}\psi)_{k_3} \cdots (\bar{\psi}\psi)_{k_r}$$

$$= \sum_{k_3 \cdots k_r \in \mathcal{M}_{st}} B_{tk_3} \cdots B_{k_r s} (\bar{\psi}\psi)_s (\bar{\psi}\psi)_t (\bar{\psi}\psi)_{k_3} \cdots (\bar{\psi}\psi)_{k_r}$$

If $k_2 = s$ then k_3 has to be equal to t and one gets the same contribution. Thus

$$\sum_{k_1 \cdots k_r \in \mathcal{M}} \tilde{B}_{k_1 k_2}^{(s,t)} \cdots \tilde{B}_{k_r k_1}^{(s,t)} (\bar{\psi}\psi)_{k_1} \cdots (\bar{\psi}\psi)_{k_r}$$

$$= r \sum_{k_3 \cdots k_r \in \mathcal{M}_{st}} B_{tk_3} \cdots B_{k_r s} (\bar{\psi}\psi)_s (\bar{\psi}\psi)_t (\bar{\psi}\psi)_{k_3} \cdots (\bar{\psi}\psi)_{k_r}$$

and one obtains

$$Z^{(s,t)} = \tag{10.39}$$

$$\int \left\{ e^{\sum_{k \in \mathcal{M}_{st}} B_{kk} (\bar{\psi}\psi)_k + \sum_{r=2}^{N} \frac{(-1)^{r+1}}{r} \sum_{k_1 \cdots k_r \in \mathcal{M}_{st}} B_{k_1 k_2} \cdots B_{k_r k_1} (\bar{\psi}\psi)_{k_1} \cdots (\bar{\psi}\psi)_{k_r}} \times \right.$$

$$\int e^{\sum_{r=2}^{N} (-1)^{r+1} \sum_{k_3 \cdots k_r \in \mathcal{M}_{st}} B_{tk_3} \cdots B_{k_r s} (\bar{\psi}\psi)_s (\bar{\psi}\psi)_t (\bar{\psi}\psi)_{k_3} \cdots (\bar{\psi}\psi)_{k_r}} d\bar{\psi}_s d\psi_s \times$$

$$\left. d\bar{\psi}_t d\psi_t \right\} e^{\sum_{k \in \mathcal{M}_{st}} (\bar{\psi}\psi)_k} \prod_{k \in \mathcal{M}_{st}} d\bar{\psi}_k d\psi_k$$

Since

$$\int e^{A_{st}(\bar\psi\psi)_s(\bar\psi\psi)_t} d\bar\psi_s d\psi_s d\bar\psi_t d\psi_t = A_{st}$$

one arrives at

$$Z^{(s,t)} = \sum_{r=2}^{N}(-1)^{r+1} \sum_{k_3\cdots k_r\in\mathcal{M}_{st}} B_{tk_3}\cdots B_{k_r s}\langle(\bar\psi\psi)_{k_3}\cdots(\bar\psi\psi)_{k_r}\rangle_{st} Z_{st}$$

$$= \sum_{r=2}^{N}(-1)^{r+1} \sum_{k_3\cdots k_r\in\mathcal{M}_{st}} B_{tk_3}\cdots B_{k_r s}Z_{stk_3\cdots k_r} \qquad (10.40)$$

such that

$$[Id + B]_{st}^{-1} = \frac{Z^{(s,t)}}{Z} = \sum_{r=2}^{N}(-1)^{r+1} \sum_{k_3\cdots k_r\in\mathcal{M}_{st}} B_{tk_3}\cdots B_{k_r s}\frac{Z_{stk_3\cdots k_r}}{Z}$$

$$= \sum_{r=2}^{N}(-1)^{r+1} \sum_{k_3\cdots k_r\in\mathcal{M}_{st}} B_{tk_3}\cdots B_{k_r s}\langle\bar\psi_s\psi_s\rangle\langle\bar\psi_t\psi_t\rangle_s \times$$

$$\langle\bar\psi_{k_3}\psi_{k_3}\rangle_{st}\cdots\langle\bar\psi_{k_r}\psi_{k_r}\rangle_{stk_3\cdots k_{r-1}}$$

This is identical to formula (10.8).

10.3 The Averaged Green Function of the Anderson Model

Let coordinate space be a lattice with unit lattice spacing and finite volume $[0, L]^d$ with periodic boundary conditions:

$$\Gamma = \{x = (n_1,\cdots,n_d) \mid 0 \le n_i \le L-1\} = \mathbb{Z}^d/(L\mathbb{Z})^d \qquad (10.41)$$

Momentum space is given by

$$\mathcal{M} := \Gamma^\sharp = \{k = \tfrac{2\pi}{L}(m_1,\cdots,m_d) \mid 0 \le m_i \le L-1\}$$

$$= \left(\tfrac{2\pi}{L}\mathbb{Z}\right)^d/(2\pi\mathbb{Z})^d \qquad (10.42)$$

We consider the averaged Green function of the Anderson model given by

$$\langle G\rangle(x, x') := \int [-\Delta - z + \lambda V]_{x,x'}^{-1}\, dP(V) \qquad (10.43)$$

where the random potential is Gaussian,

$$dP(V) = \prod_{x\in\Gamma} e^{-\frac{V_x^2}{2}}\frac{dV_x}{\sqrt{2\pi}}. \qquad (10.44)$$

Here $z = E + i\varepsilon$ and Δ is the discrete Laplacian for unit lattice spacing,

$$[-\Delta - z + \lambda V]_{x,x'} = \tag{10.45}$$

$$-\sum_{i=1}^{d} (\delta_{x',x+e_i} + \delta_{x',x-e_i} - 2\delta_{x',x}) - z\,\delta_{x,x'} + \lambda V_x\,\delta_{x,x'}$$

By taking the Fourier transform, one obtains

$$\langle G\rangle(x,x') = \tfrac{1}{L^d}\sum_{k\in\mathcal{M}} e^{ik(x'-x)}\langle G\rangle(k) \tag{10.46}$$

$$\langle G\rangle(k) = \int_{\mathbb{R}^{L^d}} \left[\left(a_p\delta_{p,p'} + \tfrac{\lambda}{\sqrt{L^d}}v_{p-p'}\right)_{p,p'\in\mathcal{M}}\right]^{-1}_{k,k} dP(v) \tag{10.47}$$

where $L^d = |\Gamma| = |\mathcal{M}|$, $dP(v)$ is given by (10.50) or (10.51) below, depending on whether L^d is even or odd, and

$$a_k = 4\sum_{i=1}^{d} \sin^2\left[\tfrac{k_i}{2}\right] - E - i\varepsilon \tag{10.48}$$

The rigorous control of $\langle G\rangle(k)$ for small disorder λ and energies inside the spectrum of the unperturbed Laplacian, $E \in [0, 4d]$, in which case a_k has a root if $\varepsilon \to 0$, is still an open problem [2, 41, 48, 52, 60]. It is expected that $\lim_{\varepsilon\searrow 0}\lim_{L\to\infty}\langle G\rangle(k) = 1/(a_k - \sigma_k)$ where $\mathrm{Im}\sigma = O(\lambda^2)$.

The integration variables v_q in (10.47) are given by the discrete Fourier transform of the V_x. In particular, observe that, if F denotes the unitary matrix of discrete Fourier transform, the variables

$$v_q \equiv (FV)_q = \tfrac{1}{\sqrt{L^d}}\sum_{x\in\Gamma} e^{-iqx}V_x \equiv \tfrac{1}{\sqrt{L^d}}\hat{V}_q \tag{10.49}$$

would not have a limit if V_x would be deterministic and cutoffs are removed, since the $\hat{V}_q$ are the quantities which have a limit in that case. But since the V_x are integration variables, we choose a unitary transform to keep the integration measure invariant. Observe also that v_q is complex, $v_q = u_q + i w_q$. Since V_x is real, $u_{-q} = u_q$ and $w_{-q} = -w_q$. In order to transform $dP(V)$ to momentum space, we have to choose a set $\mathcal{M}^+ \subset \mathcal{M}$ such that either $q \in \mathcal{M}^+$ or $-q \in \mathcal{M}^+$. If L is odd, the only momentum with $q = -q$ or $w_q = 0$ is $q = 0$. In that case $dP(V)$ becomes

$$dP(v) = e^{-\frac{u_0^2}{2}}\frac{du_0}{\sqrt{2\pi}}\prod_{q\in\mathcal{M}^+} e^{-(u_q^2+w_q^2)}\frac{du_q\,dw_q}{\pi} \tag{10.50}$$

For even L we get

$$dP(v) = e^{-\frac{1}{2}(u_0^2+u_{q_0}^2)}\frac{du_0\,du_{q_0}}{2\pi}\prod_{q\in\mathcal{M}^+} e^{-(u_q^2+w_q^2)}\frac{du_q\,dw_q}{\pi} \tag{10.51}$$

where $q_0 = \frac{2\pi m}{L}$ is the unique nonzero momentum for which $\frac{2\pi}{L}m = 2\pi(1,\cdots,1) - \frac{2\pi}{L}m$.

10.3.1   Two-Loop Approximation

Now we apply the inversion formula (10.7) to the inverse matrix element in (10.47),

$$\langle G \rangle(k) = \int_{\mathbb{R}^{L^d}} \left[a_p \delta_{p,p'} + \frac{\lambda}{\sqrt{L^d}} v_{p-p'} \right]_{k,k}^{-1} dP(v)$$

We start with the 'two-loop approximation', which we define by retaining only the $r = 2$ term in the denominator of the right hand side of (10.7),

$$G(k) \approx \frac{1}{B_{kk} - \sum_{p \in \mathcal{M} \setminus \{k\}} B_{kp} G_k(p) B_{pk}} \tag{10.52}$$

Thus, let

$$G(k) := \left[a_p \delta_{p,p'} + \frac{\lambda}{\sqrt{L^d}} v_{p-p'} \right]_{k,k}^{-1} = G(k; v, z) \tag{10.53}$$

In the infinite volume limit the spacing $2\pi/L$ of the momenta becomes arbitrarily small. Hence, in computing an inverse matrix element, it should not matter whether a single column and row labelled by some momentum t is absent or not. In other words, in the infinite volume limit one should have

$$G_t(p) = G(p) \quad \text{for } L \to \infty \tag{10.54}$$

and similarly $G_{t_1 \cdots t_j}(p) = G(p)$ as long as j is independent of the volume. We remark however that if the matrix has a block structure, say $B = (B_{k\sigma, p\tau})$ with $\sigma, \tau \in \{\uparrow, \downarrow\}$ some spin variables, this structure has to be respected. That is, for a given momentum k all rows and columns labelled by $k\uparrow$, $k\downarrow$ have to be eliminated, since otherwise (10.54) may not be true.

Thus the two-loop approximation gives

$$G(k) = \frac{1}{a_k + \frac{\lambda}{\sqrt{L^d}} v_0 - \frac{\lambda^2}{L^d} \sum_{p \neq k} v_{k-p} G(p) v_{p-k}} \tag{10.55}$$

For large L, we can disregard the $\frac{\lambda}{\sqrt{L^d}} v_0$ term. Introducing $\sigma_k = \sigma_k(v, z)$ according to

$$G(k) =: \frac{1}{a_k - \sigma_k}, \tag{10.56}$$

we get

$$\sigma_k = \frac{\lambda^2}{L^d} \sum_{p \neq k} \frac{|v_{k-p}|^2}{a_p - \sigma_p} \approx \frac{\lambda^2}{L^d} \sum_p \frac{|v_{k-p}|^2}{a_p - \sigma_p} \tag{10.57}$$

214

and arrive at

$$\langle G \rangle(k) = \int \frac{1}{a_k - \frac{\lambda^2}{L^d} \sum_p \frac{|v_{k-p}|^2}{a_p - \frac{\lambda^2}{L^d} \sum_t \frac{|v_{p-t}|^2}{a_t - \frac{\lambda^2}{L^d} \Sigma \cdots}}} \, dP(v) \qquad (10.58)$$

Now consider the infinite volume limit $L \to \infty$. By the central limit theorem of probability $\frac{1}{\sqrt{L^d}} \sum_q (|v_q|^2 - \langle |v_q|^2 \rangle)$ is, as a sum of independent random variables, normally distributed. Note that only a prefactor of $1/\sqrt{L^d}$ is required for that property. In particular, if F is some bounded function independent of L, sums which come with a prefactor of $1/L^d$ like $\frac{1}{L^d} \sum_q c_q |v_q|^2$ can be substituted by their expectation value,

$$\lim_{L \to \infty} \int F\left(\frac{1}{L^d} \sum_k c_k |v_k|^2\right) dP(v) = F\left(\lim_{L \to \infty} \frac{1}{L^d} \sum_k c_k \langle |v_k|^2 \rangle\right) \quad (10.59)$$

Therefore, in the two-loop approximation, one obtains in the infinite volume limit

$$\langle G \rangle(k) = \frac{1}{a_k - \frac{\lambda^2}{L^d} \sum_p \frac{\langle |v_{k-p}|^2 \rangle}{a_p - \frac{\lambda^2}{L^d} \sum_t \frac{\langle |v_{p-t}|^2 \rangle}{a_t - \frac{\lambda^2}{L^d} \Sigma \cdots}}} =: \frac{1}{a_k - \langle \sigma_k \rangle} \qquad (10.60)$$

where the quantity $\langle \sigma_k \rangle$ satisfies the integral equation

$$\langle \sigma_k \rangle = \frac{\lambda^2}{L^d} \sum_p \frac{\langle |v_{k-p}|^2 \rangle}{a_p - \langle \sigma_p \rangle} \overset{L \to \infty}{\longrightarrow} \lambda^2 \int_{[0,2\pi]^d} \frac{d^d p}{(2\pi)^d} \frac{\langle |v_{k-p}|^2 \rangle}{a_p - \langle \sigma_p \rangle} \qquad (10.61)$$

For a Gaussian distribution $\langle |v_q|^2 \rangle = 1$ for all q such that $\langle \sigma_k \rangle = \langle \sigma \rangle$ becomes independent of k. Thus we end up with

$$\langle G \rangle(k) = \frac{1}{4 \sum_{i=1}^d \sin^2\left[\frac{k_i}{2}\right] - E - i\varepsilon - \langle \sigma \rangle} \qquad (10.62)$$

where $\langle \sigma \rangle$ is a solution of

$$\langle \sigma \rangle = \lambda^2 \int_{[0,2\pi]^d} \frac{d^d p}{(2\pi)^d} \frac{1}{4 \sum_{i=1}^d \sin^2\left[\frac{p_i}{2}\right] - E - i\varepsilon - \langle \sigma \rangle} \qquad (10.63)$$

This equation is well known and one deduces from it that it generates a small imaginary part $\text{Im}\,\sigma = O(\lambda^2)$ for $\varepsilon \searrow 0$ if the energy E is within the spectrum of $-\Delta$, $E \in (0, 4d)$.

10.3.2 Higher Orders

We now add the higher loop terms (the terms for $r > 2$ in the denominator of (10.7)) to our discussion and give an interpretation in terms of Feynman diagrams. For the Anderson model, Feynman graphs may be obtained by brutally expanding $(C = (1/a_p \, \delta_{p,p'}), V = (\frac{\lambda}{\sqrt{L^d}} v_{p-p'})$ in the next line)

$$\int [a_k \delta_{k,p} + \tfrac{\lambda}{\sqrt{L^d}} v_{k-p}]^{-1}_{k,k} \, dP(v) \sim \sum_{r=0}^{\infty} \int (C[VC]^r)_{kk} \, dP(v)$$

$$= \sum_{r=0}^{\infty} \frac{(-\lambda)^r}{\sqrt{L^d}^r} \sum_{p_2 \cdots p_r} \frac{1}{a_k a_{p_2} \cdots a_{p_r} a_k} \int v_{k-p_2} v_{p_2-p_3} \cdots v_{p_r-k} \, dP(v) \quad (10.64)$$

For a given r, this may be represented as in figure 10.1 ($c_k := 1/a_k$).

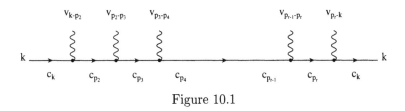

Figure 10.1

The integral over the v gives a sum of $(r-1)!!$ terms where each term is a product of $r/2$ Kroenecker-delta's, the terms for odd r vanish. If this is substituted in (10.64), the number of independent momentum sums is cut down to $r/2$ and each of the $(r-1)!!$ terms may be represented by a diagram as in figure 10.2.

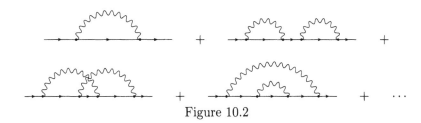

Figure 10.2

Here, the value of the, say, third diagram is given by

$$\frac{\lambda^4}{L^{2d}} \sum_{p_1, p_2} \frac{1}{a_k a_{k+p_1} a_{k+p_1+p_2} a_k}$$

For short:

$$\langle G \rangle(k) = \text{sum of all two-legged diagrams.} \quad (10.65)$$

216

Since the value of a diagram depends on its subgraph structure, one distinguishes, in the easiest case, two types of diagrams: diagrams with or without two-legged subdiagrams. Those diagrams with two-legged subgraphs usually produce anomalously large contributions. They are divided further into the one-particle irreducible ones and the reducible ones. Thereby a diagram is called one-particle reducible if it can be made disconnected by cutting one solid or 'particle' line (no squiggle), like the first diagram in figure 10.3.

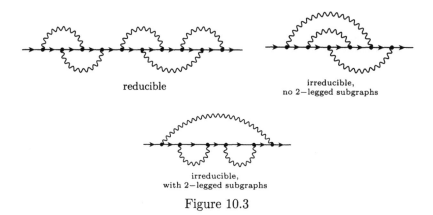

reducible

irreducible,
no 2−legged subgraphs

irreducible,
with 2−legged subgraphs

Figure 10.3

The reason for introducing reducible and irreducible diagrams is that the reducible ones can be easily resummed by writing down the Schwinger-Dyson equation which states that if the self energy Σ_k is defined through

$$\langle G \rangle(k) = \frac{1}{a_k - \Sigma_k(G_0)} \tag{10.66}$$

then $\Sigma_k(G_0)$ is the sum of all amputated (no $1/a_k$'s at the ends) one-particle irreducible diagrams. The lowest order contributions are drawn in figure 10.4. Observe that the second diagram of figure 10.2 is now absent, but the fourth diagram, which contains a two-legged subgraph, is still present in $\Sigma_k(G_0)$.

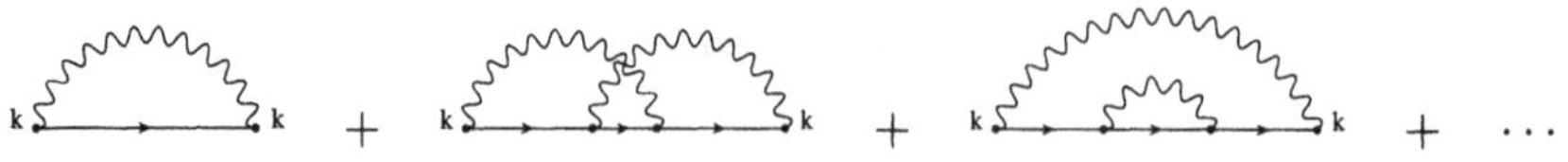

Figure 10.4

Here we wrote $\Sigma_k(G_0)$ to indicate that the factors ('propagators') assigned to the solid lines of the diagrams contributing to Σ_k are given by the free two-point function $G_0(p) = \frac{1}{a_p}$. The diagrams contributing to $\Sigma_k(G_0)$ still contain anomalously large contributions, namely irreducible diagrams which contain two-legged subdiagrams.

In the following we show, using the inversion formula (10.7) including all higher loop terms, that all graphs with two-legged subgraphs can be eliminated or resummed by writing down the following integral equation for $\langle G \rangle$:

$$\langle G \rangle(k) = \frac{1}{a_k - \sigma_k(\langle G \rangle)} \tag{10.67}$$

where

$\sigma_k(\langle G \rangle)$ is the sum of all amputated two-legged diagrams which do not contain any two-legged subdiagrams, but now with propagators $\langle G \rangle(k) = \frac{1}{a_k - \sigma_k}$ instead of $G_0 = \frac{1}{a_k}$

which may be formalized as in (10.76) below. The lowest order contributions to $\sigma_k(\langle G \rangle)$ are drawn in figure 10.5.

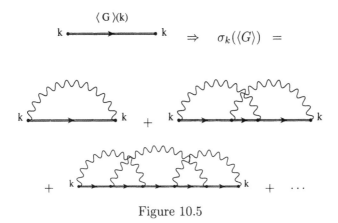

Figure 10.5

Observe that the third diagram of figure 10.4, that one with the two-legged subdiagram, has now disappeared. Thus the advantage of (10.67) compared to (10.66) is that the series for $\sigma_k(\langle G \rangle)$ can be expected to be asymptotic, that is, its lowest order contributions are a good approximation if the coupling is small, but, usually, the series for $\Sigma_k(G_0)$ is not asymptotic. Therefore renormalization in this case is nothing else that a rearrangement of the perturbation series,

$$\Sigma_k(G_0) = \sigma_k(\langle G \rangle), \tag{10.68}$$

the lowest order terms to $\Sigma_k(G_0)$ are not a good approximation but the lowest order terms of $\sigma_k(\langle G \rangle)$ are.

Equations of the type (10.67) can be found in the literature [47, 15], but usually they are derived on a more heuristic level without using the inversion formula (10.7). In section 10.6 we discuss more closely how our method is related to the integral equations which can be found in the literature.

We now show (10.67) for the Anderson model. For fixed v one has

$$G(k, v) = \frac{1}{a_k - \sigma_k(v)} \tag{10.69}$$

where

$$\sigma_k(v) = \sum_{r=2}^{L^d} (-1)^r \sum_{\substack{p_2 \cdots p_r \\ p_i \neq p_j, \, p_i \neq k}} \left(\frac{\lambda}{\sqrt{L^d}}\right)^r G_k(p_2) \cdots G_{kp_2 \cdots p_{r-1}}(p_r) \, v_{k-p_2} \cdots v_{p_r - k} \tag{10.70}$$

We cut off the r-sum in (10.70) at some arbitrary but fixed order $\ell < L^d$ where ℓ is chosen to be independent of the volume. Furthermore we substitute $G_{kp_2 \cdots p_j}(p)$ by $G(p)$ since we are interested in the infinite volume limit $L \to \infty$. Thus

$$\langle G \rangle(k) = \left\langle \frac{1}{a_k - \sum_{r=2}^{\ell} \sigma_k^r(v)} \right\rangle \tag{10.71}$$

where

$$\sigma_k^r(v) = (-1)^r \left(\frac{\lambda}{\sqrt{L^d}}\right)^r \sum_{\substack{p_2 \cdots p_r \\ p_i \neq p_j, \, p_i \neq k}} G(p_2) \cdots G(p_r) \, v_{k-p_2} \cdots v_{p_r - k} \tag{10.72}$$

Consider first two strings $s_k^{r_1}$, $s_k^{r_2}$ where

$$s_k^r(v) = \frac{\lambda^r}{\sqrt{L^{d^r}}} \sum_{\substack{p_2 \cdots p_r \\ p_i \neq p_j, \, p_i \neq k}} c_{kp_2 \cdots p_r}^r \, v_{k-p_2} \cdots v_{p_r - k} \tag{10.73}$$

and the $c_{kp_2 \cdots p_r}^r$ are some numbers. Then $\langle s_k^{r_1} s_k^{r_2} \rangle = \langle s_k^{r_1} \rangle \langle s_k^{r_2} \rangle + O(1/L^d)$ which means that in the infinite volume limit

$$\langle s_k^{r_1} s_k^{r_2} \rangle = \langle s_k^{r_1} \rangle \langle s_k^{r_2} \rangle \tag{10.74}$$

This holds because all pairings which connect the two strings have an extra volume factor $1/L^d$. Namely, if the two strings are disconnected, there are $(r_1 + r_2)/2$ loops and a volume factor of $1/\sqrt{L^d}^{(r_1 + r_2)}$ giving $(r_1 + r_2)/2$ Riemannian sums. If the two strings are connected, there are only $(r_1 + r_2 - 2)/2$ loops, leaving an extra factor of $1/L^d$. This is visualized in the following figure 10.6 for two strings with $r = 2$ and $r = 4$ squiggles. The disconnected contributions have exactly three loops, the connected contributions have exactly two loops, the prefactor is $1/L^{3d}$.

$$\int \text{(diagram)} = \text{(diagram)} = O(1) + \text{(diagram)} = O(1/L^d) + \cdots$$

Figure 10.6

By the same argument one has in the infinite volume limit

$$\langle (s_k^{r_1})^{n_1} \cdots (s_k^{r_m})^{n_m} \rangle = \langle s_k^{r_1} \rangle^{n_1} \cdots \langle s_k^{r_m} \rangle^{n_m} \tag{10.75}$$

which results in

$$\langle G \rangle(k) = \frac{1}{a_k - \sum\limits_{r=2}^{\ell} \frac{(-\lambda)^r}{\sqrt{L^d}^r} \sum\limits_{\substack{p_2 \cdots p_r \\ p_i \neq p_j, \, p_i \neq k}} \langle G \rangle(p_2) \cdots \langle G \rangle(p_r) \langle v_{k-p_2} \cdots v_{p_r-k} \rangle} \tag{10.76}$$

Now, the condition $p_2, \cdots, p_r \neq k$ and $p_i \neq p_j$ in (10.76) means exactly that two-legged subgraphs are forbidden. Namely, for a two-legged subdiagram like the one in the third diagram in figure 10.4, the incoming and outgoing momenta p_i, p_j, to which are assigned propagators $\langle G \rangle(p_i)$, $\langle G \rangle(p_j)$, must be equal which is forbidden by the condition $p_i \neq p_j$ in (10.76). Figure 10.7 may be helpful.

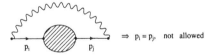

$$\Rightarrow \; p_i = p_j, \quad \text{not allowed}$$

Figure 10.7a

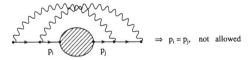

$$\Rightarrow \; p_i = p_j, \quad \text{not allowed}$$

Figure 10.7b

However, we cannot take the limit $\ell \to \infty$ in (10.76) since the series in the denominator of (10.76) is only an asymptotic one. To see this a bit more clearly suppose for the moment that there were no restrictions on the momentum sums. Then, if $V = (\frac{\lambda}{\sqrt{L^d}} v_{k-p})_{k,p}$ and $\langle G \rangle = (\langle G \rangle(k)\, \delta_{k,p})_{k,p}$,

$$\left(\frac{\lambda}{\sqrt{L^d}}\right)^r \sum_{p_2 \cdots p_r} \langle G \rangle(p_2) \cdots \langle G \rangle(p_r) \langle v_{k-p_2} \cdots v_{p_r-k} \rangle = \langle (V[\langle G \rangle V]^{r-1})_{kk} \rangle$$

(10.77)

and for $\ell \to \infty$ we would get

$$\langle G \rangle(k) = \frac{1}{a_k - \langle (V \frac{\langle G \rangle V}{Id + \langle G \rangle V})_{kk} \rangle} = \frac{1}{a_k - \langle (V \frac{1}{\langle G \rangle^{-1} + V} V)_{kk} \rangle}$$

(10.78)

That is, the factorials produced by the number of diagrams in the denominator of (10.76) are basically the same as those in the expansion

$$\int_{\mathbb{R}} \frac{x^2}{z + \lambda x} e^{-\frac{x^2}{2}} \frac{dx}{\sqrt{2\pi}} = \sum_{r=0}^{\ell} \frac{\lambda^{2r}}{z^{2r+1}} (2r+1)!! + R_{\ell+1}(\lambda)$$

(10.79)

where the remainder satisfies the bound $|R_{\ell+1}(\lambda)| \le \ell!\, const_z^\ell\, \lambda^\ell$.

We close this section with two further remarks. So far the computations were done in momentum space. One may wonder what one gets if the inversion formula (10.7) is applied to $[-\Delta + z + \lambda V]^{-1}$ in coordinate space. Whereas a geometric series expansion of $[-\Delta + z + \lambda V]^{-1}$ gives a representation in terms of the simple random walk [59], application of (10.7) results in a representation in terms of the self-avoiding walk:

$$[-\Delta + z + \lambda V]^{-1}_{0,x} = \sum_{\substack{\gamma:0 \to x \\ \gamma\ self-avoiding}} \frac{\det\left[(-\Delta + z + \lambda V)_{y,y' \in \Gamma \backslash \gamma}\right]}{\det\left[(-\Delta + z + \lambda V)_{y,y' \in \Gamma}\right]}$$

(10.80)

where Γ is the lattice in coordinate space. Namely, if $|x| > 1$, the inversion formula (10.8) for the off-diagonal elements gives

$$[-\Delta + \lambda V]^{-1}_{0,x} =$$

$$\sum_{r=3}^{L^d} (-1)^{r+1} \sum_{\substack{x_3 \cdots x_r \in \Gamma \backslash \{0,x\} \\ x_i \ne x_j}} G(0) G_0(x_3) \cdots G_{0 x_3 \cdots x_r}(x) (-\Delta)_{0 x_3} \cdots (-\Delta)_{x_r x}$$

$$= \sum_{r=3}^{L^d} \sum_{\substack{x_2=0, x_3, \cdots, x_r, x_{r+1}=x \in \Gamma \\ |x_i - x_{i+1}|=1\ \forall i=2 \cdots r}} \frac{\det\left[(-\Delta + \lambda V)_{y,y' \in \Gamma \backslash \{0\}}\right]}{\det\left[(-\Delta + \lambda V)_{y,y' \in \Gamma}\right]} \times \cdots$$

$$\cdots \times \frac{\det\left[(-\Delta + \lambda V)_{y,y' \in \Gamma \backslash \{0,x_3 \cdots x_r, x\}}\right]}{\det\left[(-\Delta + \lambda V)_{y,y' \in \Gamma \backslash \{0,x_3 \cdots x_r\}}\right]}$$

which coincides with (10.80).

Finally we remark that, while the argument following (10.73) leads to a factorization property for on-diagonal elements in momentum space, $\langle G(k)\,G(p)\rangle$ equals $\langle G(k)\rangle\,\langle G(p)\rangle$, there is no such property for products of off-diagonal elements which appear in a quantity like

$$\Lambda(q) = \tfrac{1}{L^d} \sum_{k,p} \left\langle \left[a_k \delta_{k,p} + \tfrac{\lambda}{\sqrt{L}^d} v_{k-p} \right]_{k,p}^{-1} \left[\bar{a}_k \delta_{k,p} + \tfrac{\lambda}{\sqrt{L}^d} \bar{v}_{k-p} \right]_{k-q,p-q}^{-1} \right\rangle \quad (10.81)$$

which is the Fourier transform of $\langle |[-\Delta + z + \lambda V]_{x,y}^{-1}|^2 \rangle$. (Each off-diagonal inverse matrix element is proportional to $1/\sqrt{L^d}$, therefore the prefactor of $1/L^d$ in (10.81) is correct.)

10.4 The Many-Electron System with Attractive Delta Interaction

10.4.1 The Integral Equations in Two-Loop Approximation

The goal in this section is to apply the inversion formula (10.7) to the inverse matrix element in (10.2) and to compute the two point function $\langle \bar\psi_{k\sigma} \psi_{k\sigma} \rangle$ for the many-electron system with delta interaction, or more precisely, to derive a closed set of integral equations for it. We briefly recall the setup.

We consider the many-electron system in the grand canonical ensemble in finite volume $[0, L]^d$ and at some small but positive temperature $T = 1/\beta > 0$ with attractive delta interaction ($\lambda > 0$) given by the Hamiltonian

$$\begin{aligned}
H &= H_0 - \lambda H_{\text{int}} \\
&= \tfrac{1}{L^d} \sum_{k\sigma} (\tfrac{k^2}{2m} - \mu) a_{k\sigma}^+ a_{k\sigma} - \tfrac{\lambda}{L^{3d}} \sum_{kpq} a_{k\uparrow}^+ a_{q-k\downarrow}^+ a_{q-p\downarrow} a_{p\uparrow} \quad (10.82)
\end{aligned}$$

Our normalization conventions concerning the volume factors are such that the canonical anticommutation relations read $\{a_{k\sigma}, a_{p\tau}^+\} = L^d \delta_{k,p} \delta_{\sigma,\tau}$. The momentum sums range over some subset of $\left(\tfrac{2\pi}{L}\mathbb{Z}\right)^d$, say $\mathcal{M} = \{k \in \left(\tfrac{2\pi}{L}\mathbb{Z}\right)^d \,||e_k| \leq 1\}$, $e_k = k^2/2m - \mu$, and $q \in \{k - p\,|k, p \in \mathcal{M}\}$. We are interested in the momentum distribution

$$\langle a_{k\sigma}^+ a_{k\sigma} \rangle = Tr[e^{-\beta H} a_{k\sigma}^+ a_{k\sigma}]/Tr\, e^{-\beta H} \quad (10.83)$$

and in the expectation value of the energy

$$\langle H_{\text{int}} \rangle = \sum_q \Lambda(q) \quad (10.84)$$

where

$$\Lambda(\mathbf{q}) = \frac{\lambda}{L^{3d}} \sum_{\mathbf{k},\mathbf{p}} Tr[e^{-\beta H} a^+_{\mathbf{k}\uparrow} a^+_{\mathbf{q}-\mathbf{k}\downarrow} a_{\mathbf{q}-\mathbf{p}\downarrow} a_{\mathbf{p}\uparrow}]/Tr\, e^{-\beta H} \qquad (10.85)$$

By writing down the perturbation series for the partition function, rewriting it as a Grassmann integral

$$\frac{Tr\, e^{-\beta(H_0-\lambda H_{\text{int}})}}{Tr\, e^{-\beta H_0}} = \int e^{\frac{\lambda}{(\beta L^d)^3} \sum_{kpq} \bar{\psi}_{k\uparrow} \bar{\psi}_{q-k\downarrow} \psi_{q-p\downarrow} \psi_{p\uparrow}} d\mu_C(\psi, \bar{\psi}) \qquad (10.86)$$

$$d\mu_C = \prod_{k\sigma} \frac{\beta L^d}{ik_0 - e_{\mathbf{k}}} e^{-\frac{1}{\beta L^d} \sum_{k\sigma} (ik_0 - e_{\mathbf{k}}) \bar{\psi}_{k\sigma} \psi_{k\sigma}} \prod_{k\sigma} d\psi_{k\sigma} d\bar{\psi}_{k\sigma}$$

performing a Hubbard-Stratonovich transformation ($\phi_q = u_q + iv_q$, $d\phi_q d\bar{\phi}_q :=du_q dv_q$)

$$e^{-\sum_q a_q b_q} = \int e^{i \sum_q (a_q \phi_q + b_q \bar{\phi}_q)} e^{-\sum_q |\phi_q|^2} \prod_q \frac{d\phi_q d\bar{\phi}_q}{\pi} \qquad (10.87)$$

with

$$a_q = \frac{\lambda^{\frac{1}{2}}}{(\beta L^d)^{\frac{3}{2}}} \sum_k \bar{\psi}_{k\uparrow} \bar{\psi}_{q-k\downarrow}, \quad b_q = \frac{\lambda^{\frac{1}{2}}}{(\beta L^d)^{\frac{3}{2}}} \sum_p \psi_{p\uparrow} \psi_{q-p\downarrow} \qquad (10.88)$$

and then integrating out the $\psi, \bar{\psi}$ variables, one arrives at the following representation which is the starting point for our analysis:

$$\frac{1}{L^d} \langle a^+_{\mathbf{k}\sigma} a_{\mathbf{k}\sigma} \rangle = \frac{1}{\beta L^d} \frac{1}{\beta} \sum_{k_0 \in \frac{\pi}{\beta}(2\mathbb{Z}+1)} \langle \psi^+_{\mathbf{k}k_0\sigma} \psi_{\mathbf{k}k_0\sigma} \rangle \qquad (10.89)$$

where, abbreviating $k = (\mathbf{k}, k_0)$, $\kappa = \beta L^d$, $a_k = ik_0 - e_{\mathbf{k}}$, $g = \lambda^{\frac{1}{2}}$,

$$\frac{1}{\kappa} \langle \bar{\psi}_{t\sigma} \psi_{t\sigma} \rangle = \int \left[\begin{array}{cc} a_k \delta_{k,p} & \frac{ig}{\sqrt{\kappa}} \bar{\phi}_{p-k} \\ \frac{ig}{\sqrt{\kappa}} \phi_{k-p} & a_{-k} \delta_{k,p} \end{array} \right]^{-1}_{t\sigma, t\sigma} dP(\phi) \qquad (10.90)$$

and $dP(\phi)$ is the normalized measure

$$dP(\phi) = \frac{1}{Z} \det \left[\begin{array}{cc} a_k \delta_{k,p} & \frac{ig}{\sqrt{\kappa}} \bar{\phi}_{p-k} \\ \frac{ig}{\sqrt{\kappa}} \phi_{k-p} & a_{-k} \delta_{k,p} \end{array} \right] e^{-\sum_q |\phi_q|^2} \prod_q d\phi_q d\bar{\phi}_q \qquad (10.91)$$

Furthermore

$$\Lambda(\mathbf{q}) = \frac{1}{\beta} \sum_{q_0 \in \frac{2\pi}{\beta} \mathbb{Z}} \Lambda(\mathbf{q}, q_0) \qquad (10.92)$$

where

$$\Lambda(q) = \frac{\lambda}{(\beta L^d)^3} \sum_{k,p} \langle \bar{\psi}_{k\uparrow} \bar{\psi}_{q-k\downarrow} \psi_{q-p\downarrow} \psi_{p\uparrow} \rangle$$

$$= \langle |\phi_q|^2 \rangle - 1 \qquad (10.93)$$

and the expectation in the last line is integration with respect to $dP(\phi)$. The expectation on the ψ variables in $\langle \bar{\psi}_{k\sigma} \psi_{k\sigma} \rangle$ is Grassmann integration, $\langle \bar{\psi}_{k\sigma} \psi_{k\sigma} \rangle = \frac{1}{2} \int \bar{\psi}_{k\sigma} \psi_{k\sigma} e^{\frac{\lambda}{\kappa^3} \sum_{k,p,q} \bar{\psi}_{k\uparrow} \bar{\psi}_{q-k\downarrow} \psi_{q-p\downarrow} \psi_{p\uparrow}} d\mu_C$, but these representations are not used in the following. The matrix and the integral in (10.90) become finite dimensional if we choose some cutoff on the k_0 variables which is removed in the end. The set $\mathcal{M}$ for the spatial momenta is already finite since we have chosen a fixed UV-cutoff $|e_{\mathbf{k}}| = |\mathbf{k}^2/2m - \mu| \leq 1$ which will not be removed in the end since we are interested in the infrared properties at $\mathbf{k}^2/2m = \mu$.

Our goal is to apply the inversion formula to the inverse matrix element in (10.90). Instead of writing the matrix in terms of four $N \times N$ blocks $(a_k \delta_{k,p})_{k,p}$, $(\bar{\phi}_{p-k})_{k,p}$, $(\phi_{k-p})_{k,p}$ and $(a_{-k} \delta_{k,p})_{k,p}$ where N is the number of the $d+1$-dimensional momenta k, p, we interchange rows and columns to rewrite it in terms of N blocks of size 2×2 (the matrix U in the next line interchanges the rows and columns):

$$U \begin{bmatrix} a_k \delta_{k,p} & \frac{ig}{\sqrt{\kappa}} \bar{\phi}_{p-k} \\ \frac{ig}{\sqrt{\kappa}} \phi_{k-p} & a_{-k} \delta_{k,p} \end{bmatrix} U^{-1} = B = (B_{kp})_{k,p}$$

where the 2×2 blocks B_{kp} are given by

$$B_{kk} = \begin{pmatrix} a_k & \frac{ig}{\sqrt{\kappa}} \bar{\phi}_0 \\ \frac{ig}{\sqrt{\kappa}} \phi_0 & a_{-k} \end{pmatrix}, \quad B_{kp} = \frac{ig}{\sqrt{\kappa}} \begin{pmatrix} 0 & \bar{\phi}_{p-k} \\ \phi_{k-p} & 0 \end{pmatrix} \text{ if } k \neq p \qquad (10.94)$$

We want to compute the 2×2 matrix

$$\langle G \rangle(k) = \int G(k) \, dP(\phi) \qquad (10.95)$$

where

$$G(k) = [B^{-1}]_{kk} \qquad (10.96)$$

As for the Anderson model, we start again with the two-loop approximation which retains only the $r = 2$ term in the denominator of (10.7). The result will be equation (10.101) below where the quantities $\langle \sigma_k \rangle$ and $\langle |\phi_0|^2 \rangle$ appearing in (10.101) have to satisfy the equations (10.102) and (10.105) which have to be solved in conjunction with (10.110). The solution to these equations is discussed below (10.111).

We first derive (10.101). In the two loop approximation,

$$G(k) \approx \left[B_{kk} - \sum_{p \neq k} B_{kp} \, G_k(p) \, B_{pk} \right]^{-1}$$

$$= \left[\begin{pmatrix} a_k & \frac{ig}{\sqrt{\kappa}} \, \bar{\phi}_0 \\ \frac{ig}{\sqrt{\kappa}} \, \phi_0 & a_{-k} \end{pmatrix} + \frac{\lambda}{\kappa} \sum_{p \neq k} \begin{pmatrix} & \bar{\phi}_{p-k} \\ \phi_{k-p} & \end{pmatrix} G_k(p) \begin{pmatrix} & \bar{\phi}_{k-p} \\ \phi_{p-k} & \end{pmatrix} \right]^{-1}$$

$$=: \left[\begin{pmatrix} a_k & \frac{ig}{\sqrt{\kappa}} \, \bar{\phi}_0 \\ \frac{ig}{\sqrt{\kappa}} \, \phi_0 & \bar{a}_k \end{pmatrix} + \Sigma(k) \right]^{-1} \tag{10.97}$$

where, substituting again $G_k(p)$ by $G(p)$ in the infinite volume limit,

$$\Sigma(k) = \frac{\lambda}{\kappa} \sum_{p \neq k} \begin{pmatrix} & \bar{\phi}_{p-k} \\ \phi_{k-p} & \end{pmatrix} \left[\begin{pmatrix} a_p & \frac{ig}{\sqrt{\kappa}} \, \bar{\phi}_0 \\ \frac{ig}{\sqrt{\kappa}} \, \phi_0 & \bar{a}_p \end{pmatrix} + \Sigma(p) \right]^{-1} \begin{pmatrix} & \bar{\phi}_{k-p} \\ \phi_{p-k} & \end{pmatrix} \tag{10.98}$$

Anticipating the fact that the off-diagonal elements of $\langle \Sigma \rangle(k)$ will be zero (for 'zero external field'), we make the ansatz

$$\Sigma(k) = \begin{pmatrix} \sigma_k & \\ & \bar{\sigma}_k \end{pmatrix} \tag{10.99}$$

and obtain

$$\begin{pmatrix} \sigma_k & \\ & \bar{\sigma}_k \end{pmatrix} = \tag{10.100}$$

$$\frac{\lambda}{\kappa} \sum_{p \neq k} \frac{1}{(a_p + \sigma_p)(\bar{a}_p + \bar{\sigma}_p) + \frac{\lambda}{\kappa} |\phi_0|^2} \begin{pmatrix} (a_k + \sigma_k)|\phi_{p-k}|^2 & -\frac{ig}{\sqrt{\kappa}} \, \phi_0 \bar{\phi}_{k-p} \bar{\phi}_{p-k} \\ -\frac{ig}{\sqrt{\kappa}} \, \bar{\phi}_0 \phi_{k-p} \phi_{p-k} & (\bar{a}_k + \bar{\sigma}_k)|\phi_{k-p}|^2 \end{pmatrix}$$

As for the Anderson model, we perform the functional integral by substituting the quantities $|\phi_q|^2$ by their expectation values $\langle |\phi_q|^2 \rangle$. Apparently this is less obvious in this case since $dP(\phi)$ is no longer Gaussian and the $|\phi_q|^2$ are no longer identically, independently distributed. We will comment on this after (10.122) below and in section 10.6 by reinterpreting this procedure as a resummation of diagrams. For now, we simply continue in this way. Then

$$\langle G \rangle(k) = \frac{1}{|a_k + \langle \sigma_k \rangle|^2 + \frac{\lambda}{\kappa} \langle |\phi_0|^2 \rangle} \begin{pmatrix} \bar{a}_k + \langle \bar{\sigma}_k \rangle & -\frac{ig}{\sqrt{\kappa}} \, \langle \bar{\phi}_0 \rangle \\ -\frac{ig}{\sqrt{\kappa}} \, \langle \phi_0 \rangle & a_k + \langle \sigma_k \rangle \end{pmatrix} \tag{10.101}$$

where the quantity $\langle \sigma_k \rangle$ has to satisfy the equation

$$\langle \sigma_k \rangle = \frac{\lambda}{\kappa} \sum_{p \neq k} \frac{\bar{a}_p + \langle \bar{\sigma}_p \rangle}{|a_p + \langle \sigma_p \rangle|^2 + \frac{\lambda}{\kappa} \langle |\phi_0|^2 \rangle} \langle |\phi_{p-k}|^2 \rangle \tag{10.102}$$

Since $dP(\phi)$ is not Gaussian, we do not know the expectations $\langle |\phi_q|^2 \rangle$. However, by partial integration, we obtain

$$\langle |\phi_q|^2 \rangle = 1 + \tfrac{ig}{\sqrt{\kappa}} \sum_p \int \phi_q \, [B^{-1}(\phi)]_{p\uparrow,p+q\downarrow} \, dP(\phi) \qquad (10.103)$$

Namely,

$$\langle |\phi_q|^2 \rangle = \tfrac{1}{Z} \int \phi_q \bar{\phi}_q \, \det\left[\{B_{kp}(\phi)\}_{k,p}\right] e^{-\sum_q |\phi_q|^2} d\phi_q d\bar{\phi}_q$$

$$= 1 + \tfrac{1}{Z} \int \phi_q \left(\tfrac{\partial}{\partial \phi_q} \det\left[\{B_{kp}(\phi)\}_{k,p}\right]\right) e^{-\sum_q |\phi_q|^2} d\phi_q d\bar{\phi}_q$$

$$= 1 + \tfrac{1}{Z} \int \phi_q \sum_{p,\tau} \det \left[\; \bigg| \; B_{k\sigma,p'\tau'} \; \bigg| \; \frac{\partial B_{k\sigma,p\tau}}{\partial \phi_q} \; \bigg| \; B_{k\sigma,p''\tau''} \; \bigg| \; \right] e^{-\sum_q |\phi_q|^2} d\phi_q d\bar{\phi}_q$$

Since

$$\frac{\partial}{\partial \phi_q} B_{kp} = \tfrac{ig}{\sqrt{\kappa}} \begin{pmatrix} 0 & 0 \\ 1 & 0 \end{pmatrix} \delta_{k-p,q}$$

we have

$$\det \left[\; \bigg| \; B_{k\sigma,p'\tau'} \; \bigg| \; \frac{\partial B_{k\sigma,p\tau}}{\partial \phi_q} \; \bigg| \; B_{k\sigma,p''\tau''} \; \bigg| \; \right] \bigg/ \det\left[\{B_{kp}\}_{k,p}\right] = \begin{cases} 0 & \text{if } \tau = \downarrow \\[2mm] \tfrac{ig}{\sqrt{\kappa}} [B^{-1}]_{p\uparrow,p+q\downarrow} & \text{if } \tau = \uparrow \end{cases}$$

which results in (10.103).

The inverse matrix element in (10.103) we compute again with (10.7,10.8) in the two-loop approximation. Consider first the case $q = 0$. Then one gets

$$\langle |\phi_0|^2 \rangle = 1 + \tfrac{ig}{\sqrt{\kappa}} \sum_p \int \phi_0 \, G(p)_{\uparrow\downarrow} \, dP(\phi)$$

$$= 1 + \tfrac{ig}{\sqrt{\kappa}} \sum_p \int \phi_0 \frac{1}{|a_p + \sigma_p|^2 + \tfrac{\lambda}{\kappa}|\phi_0|^2} \begin{pmatrix} \bar{a}_p + \bar{\sigma}_p & -\tfrac{ig}{\sqrt{\kappa}} \bar{\phi}_0 \\ -\tfrac{ig}{\sqrt{\kappa}} \phi_0 & a_p + \sigma_p \end{pmatrix}_{\uparrow\downarrow} dP(\phi)$$

$$= 1 + \tfrac{\lambda}{\kappa} \sum_p \int \phi_0 \frac{\bar{\phi}_0}{|a_p + \sigma_p|^2 + \tfrac{\lambda}{\kappa}|\phi_0|^2} \, dP(\phi) \qquad (10.104)$$

Performing the functional integral by substitution of expectation values gives

$$\langle |\phi_0|^2 \rangle = 1 + \tfrac{\lambda}{\kappa} \sum_p \langle |\phi_0|^2 \rangle \frac{1}{|a_p + \langle \sigma_p \rangle|^2 + \tfrac{\lambda}{\kappa}\langle |\phi_0|^2 \rangle}$$

or

$$\langle |\phi_0|^2 \rangle = \frac{1}{1 - \tfrac{\lambda}{\kappa} \sum_p \dfrac{1}{|a_p + \langle \sigma_p \rangle|^2 + \tfrac{\lambda}{\kappa}\langle |\phi_0|^2 \rangle}} \qquad (10.105)$$

Before we discuss (10.105), we write down the equation for $q \neq 0$. In that case we use (10.8) to compute $[B^{-1}(\phi)]_{p\uparrow, p+q\downarrow}$ in the two-loop approximation. We obtain

$$[B^{-1}(\phi)]_{p\uparrow, p+q\downarrow} \approx -[G(p)B_{p,p+q}G(p+q)]_{\uparrow\downarrow}$$

$$= -\frac{1}{|a_p + \sigma_p|^2 + \frac{\lambda}{\kappa}|\phi_0|^2} \frac{1}{|a_{p+q} + \sigma_{p+q}|^2 + \frac{\lambda}{\kappa}|\phi_0|^2} \frac{ig}{\sqrt{\kappa}} \times$$

$$\begin{pmatrix} -\frac{ig}{\sqrt{\kappa}}[(\bar{a}+\bar{\sigma})_{p+q}\bar{\phi}_0\phi_{-q} + (\bar{a}+\bar{\sigma})_p\phi_0\bar{\phi}_q] & (\bar{a}+\bar{\sigma})_p(a+\sigma)_{p+q}\bar{\phi}_q - \frac{\lambda}{\kappa}\bar{\phi}_0^2\phi_{-q} \\ (a+\sigma)_p(\bar{a}+\bar{\sigma})_{p+q}\phi_{-q} - \frac{\lambda}{\kappa}\phi_0^2\bar{\phi}_q & -\frac{ig}{\sqrt{\kappa}}[(a+\sigma)_{p+q}\phi_0\bar{\phi}_q + (a+\sigma)_p\bar{\phi}_0\phi_{-q}] \end{pmatrix}_{\uparrow\downarrow}$$

$$= -\frac{ig}{\sqrt{\kappa}} \frac{(\bar{a}+\bar{\sigma})_p(a+\sigma)_{p+q}\bar{\phi}_q - \frac{\lambda}{\kappa}\bar{\phi}_0^2\phi_{-q}}{(|a_p+\sigma_p|^2 + \frac{\lambda}{\kappa}|\phi_0|^2)(|a_{p+q}+\sigma_{p+q}|^2 + \frac{\lambda}{\kappa}|\phi_0|^2)} \tag{10.106}$$

which gives

$$\langle |\phi_q|^2 \rangle = 1 + \frac{\lambda}{\kappa} \sum_p \int \phi_q \frac{(\bar{a}+\bar{\sigma})_p(a+\sigma)_{p+q}\bar{\phi}_q - \frac{\lambda}{\kappa}\bar{\phi}_0^2\phi_{-q}}{(|a_p+\sigma_p|^2 + \frac{\lambda}{\kappa}|\phi_0|^2)(|a_{p+q}+\sigma_{p+q}|^2 + \frac{\lambda}{\kappa}|\phi_0|^2)} \, dP(\phi)$$

$$= 1 + \frac{\lambda}{\kappa} \sum_p \frac{(\bar{a}_p + \langle \bar{\sigma}_p \rangle)(a_{p+q} + \langle \sigma_{p+q} \rangle)\langle |\phi_q|^2 \rangle - \frac{\lambda}{\kappa}\langle \bar{\phi}_0^2\phi_q\phi_{-q} \rangle}{(|a_p + \langle \sigma_p \rangle|^2 + \frac{\lambda}{\kappa}\langle |\phi_0|^2 \rangle)(|a_{p+q} + \langle \sigma_{p+q} \rangle|^2 + \frac{\lambda}{\kappa}\langle |\phi_0|^2 \rangle)} \tag{10.107}$$

The expectation $\langle \bar{\phi}_0^2\phi_q\phi_{-q} \rangle$ can be computed again by partial integration:

$$\langle \bar{\phi}_0^2\phi_q\phi_{-q} \rangle = \frac{1}{Z} \int \bar{\phi}_0^2\phi_q\phi_{-q} \det[\{B_{kp}(\phi)\}_{k,p}] \, e^{-\sum_q |\phi_q|^2} d\phi_q d\bar{\phi}_q$$

$$= \frac{1}{Z} \int \bar{\phi}_0^2\phi_q \left(\frac{\partial}{\partial \phi_{-q}} \det[\{B_{kp}(\phi)\}_{k,p}] \right) e^{-\sum_q |\phi_q|^2} d\phi_q d\bar{\phi}_q$$

$$= \frac{1}{Z} \int \bar{\phi}_0^2\phi_q \sum_{p,\tau} \det\left[B_{k\sigma,p'\tau'} \,\middle|\, \frac{\partial B_{k\sigma,p\tau}}{\partial \phi_{-q}} \,\middle|\, B_{k\sigma,p''\tau''} \right] e^{-\sum_q |\phi_q|^2} d\phi_q d\bar{\phi}_q$$

The above determinant is multiplied and divided by $\det[\{B_{kp}\}_{k,p}]$ to give

$$\det\left[B_{k\sigma,p'\tau'} \,\middle|\, \frac{\partial B_{k\sigma,p\tau}}{\partial \phi_{-q}} \,\middle|\, B_{k\sigma,p''\tau''} \right] \middle/ \det[\{B_{kp}\}_{k,p}] = \begin{cases} 0 & \text{if } \tau = \uparrow \\ \frac{ig}{\sqrt{\kappa}}[B^{-1}]_{p\downarrow, p+q\uparrow} & \text{if } \tau = \downarrow \end{cases}$$

Computing the inverse matrix element again in the two-loop approximation (10.106), we arrive at

$$\langle \bar{\phi}_0^2\phi_q\phi_{-q} \rangle = \frac{\lambda}{\kappa} \sum_p \left\langle \frac{(a_p+\sigma_p)(\bar{a}_{p+q}+\bar{\sigma}_{p+q})\bar{\phi}_0^2\phi_q\phi_{-q} - \frac{\lambda}{\kappa}\bar{\phi}_0^2\phi_0^2\phi_q\bar{\phi}_q}{(|a_p+\sigma_p|^2 + \frac{\lambda}{\kappa}|\phi_0|^2)(|a_{p+q}+\sigma_{p+q}|^2 + \frac{\lambda}{\kappa}|\phi_0|^2)} \right\rangle$$

Abbreviating

$$g_p = \frac{a_p + \langle \sigma_p \rangle}{|a_p + \langle \sigma_p \rangle|^2 + \frac{\lambda}{\kappa}\langle |\phi_0|^2 \rangle}, \qquad f_p = \frac{\sqrt{\frac{\lambda}{\kappa}\langle |\phi_0|^2 \rangle}}{|a_p + \langle \sigma_p \rangle|^2 + \frac{\lambda}{\kappa}\langle |\phi_0|^2 \rangle} \tag{10.108}$$

this gives

$$\tfrac{\lambda}{\kappa} \langle \bar{\phi}_0^2 \phi_q \phi_{-q} \rangle = \tfrac{\lambda}{\kappa} \sum_p g_p \bar{g}_{p+q} \, \tfrac{\lambda}{\kappa} \langle \bar{\phi}_0^2 \phi_q \phi_{-q} \rangle - \tfrac{\lambda}{\kappa} \sum_p f_p f_{p+q} \, \tfrac{\lambda}{\kappa} \langle |\phi_0|^2 \rangle \langle |\phi_q|^2 \rangle$$

or

$$\tfrac{\lambda}{\kappa} \langle \bar{\phi}_0^2 \phi_q \phi_{-q} \rangle = \frac{ -\tfrac{\lambda}{\kappa} \sum_p f_p f_{p+q} \, \tfrac{\lambda}{\kappa} \langle |\phi_0|^2 \rangle }{ 1 - \tfrac{\lambda}{\kappa} \sum_p g_p \bar{g}_{p+q} } \langle |\phi_q|^2 \rangle \qquad (10.109)$$

Substituting this in (10.107), we finally arrive at

$$\langle |\phi_q|^2 \rangle = \frac{ 1 - \tfrac{\lambda}{\kappa} \sum_p g_p \bar{g}_{p+q} }{ \left| 1 - \tfrac{\lambda}{\kappa} \sum_p g_p \bar{g}_{p+q} \right|^2 - \left(\tfrac{\lambda}{\kappa} \sum_p f_p f_{p+q} \right)^2 } \qquad (10.110)$$

where g_p, f_p are given by (10.108). Observe that, since $dP(\phi)$ is complex, also $\langle |\phi_q|^2 \rangle$ is in general complex. Only after summation over the q_0 variables we obtain necessarily a real quantity which is given by (10.85) and (10.92).

10.4.2 Discussion

We now discuss the solutions to (10.105) and (10.110). We assume that the solution $\langle \sigma_k \rangle$ of (10.102) is sufficiently small such that the BCS equation

$$\frac{\lambda}{\kappa} \sum_p \frac{1}{|a_p + \langle \sigma_p \rangle|^2 + |\Delta|^2} = 1 \qquad (10.111)$$

has a nonzero solution $\Delta \neq 0$ (in particular this excludes large corrections like $\langle \sigma_p \rangle \sim p_0^\alpha$, $\alpha \leq 1/2$, which one may expect in the case of Luttinger liquid behavior, for $d = 1$ one should make a separate analysis), and make the ansatz

$$\lambda \langle |\phi_0|^2 \rangle = \beta L^d |\Delta|^2 + \eta \qquad (10.112)$$

where η is independent of the volume. Then

$$\frac{\lambda}{\kappa} \sum_p \frac{1}{|a_p + \langle \sigma_p \rangle|^2 + \tfrac{\lambda}{\kappa} \langle |\phi_0|^2 \rangle} = \frac{\lambda}{\kappa} \sum_p \frac{1}{|a_p + \langle \sigma_p \rangle|^2 + |\Delta|^2 + \tfrac{\eta}{\kappa}}$$

$$= \frac{\lambda}{\kappa} \sum_p \frac{1}{|a_p + \langle \sigma_p \rangle|^2 + |\Delta|^2} - \frac{\lambda}{\kappa} \sum_p \frac{\eta/\kappa}{(|a_p + \langle \sigma_p \rangle|^2 + |\Delta|^2)^2} + O\left((\tfrac{\eta}{\kappa})^2 \right)$$

$$= 1 - c_\Delta \tfrac{\eta}{\kappa} + O\left((\tfrac{\eta}{\kappa})^2 \right) \qquad (10.113)$$

where we put $c_\Delta = \tfrac{\lambda}{\kappa} \sum_p \frac{1}{(|a_p + \langle \sigma_p \rangle|^2 + |\Delta|^2)^2}$ and used the BCS equation (10.111) in the last line. Equation (10.105) becomes

$$\kappa |\Delta|^2 + \eta = \frac{\lambda}{c_\Delta \tfrac{\eta}{\kappa} + O\left((\tfrac{\eta}{\kappa})^2 \right)} = \kappa \frac{\lambda}{c_\Delta \eta} + O(1) \qquad (10.114)$$

and has a solution $\eta = \lambda/(c_\Delta |\Delta|^2)$.

Now consider $\langle|\phi_q|^2\rangle$ for small but nonzero q. In the limit $q \to 0$ the denominator in (10.110) vanishes, or more precisely, is of order $O(1/\kappa)$ since

$$1 - \tfrac{\lambda}{\kappa}\sum_p g_p \bar{g}_p - \tfrac{\lambda}{\kappa}\sum_p f_p f_p =$$

$$1 - \tfrac{\lambda}{\kappa}\sum_p \frac{1}{|a_p + \langle\sigma_p\rangle|^2 + \tfrac{\lambda}{\kappa}\langle|\phi_0|^2\rangle} = O(1/\kappa) \qquad (10.115)$$

because of (10.113). If we assume the second derivatives of $\langle\sigma_k\rangle$ to be integrable (which should be the case for $d = 3$ and $\langle|\phi_q|^2\rangle \sim 1/q^2$ by virtue of (10.102)), then, since the denominator in (10.110) is an even function of q, the small q behavior of $\langle|\phi_q|^2\rangle$ is $1/q^2$. This agrees with the common expectations [25, 14, 11]. Usually the behavior of $\langle|\phi_q|^2\rangle$ is inferred from the second order Taylor expansion of the effective potential

$$V_{\text{eff}}(\{\phi_q\}) = \sum_q |\phi_q|^2 - \log\det\begin{bmatrix} \delta_{k,p} & \frac{ig}{\sqrt{\kappa}}\frac{\bar{\phi}_{p-k}}{a_k} \\ \frac{ig}{\sqrt{\kappa}}\frac{\phi_{k-p}}{a_{-k}} & \delta_{k,p} \end{bmatrix} \qquad (10.116)$$

around its global minimum (see section 5.2)

$$\phi_q^{\min} = \sqrt{\beta L^d}\, \tfrac{|\Delta|}{\sqrt{\lambda}}\, \delta_{q,0}\, e^{i\theta_0} \qquad (10.117)$$

where the phase θ_0 of ϕ_0 is arbitrary. If one expands V_{eff} up to second order in

$$\xi_q = \phi_q - \delta_{q,0}\sqrt{\beta L^d}\, \tfrac{|\Delta|}{\sqrt{\lambda}}\, e^{i\theta_0} = \begin{cases} \left(\rho_0 - \sqrt{\beta L^d}\,\tfrac{|\Delta|}{\sqrt{\lambda}}\right) e^{i\theta_0} & \text{for } q = 0 \\ \rho_q\, e^{i\theta_q} & \text{for } q \neq 0 \end{cases} \qquad (10.118)$$

one obtains (see section 5.2)

$$V_{\text{eff}}(\{\phi_q\}) = V_{\min} + 2\beta_0\left(\rho_0 - \sqrt{\beta L^d}\,\tfrac{|\Delta|}{\sqrt{\lambda}}\right)^2 + \sum_{q\neq 0}(\alpha_q + i\gamma_q)\rho_q^2$$

$$+ \tfrac{1}{2}\sum_{q\neq 0}\beta_q\,|e^{-i\theta_0}\phi_q + e^{i\theta_0}\bar{\phi}_{-q}|^2 + O(\xi^3) \qquad (10.119)$$

where for small q one has $\alpha_q, \gamma_q \sim q^2$. Hence, if V_{eff} is substituted by the right hand side of (10.119) one obtains $\langle|\phi_q|^2\rangle \sim 1/q^2$.

For $d = 3$, this seems to be the right answer, but in lower dimensions one would expect an integrable singularity due to (10.102) and (10.84), (10.85) and (10.92). In particular, we think it would be a very interesting problem to solve the integral equations (10.102), (10.105) and (10.110) for $d = 1$ and to check the result for Luttinger liquid behavior. A good warm-up exercise would be to consider the $0+1$ dimensional problem, that is, we only have the k_0, p_0, q_0-variables. In that case the 'bare BCS equation'

$$\frac{\lambda}{\beta} \sum_{p_0 \in \frac{\pi}{\beta}(2\mathbb{Z}+1)} \frac{1}{p_0^2 + |\Delta|^2} = 1 \tag{10.120}$$

still has a nonzero solution Δ for sufficiently small $T = 1/\beta$ and the question would be whether the correction $\langle \sigma_{p_0} \rangle$ is sufficiently big to destroy the gap. That is, does the 'renormalized BCS equation'

$$\frac{\lambda}{\beta} \sum_{p_0 \in \frac{\pi}{\beta}(2\mathbb{Z}+1)} \frac{1}{|p_0 + \langle \sigma_{p_0} \rangle|^2 + |\Delta|^2} = 1 \tag{10.121}$$

$\langle \sigma_{p_0} \rangle$ being the solution to (10.102), (10.105) and (10.110), still have a nonzero solution? We remark that, if the gap vanishes (for arbitrary dimension), then also the singularity of $\langle |\phi_q|^2 \rangle$ disappears. Namely, if the gap equation has no solution, that is, if $\frac{1}{\kappa} \sum_p \frac{1}{|a_p + \langle \sigma_p \rangle|^2} < \infty$, then $\langle |\phi_0|^2 \rangle$ given by (10.105) is no longer macroscopic (for sufficiently small coupling) and $\frac{\lambda}{\kappa} \langle |\phi_0|^2 \rangle$ vanishes in the infinite volume limit. And the denominator in (10.110) becomes for $q \to 0$

$$1 - \frac{\lambda}{\kappa} \sum_p \frac{1}{|a_p + \langle \sigma_p \rangle|^2}$$

which would be nonzero (for sufficiently small coupling).

Finally we argue why it is reasonable to substitute $|\phi_0|^2$ by its expectation value while performing the functional integral. We may write the effective potential (10.116) as

$$V_{\text{eff}}(\{\phi_q\}) = V_1(\phi_0) + V_2(\{\phi_q\}) \tag{10.122}$$

where

$$V_1(\phi_0) = |\phi_0|^2 - \sum_k \log \left[1 + \frac{\lambda}{\kappa} \frac{|\phi_0|^2}{k_0^2 + e_{\mathbf{k}}^2} \right]$$

$$= \kappa \left(\frac{|\phi_0|^2}{\kappa} - \frac{1}{\kappa} \sum_k \log \left[1 + \frac{\frac{\lambda}{\kappa} |\phi_0|^2}{k_0^2 + e_{\mathbf{k}}^2} \right] \right) \equiv \kappa V_{\text{BCS}} \left(\frac{|\phi_0|}{\sqrt{\kappa}} \right) \tag{10.123}$$

and

$$V_2(\{\phi_q\}) = \tag{10.124}$$

$$\sum_{q \neq 0} |\phi_q|^2 - \log \det \left[\begin{pmatrix} \delta_{k,p} & \frac{ig}{\sqrt{\kappa}} \frac{\bar{\phi}_0}{a_k} \delta_{k,p} \\ \frac{ig}{\sqrt{\kappa}} \frac{\phi_0}{a_{-k}} \delta_{k,p} & \delta_{k,p} \end{pmatrix}^{-1} \begin{pmatrix} \delta_{k,p} & \frac{ig}{\sqrt{\kappa}} \frac{\bar{\phi}_{p-k}}{a_k} \\ \frac{ig}{\sqrt{\kappa}} \frac{\phi_{k-p}}{a_{-k}} & \delta_{k,p} \end{pmatrix} \right]$$

If we ignore the ϕ_0-dependence of V_2, then the ϕ_0-integral

$$\frac{\int F\left(\frac{1}{\kappa}|\phi_0|^2\right) e^{-V_1(\phi_0)} d\phi_0 d\bar{\phi}_0}{\int e^{-V_1(\phi_0)} d\phi_0 d\bar{\phi}_0} = \tag{10.125}$$

$$\frac{\int F\left(\rho^2\right) e^{-\kappa V_{\text{BCS}}(\rho)} \rho \, d\rho}{\int e^{-\kappa V_{\text{BCS}}(\rho)} \rho \, d\rho} \overset{\kappa \to \infty}{\longrightarrow} F(\rho_{\min}^2) = F\left(\frac{1}{\kappa}\langle|\phi_0|^2\rangle\right) \tag{10.126}$$

simply puts $|\phi_0|^2$ at the global minimum of the (BCS) effective potential.

10.5 Application to Bosonic Models

10.5.1 The φ^4-Model

In this section we choose the φ^4-model as a typical bosonic model to demonstrate our resummation method. As usual, we start in finite volume $[0, L]^d$ on a lattice with lattice spacing $1/M$. The two point function is given by

$$S(x, y) = \langle \varphi_x \varphi_y \rangle \tag{10.127}$$

$$:= \frac{\int_{\mathbb{R}^{N^d}} \varphi_x \varphi_y \, e^{-\frac{g^2}{2} \frac{1}{M^d} \sum_x \varphi_x^4} e^{-\frac{1}{M^{2d}} \sum_{x,y} (-\Delta + m^2)_{x,y} \varphi_x \varphi_y} \prod_x d\varphi_x}{\int_{\mathbb{R}^{N^d}} e^{-\frac{g^2}{2} \frac{1}{M^d} \sum_x \varphi_x^4} e^{-\frac{1}{M^{2d}} \sum_{x,y} (-\Delta + m^2)_{x,y} \varphi_x \varphi_y} \prod_x d\varphi_x}$$

where

$$(-\Delta + m^2)_{x,y} = M^d \left[-M^2 \sum_{i=1}^d (\delta_{x,y-e_i/M} + \delta_{x,y+e_i/M} - 2\delta_{x,y}) + m^2 \delta_{x,y} \right] \tag{10.128}$$

First we have to bring this into the form $\int [P + Q]_{x,y}^{-1} d\mu$, P diagonal in momentum space, Q diagonal in coordinate space. This is done again by making a Hubbard-Stratonovich transformation which in this case reads

$$e^{-\frac{1}{2}\sum_x a_x^2} = \int e^{i\sum_x a_x u_x} e^{-\frac{1}{2}\sum_x u_x^2} \prod_x \frac{du_x}{\sqrt{2\pi}} \tag{10.129}$$

with

$$a_x = \frac{g}{\sqrt{M^d}} \varphi_x^2 \tag{10.130}$$

The result is Gaussian in the φ_x-variables and the integral over these variables gives

$$S(x,y) = \int_{\mathbb{R}^{N^d}} \left[\tfrac{1}{M^{2d}}(-\Delta + m^2)_{x,y} - \tfrac{ig}{\sqrt{M^d}} u_x \delta_{x,y} \right]^{-1}_{x,y} dP(u) \qquad (10.131)$$

where

$$dP(u) = \tfrac{1}{Z} \det \left[\tfrac{1}{M^{2d}}(-\Delta + m^2)_{x,y} - \tfrac{ig}{\sqrt{M^d}} u_x \delta_{x,y} \right]^{-\frac{1}{2}} e^{-\frac{1}{2}\sum_x u_x^2} \prod_x du_x$$

Since we have bosons, the determinant comes with a power of $-1/2$ which is the only difference compared to a fermionic system. In momentum space this reads

$$S(x-y) = \tfrac{1}{L^d} \sum_k e^{ik(x-y)} \langle G \rangle(k) \qquad (10.132)$$

where ($\gamma_q = v_q + iw_q$, $\gamma_{-q} = \bar{\gamma}_q$, $d\gamma_q d\bar{\gamma}_q := dv_q dw_q$)

$$\langle G \rangle(k) = \int_{\mathbb{R}^{N^d}} \left[a_k \delta_{k,p} - \tfrac{ig}{\sqrt{L^d}} \gamma_{k-p} \right]^{-1}_{kk} dP(\gamma) \qquad (10.133)$$

and

$$dP(\gamma) = \tfrac{1}{Z} \det \left[a_k \delta_{k,p} - \tfrac{ig}{\sqrt{L^d}} \gamma_{k-p} \right]^{-\frac{1}{2}} e^{-\frac{1}{2} v_0^2} dv_0 \prod_{q \in \mathcal{M}^+} e^{-|\gamma_q|^2} d\gamma_q d\bar{\gamma}_q$$

and $\mathcal{M}^+$ again is a set such that either $q \in \mathcal{M}^+$ or $-q \in \mathcal{M}^+$. Furthermore

$$a_k = 4M^2 \sum_{i=1}^{d} \sin^2 \left[\tfrac{k_i}{2M} \right] + m^2 \qquad (10.134)$$

Equation (10.133) is our starting point. We apply (10.7) to the inverse matrix element in (10.133). In the two loop approximation one obtains ($\gamma_0 = v_0 \in \mathbb{R}$)

$$\left[a_k \delta_{k,p} - \tfrac{ig}{\sqrt{L^d}} \gamma_{k-p} \right]^{-1}_{kk} \approx \frac{1}{a_k - \tfrac{igv_0}{\sqrt{L^d}} + \tfrac{g^2}{L^d} \sum_{p \neq k} G_k(p) |\gamma_{k-p}|^2}$$

$$=: \frac{1}{a_k + \sigma_k} \qquad (10.135)$$

where

$$\sigma_k = -\tfrac{ig}{\sqrt{L^d}} v_0 + \tfrac{g^2}{L^d} \sum_{p \neq k} \frac{|\gamma_{k-p}|^2}{a_p - \tfrac{igv_0}{\sqrt{L^d}} + \sigma_p} \qquad (10.136)$$

which results in

$$\langle G \rangle (k) = \frac{1}{a_k + \langle \sigma_k \rangle} \tag{10.137}$$

where $\langle \sigma_k \rangle$ has to satisfy the equation

$$\langle \sigma_k \rangle = -\frac{ig}{\sqrt{L^d}} \langle v_0 \rangle + \frac{g^2}{L^d} \sum_{p \neq k} \frac{\langle |\gamma_{k-p}|^2 \rangle}{a_p + \langle \sigma_p \rangle} \tag{10.138}$$

$$= \frac{g^2}{2L^d} \sum_p \langle G \rangle (p) + \frac{g^2}{L^d} \sum_{p \neq k} \frac{\langle |\gamma_{k-p}|^2 \rangle}{a_p + \langle \sigma_p \rangle} = \frac{g^2}{L^d} \sum_{p \neq k} \frac{\langle |\gamma_{k-p}|^2 \rangle + \frac{1}{2}}{a_p + \langle \sigma_p \rangle}$$

where the last line is due to

$$\langle v_0 \rangle = \frac{1}{Z} \int v_0 \det \left[a_k \delta_{k,p} - \frac{ig}{\sqrt{L^d}} \gamma_{k-p} \right]^{-\frac{1}{2}} e^{-\frac{1}{2} v_0^2} dv_0 \prod_{q \in \mathcal{M}^+} e^{-|\gamma_q|^2} d\gamma_q d\bar{\gamma}_q$$

$$= \frac{1}{Z} \int \left\{ \frac{\partial}{\partial v_0} \det \left[a_k \delta_{k,p} - \frac{ig}{\sqrt{L^d}} \gamma_{k-p} \right]^{-\frac{1}{2}} \right\} e^{-\frac{1}{2} v_0^2} dv_0 \prod_{q \in \mathcal{M}^+} e^{-|\gamma_q|^2} d\gamma_q d\bar{\gamma}_q$$

$$= -\frac{1}{2} \sum_p \left(-\frac{ig}{\sqrt{L^d}} \right) \int \left[a_k \delta_{k,p} - \frac{ig}{\sqrt{L^d}} \gamma_{k-p} \right]^{-1}_{pp} dP(\gamma) \tag{10.139}$$

As for the many-electron system, we can derive an equation for $\langle |\gamma_q|^2 \rangle$ by partial integration:

$$\langle |\gamma_q|^2 \rangle = \frac{1}{Z} \int \gamma_q \bar{\gamma}_q \det \left[a_k \delta_{k,p} - \frac{ig}{\sqrt{L^d}} \gamma_{k-p} \right]^{-\frac{1}{2}} e^{-\frac{v_0^2}{2}} dv_0 \prod_q e^{-|\gamma_q|^2} d\gamma_q d\bar{\gamma}_q$$

$$= 1 + \frac{1}{Z} \int \gamma_q \frac{\partial}{\partial \gamma_q} \left\{ \det \left[a_k \delta_{k,p} - \frac{ig}{\sqrt{L^d}} \gamma_{k-p} \right]^{-\frac{1}{2}} \right\} e^{-\frac{v_0^2}{2}} dv_0 \prod_q e^{-\frac{1}{2} |\gamma_q|^2} d\gamma_q d\bar{\gamma}_q$$

$$= 1 - \frac{1}{2} \int \gamma_q \frac{\frac{\partial}{\partial \gamma_q} \det \left[a_k \delta_{k,p} - \frac{ig}{\sqrt{L^d}} \gamma_{k-p} \right]}{\det \left[a_k \delta_{k,p} - \frac{ig}{\sqrt{L^d}} \gamma_{k-p} \right]} dP(\gamma)$$

$$= 1 - \frac{1}{2} \sum_p \frac{-ig}{\sqrt{L^d}} \int \gamma_q \left[a_k \delta_{k,p} - \frac{ig}{\sqrt{L^d}} \gamma_{k-p} \right]^{-1}_{p,p+q} dP(\gamma) \tag{10.140}$$

Computing the inverse matrix element in (10.140) again with (10.8) in the two-loop approximation, one arrives at

$$\langle |\gamma_q|^2 \rangle = 1 - \langle |\gamma_q|^2 \rangle \frac{g^2}{2L^d} \sum_p \frac{1}{(a_p + \langle \sigma_p \rangle)(a_{p+q} + \langle \sigma_{p+q} \rangle)}$$

or

$$\langle |\gamma_q|^2 \rangle = \frac{1}{1 + \frac{g^2}{2} \int_{[0,2\pi M]^d} \frac{d^d p}{(2\pi)^d} \frac{1}{(a_p + \langle \sigma_p \rangle)(a_{p+q} + \langle \sigma_{p+q} \rangle)}} \tag{10.141}$$

which has to be solved in conjunction with

$$\langle \sigma_k \rangle = g^2 \int_{[0,2\pi M]^d} \frac{d^d p}{(2\pi)^d} \frac{\langle |\gamma_{k-p}|^2 \rangle + \frac{1}{2}}{a_p + \langle \sigma_p \rangle} \tag{10.142}$$

Introducing the rescaled quantities

$$\langle \sigma_k \rangle = M^2 s_{\frac{p}{M}}, \quad \langle |\gamma_q|^2 \rangle = \lambda_{\frac{q}{M}}, \quad a_k = M^2 \varepsilon_{\frac{k}{M}}, \quad \varepsilon_k = \sum_{i=1}^{d} \sin^2 \frac{k_i}{2} + \frac{m^2}{M^2} \tag{10.143}$$

(10.141) and (10.142) read

$$s_k = M^{d-4} g^2 \int_{[0,2\pi]^d} \frac{d^d p}{(2\pi)^d} \frac{\lambda_{k-p} + \frac{1}{2}}{\varepsilon_p + s_p} \tag{10.144}$$

$$\lambda_q = \frac{1}{1 + M^{d-4}\frac{g^2}{2} \int_{[0,2\pi]^d} \frac{d^d p}{(2\pi)^d} \frac{1}{(\varepsilon_p + s_p)(\varepsilon_{p+q} + s_{p+q})}} \tag{10.145}$$

Unfortunately we cannot check this result with the rigorously proven triviality theorem since $\langle \sigma_k \rangle$ and $\langle |\gamma_q|^2 \rangle$ only give information on the two-point function $S(x, y)$, (10.127), and on $\frac{g^2}{M^d} \sum_x \langle \varphi(x)^4 \rangle = \sum_q \Lambda(q)$ where $\Lambda(q) = \langle |\gamma_q|^2 \rangle - 1$. However, the triviality theorem [32, 30] makes a statement on the connected four-point function $S_{4,c}(x_1, x_2, x_3, x_4)$ at noncoinciding arguments, namely that this function vanishes in the continuum limit in dimension $d > 4$.

10.5.2 Higher Orders

We now include the higher loop terms of (10.7), (10.8) and give an interpretation in terms of diagrams. The exact equations for $\langle G \rangle(k)$ and $\langle |\gamma_q|^2 \rangle$ are

$$\langle G \rangle(k) = \int \left[a_k \delta_{k,p} - \frac{ig}{\sqrt{L^d}} \gamma_{k-p} \right]_{kk}^{-1} dP(\gamma) = \left\langle \frac{1}{a_k + \sigma_k} \right\rangle \tag{10.146}$$

$$\sigma_k = -\frac{ig}{\sqrt{L^d}} v_0 + \sum_{r=2}^{N^d} \left(\frac{ig}{\sqrt{L^d}} \right)^r \sum_{\substack{p_2 \cdots p_r \neq k \\ p_i \neq p_j}} G_k(p_2) \cdots G_{kp_2 \cdots p_{r-1}}(p_r) \times$$

$$\times \gamma_{k-p_2} \gamma_{p_2-p_3} \cdots \gamma_{p_r-k}$$

and

$$\langle |\gamma_q|^2 \rangle = 1 + \frac{ig}{2\sqrt{L^d}} \sum_p \int \gamma_q \left[a_k \delta_{k,p} - \frac{ig}{\sqrt{L^d}} \gamma_{k-p} \right]_{p,p+q}^{-1} dP(\gamma) \tag{10.147}$$

$$\stackrel{p \to p_2}{=} 1 + \frac{1}{2} \sum_{r=2}^{N^d} \left(\frac{ig}{\sqrt{L^d}} \right)^r \sum_{\substack{p_2 \cdots p_r \neq p_2 + q \\ p_i \neq p_j}} \left\langle G(p_2) G_{p_2}(p_3) \times \cdots \times G_{p_2 \cdots p_{r-1}}(p_r) \times \right.$$

$$\left. \times \gamma_{p_2-p_3} \cdots \gamma_{p_{r-1}-p_r} \gamma_{p_r-p_2-q} \gamma_{p_2+q-p_2} \right\rangle$$

For $r > 2$, we obtain terms $\langle \gamma_{k_1} \cdots \gamma_{k_r} \rangle$ whose connected contributions are, in terms of the electron or φ^4-lines, are at least six-legged. Since for the many-electron system and for the φ^4-model [54] (for $d = 4$) the relevant diagrams are two- and four-legged, one may start with an approximation which ignores the connected r-loop contributions for $r > 2$. This is obtained by writing

$$\langle \gamma_{k_1} \cdots \gamma_{k_n} \rangle \approx \langle \gamma_{k_1} \cdots \gamma_{k_n} \rangle_2 \qquad (10.148)$$

where (the index '2' for 'retaining only two-loop contributions')

$$\langle \gamma_{k_1} \cdots \gamma_{k_{2n}} \rangle_2 := \sum_{\text{pairings } \sigma} \langle \gamma_{k_{\sigma 1}} \gamma_{k_{\sigma 2}} \rangle \cdots \langle \gamma_{k_{\sigma(2n-1)}} \gamma_{k_{\sigma 2n}} \rangle$$

$$= \int \gamma_{k_1} \cdots \gamma_{k_{2n}} \, dP_2(\gamma) \qquad (10.149)$$

if we define

$$dP_2(\gamma) := e^{-\sum_q \frac{|\gamma_q|^2}{\langle |\gamma_q|^2 \rangle}} \prod_q \frac{d\gamma_q \, d\bar{\gamma}_q}{\pi \langle |\gamma_q|^2 \rangle} \qquad (10.150)$$

Substituting dP by dP_2 in (10.147) and (10.148), we obtain a model which differs from the original model only by irrelevant contributions and for which we are able to write down a closed set of equations for the two-legged particle correlation function $\langle G \rangle(k)$ and the two-legged interaction correlation function $\langle |\gamma_q|^2 \rangle$ by resumming all two-legged (particle and squiggle) subdiagrams. The exact equations of this model are

$$\langle G \rangle(k) = \int \left[a_k \delta_{k,p} - \frac{ig}{\sqrt{L^d}} \gamma_{k-p} \right]_{kk}^{-1} dP_2(\gamma) \qquad (10.151)$$

$$\langle |\gamma_q|^2 \rangle = 1 + \frac{ig}{2\sqrt{L^d}} \sum_p \int \gamma_q \left[a_k \delta_{k,p} - \frac{ig}{\sqrt{L^d}} \gamma_{k-p} \right]_{p,p+q}^{-1} dP_2(\gamma) \qquad (10.152)$$

and the resummation of the two-legged particle and squiggle subdiagrams is obtained by applying the inversion formula (10.7) and (10.8) to the inverse matrix elements in (10.151) and (10.152). A discussion similar to those of section 10.3 gives the following closed set of equations for the quantities $\langle G \rangle(k)$ and $\langle |\gamma_q|^2 \rangle$:

$$\langle G \rangle(k) = \frac{1}{a_k + \langle \sigma_k \rangle}, \qquad \langle |\gamma_q|^2 \rangle = \frac{1}{1 + \langle \pi_q \rangle} \qquad (10.153)$$

where

$$\langle \sigma_k \rangle = \frac{g^2}{2L^d} \sum_p \langle G \rangle (p) + \sum_{r=2}^{\ell} \left(\frac{ig}{\sqrt{L^d}} \right)^r \sum_{\substack{p_2 \cdots p_r \neq k \\ p_i \neq p_j}} \langle G \rangle (p_2) \cdots \langle G \rangle (p_r) \times$$

$$\times \; \langle \gamma_{k-p_2} \gamma_{p_2-p_3} \cdots \gamma_{p_r-k} \rangle_2 \qquad (10.154)$$

$$\langle \pi_q \rangle = -\frac{1}{2} \sum_{r=2}^{\ell} \left(\frac{ig}{\sqrt{L^d}} \right)^r \sum_{s=3}^{r-1} \sum_{\substack{p_2 \cdots p_r \neq p_2 + q \\ p_i \neq p_j}} \left(\delta_{q, p_{s+1} - p_s} \langle G \rangle (p_2) \cdots \langle G \rangle (p_r) \times \right.$$

$$\left. \times \; \langle G \rangle (p_2 + q) \langle \gamma_{p_2 - p_3} \cdots \widehat{\gamma}_{p_s - p_{s+1}} \cdots \gamma_{p_{r-1} - p_r} \gamma_{p_r - p_2 - q} \rangle_2 \right) \qquad (10.155)$$

In the last line we used that γ_q in (10.148) cannot contract to $\gamma_{p_2-p_3}$ or to $\gamma_{p_r-p_2-q}$. If the expectations of the γ-fields on the right hand side of (10.154) and (10.155) are computed according to (10.149), one obtains the expansion into diagrams. The graphs contributing to $\langle \sigma_k \rangle$ have exactly one string of particle lines, each line having $\langle G \rangle$ as propagator, and no particle loops (up to the tadpole diagram). Each squiggle corresponds to a factor $\langle |\gamma|^2 \rangle$. The diagrams contributing to $\langle \pi \rangle$ have exactly one particle loop, the propagators being again the interacting two-point functions, $\langle G \rangle$ for the particle lines and $\langle |\gamma|^2 \rangle$ for the squiggles. In both cases there are no two-legged subdiagrams. However, although the equation $\langle |\gamma_q|^2 \rangle = \frac{1}{1+\langle \pi_q \rangle}$ resums ladder or bubble diagrams and more general four-legged particle subdiagrams if the terms for $r \geq 4$ in (10.155) are taken into account, the right hand side of (10.154) and (10.155) still contains diagrams with four-legged particle subdiagrams. Thus, the resummation of four-legged particle subdiagrams is only partially through the complete resummation of two-legged squiggle diagrams. Also observe that, in going from (10.151), (10.152) to (10.153) to (10.155), we cut off the r-sum at some fixed order ℓ independent of the volume since we can only expect that the expansions are asymptotic ones, compare the discussion in section 10.3.

10.5.3 The $\varphi^2 \psi^2$-Model

This problem was suggested to us by A. Sokal. One has two real scalar bosonic fields φ and ψ or φ_1 and φ_2 (since ψ 'looks fermionic') with free action $-\Delta + m_i^2$ and coupling $\lambda \sum_x \varphi_1^2(x) \varphi_2^2(x)$. One expects that there is mass generation in the limit $m_1 = m_2 \to 0$. In the following we present a computation using (10.7) in two-loop approximation which shows this behavior.

Let coordinate space be a lattice of finite volume with periodic boundary conditions, lattice spacing 1 and volume $[0, L]^d$:

$$\Gamma = \{ x = (n_1, \cdots, n_d) \mid 0 \leq n_i \leq L - 1 \} = \mathbb{Z}^d / (L\mathbb{Z})^d$$

Momentum space is given by

$$\mathcal{M} := \Gamma^{\sharp} = \left\{ k = \frac{2\pi}{L} (m_1, \cdots, m_d) \mid 0 \leq m_i \leq L - 1 \right\} = \left(\frac{2\pi}{L} \mathbb{Z} \right)^d / (2\pi \mathbb{Z})^d$$

We consider the two-point functions ($i \in \{1,2\}$)

$$S_i(x,y) = \frac{1}{Z} \int_{\mathbb{R}^{2L^d}} \varphi_{i,x} \varphi_{i,y} \, e^{-\lambda \sum_{x \in \Gamma} \varphi_{1,x}^2 \varphi_{2,x}^2} \times \tag{10.156}$$

$$e^{-\sum_{i=1}^{2} \sum_{x \in \Gamma} \varphi_{i,x}(-\Delta + m_i^2)\varphi_{i,x}} \prod_{x \in \Gamma} d\varphi_{1,x} d\varphi_{2,x}$$

where

$$Z = \int_{\mathbb{R}^{2L^d}} e^{-\lambda \sum_{x \in \Gamma} \varphi_{1,x}^2 \varphi_{2,x}^2} e^{-\sum_{i=1}^{2} \sum_{x \in \Gamma} \varphi_{i,x}(-\Delta + m_i^2)\varphi_{i,x}} \prod_{x \in \Gamma} d\varphi_{1,x} d\varphi_{2,x}$$
$$\tag{10.157}$$

Consider, say, S_1. We may integrate out the φ_2 field to obtain

$$S_1(x,y) = \frac{1}{Z} \int_{\mathbb{R}^{L^d}} \varphi_{1,x} \varphi_{1,y} \, \det\left[-\Delta + m_2^2 + \lambda \varphi_{1,x}^2 \delta_{x,x'}\right]^{-\frac{1}{2}} \times$$

$$e^{-\sum_x \varphi_{1,x}(-\Delta + m_1^2)\varphi_{1,x}} \prod_x d\varphi_{1,x} \tag{10.158}$$

On the other hand, integrating out the φ_1 field gives the following representation:

$$S_1(x,y) = \frac{1}{Z} \int_{\mathbb{R}^{L^d}} \left\{ \int_{\mathbb{R}^{L^d}} \varphi_{1,x} \varphi_{1,y} \, \frac{e^{-\sum_x \varphi_{1,x}(-\Delta + m_1^2 + \lambda \varphi_{2,x}^2)\varphi_{1,x}}}{\det\left[-\Delta + m_1^2 + \lambda \varphi_2^2\right]^{-\frac{1}{2}}} \prod_x d\varphi_{1,x} \right\} \times$$

$$\det\left[-\Delta + m_1^2 + \lambda \varphi_2^2\right]^{-\frac{1}{2}} e^{-\sum_x \varphi_{2,x}(-\Delta + m_2^2)\varphi_{2,x}} \prod_x d\varphi_{2,x}$$

$$= \frac{1}{Z} \int_{\mathbb{R}^{L^d}} \left[-\Delta + m_1^2 + \lambda \varphi_2^2\right]^{-1}_{x,y} dP(\varphi_2) \tag{10.159}$$

where

$$dP(\varphi_2) = \frac{1}{Z} \det\left[-\Delta + m_1^2 + \lambda \varphi_2^2\right]^{-\frac{1}{2}} e^{-\sum_x \varphi_{2,x}(-\Delta + m_2^2)\varphi_{2,x}} \prod_x d\varphi_{2,x}$$

and $\int dP(\varphi_2) = 1$ (the Z factor above differs from its definition in (10.157) by some factors of 2π, for notational simplicity we have chosen the same symbol). By taking the Fourier transform (and dropping the 'hats' and the index 2 on the Fourier transformed φ_2 field)

$$S_1(x-y) = \frac{1}{L^d} \sum_k e^{ik(x-y)} \langle G_1 \rangle(k) \tag{10.160}$$

where ($\varphi_q = v_q + iw_q$, $\varphi_{-q} = \bar{\varphi}_q$, $d\varphi_q d\bar{\varphi}_q := dv_q dw_q$)

$$\langle G_1 \rangle(k) = \int_{\mathbb{R}^{L^d}} \left[a_{1,k}\delta_{k,p} + \frac{\lambda}{\sqrt{L^d}} \widehat{(\varphi^2)}_{k-p}\right]^{-1}_{k,k} dP(\{\varphi_q\})$$

$$= \int_{\mathbb{R}^{L^d}} \left[a_{1,k}\delta_{k,p} + \frac{\lambda}{L^d} \sum_t \varphi_{k-t}\varphi_{t-p}\right]^{-1}_{k,k} dP(\{\varphi_q\}) \tag{10.161}$$

To keep track of the volume factors recall that φ_q is obtained from φ_x by applying the unitary matrix $F = (\frac{1}{\sqrt{L^d}} e^{-ikx})_{k,x}$ of discrete Fourier transform, that is

$$\varphi_q = \frac{1}{\sqrt{L^d}} \sum_x e^{-iqx} \varphi_x, \quad \varphi_x = \frac{1}{\sqrt{L^d}} \sum_q e^{iqx} \varphi_q$$

$$\left(F \left[\varphi_x^2 \delta_{x,x'} \right] F^* \right)_{k,p} = \frac{1}{L^d} \sum_x e^{-i(k-p)x} \varphi_x^2 = \frac{1}{\sqrt{L^d}} \widehat{(\varphi^2)}_{k-p}$$

and

$$\varphi_x^2 = \frac{1}{L^d} \sum_{k,p} e^{i(k+p)x} \varphi_k \varphi_p = \frac{1}{\sqrt{L^d}} \sum_q e^{iqx} \frac{1}{\sqrt{L^d}} \sum_k \varphi_k \varphi_{q-k}$$

which gives

$$\widehat{(\varphi^2)}_{k-p} = \frac{1}{\sqrt{L^d}} \sum_t \varphi_t \varphi_{k-p-t} = \frac{1}{\sqrt{L^d}} \sum_t \varphi_{t-p} \varphi_{k-t} \qquad (10.162)$$

Furthermore

$$dP(\{\varphi_q\}) = \frac{1}{Z} \det \left[a_{1,k} \delta_{k,p} + \frac{\lambda}{L^d} \sum_t \varphi_{k-t} \varphi_{t-p} \right]^{-\frac{1}{2}} \times$$

$$\times \prod_{q \in \mathcal{M}^+} e^{-(q^2+m_2^2)|\varphi_q|^2} d\varphi_q d\bar{\varphi}_q \qquad (10.163)$$

and

$$a_{1,k} = 4 \sum_{i=1}^d \sin^2 \left[\tfrac{k_i}{2} \right] + m_1^2 \qquad (10.164)$$

Equation (10.161) is the starting point for the application of the inversion formula (10.7). We have to compute $\langle G_1 \rangle(k) = \int G_1(k) \, dP(\varphi)$ where

$$G_1(k) = \left[a_{1,k} \delta_{k,p} + \frac{\lambda}{L^d} \sum_t \varphi_{k-t} \varphi_{t-p} \right]_{k,k}^{-1} \qquad (10.165)$$

In the two-loop approximation one obtains

$$G_1(k) = \frac{1}{a_{1,k} + \sigma_{1,k}} \qquad (10.166)$$

where, approximating $G_{1,k}(p)$ by $G_1(p)$ in the infinite volume limit,

$$\sigma_{1,k} = \frac{\lambda}{L^d} \sum_t \varphi_{k-t} \varphi_{t-k} - \sum_{\substack{p \\ p \neq k}} \frac{\lambda}{L^d} \sum_s \varphi_{k-s} \varphi_{s-p} \, G_1(p) \frac{\lambda}{L^d} \sum_t \varphi_{p-t} \varphi_{t-k}$$

$$= \frac{\lambda}{L^d} \sum_t |\varphi_t|^2 - \frac{\lambda^2}{L^{2d}} \sum_{\substack{p,s,t \\ p \neq k}} \varphi_{k-s} \varphi_{s-p} \varphi_{p-t} \varphi_{t-k} \, G_1(p) \qquad (10.167)$$

Thus

$$\langle G_1 \rangle(k) = \frac{1}{a_{1,k} + \langle \sigma_{1,k} \rangle} \tag{10.168}$$

where

$$\langle \sigma_{1,k} \rangle = \frac{\lambda}{L^d} \sum_t \langle |\varphi_t|^2 \rangle - \frac{\lambda^2}{L^{2d}} \sum_{\substack{p,s,t \\ p \neq k}} \langle \varphi_{k-s} \varphi_{s-p} \varphi_{p-t} \varphi_{t-k} \rangle \langle G_1 \rangle(p) \tag{10.169}$$

To obtain a closed set of integral equations, we ignore connected three and higher loops and approximate

$$\begin{aligned} \langle \varphi_{k-s} \varphi_{s-p} \varphi_{p-t} \varphi_{t-k} \rangle &\approx \langle \varphi_{k-s} \varphi_{s-p} \rangle \langle \varphi_{p-t} \varphi_{t-k} \rangle + \langle \varphi_{k-s} \varphi_{p-t} \rangle \langle \varphi_{s-p} \varphi_{t-k} \rangle \\ &\quad + \langle \varphi_{k-s} \varphi_{t-k} \rangle \langle \varphi_{s-p} \varphi_{p-t} \rangle \\ &= \delta_{k,p} \langle |\varphi_{k-s}|^2 \rangle \langle |\varphi_{t-k}|^2 \rangle + \delta_{t-p,k-s} \langle |\varphi_{k-s}|^2 \rangle \langle |\varphi_{s-p}|^2 \rangle \\ &\quad + \delta_{s,t} \langle |\varphi_{k-s}|^2 \rangle \langle |\varphi_{s-p}|^2 \rangle \\ &= \delta_{t-p,k-s} \langle |\varphi_{k-s}|^2 \rangle \langle |\varphi_{s-p}|^2 \rangle + \delta_{s,t} \langle |\varphi_{k-s}|^2 \rangle \langle |\varphi_{s-p}|^2 \rangle \tag{10.170} \end{aligned}$$

where the last line is due to the constraint $k \neq p$ in (10.169) which comes out of the inversion formula (10.7). Thus $\langle \sigma_{1,k} \rangle$ has to satisfy the equation

$$\langle \sigma_{1,k} \rangle = \frac{\lambda}{L^d} \sum_t \langle |\varphi_t|^2 \rangle - 2 \frac{\lambda^2}{L^{2d}} \sum_{\substack{p,s \\ p \neq k}} \frac{\langle |\varphi_{k-s}|^2 \rangle \langle |\varphi_{s-p}|^2 \rangle}{a_{1,p} + \langle \sigma_{1,p} \rangle} \tag{10.171}$$

To close the system of equations, one needs another equation for $\langle |\varphi_{2,q}|^2 \rangle$. This can be done again by partial integration as we did in the last section for the φ^4-model. However, for the specific model at hand, one has, by virtue of (10.158) with the indices 1 and 2 interchanged,

$$\langle |\varphi_{2,q}|^2 \rangle = \langle G_2 \rangle(q) \tag{10.172}$$

Thus one ends up with

$$\langle G_1 \rangle(k) = \frac{1}{a_{1,k} + \langle \sigma_{1,k} \rangle} \tag{10.173}$$

where

$$\langle \sigma_{1,k} \rangle = \frac{\lambda}{L^d} \sum_t \langle G_2 \rangle(t) - 2 \frac{\lambda^2}{L^{2d}} \sum_{\substack{p,s \\ p \neq k}} \langle G_2 \rangle(k-s) \langle G_2 \rangle(s-p) \langle G_1 \rangle(p) \tag{10.174}$$

In particular, for $m_1 = m_2 = m$ one has $\langle G_1 \rangle = \langle G_2 \rangle \equiv G$,

$$G(k) = \frac{1}{k^2 + m^2 + \sigma_k} \tag{10.175}$$

and σ_k has to satisfy the equation

$$\sigma_k = \lambda \int_{[-\pi,\pi]^d} \frac{d^d p}{(2\pi)^d} \, G(p) - 2\lambda^2 \int_{[-\pi,\pi]^{2d}} \frac{d^d p}{(2\pi)^d} \frac{d^d q}{(2\pi)^d} \, G(p)\,G(q)\,G(k-p-q) \tag{10.176}$$

To see how the solution looks like one may ignore the λ^2-term in which case $\sigma_k = \sigma$ becomes independent of k. In the limit $m^2 \downarrow 0$ one obtains, if we substitute $4 \sum_{i=1}^d \sin^2 [k_i/2]$ by k^2 for simplicity,

$$\sigma = \lambda \int_{[-\pi,\pi]^d} \frac{d^d p}{(2\pi)^d} \frac{1}{p^2 + \sigma} \tag{10.177}$$

which gives

$$\sigma = \begin{cases} O(\lambda) & \text{if } d \geq 3 \\ O(\lambda \log[1/\lambda]) & \text{if } d = 2 \\ O(\lambda^{\frac{2}{3}}) & \text{if } d = 1 \,. \end{cases} \tag{10.178}$$

10.6 General Structure of the Integral Equations

In the general case, without making the approximation (10.148), we expect the following picture for a generic quartic field theoretical model. Let G and G_0 be the interacting and free particle Green function (one solid line goes in, one solid line goes out), and let D and D_0 be the interacting and free interaction Green function (one wavy line goes in, one wavy line goes out). Then we expect the following closed set of integral equations for G and D:

$$G = \frac{1}{G_0^{-1} + \sigma(G,D)}, \qquad D = \frac{1}{D_0^{-1} + \pi(G,D)} \tag{10.179}$$

where σ and π are the sum of all two-legged diagrams without two-legged (particle and wavy line) subdiagrams with propagators G and D (instead of G_0, D_0). Thus (10.179) simply eliminates all two-legged insertions by substituting them by the full propagators. For the Anderson model $D = D_0 = 1$ and (10.179) reduces to (10.67) and (10.76).

A variant of equations (10.179) has been derived on a more heuristic level in [15] and [47]. Their integral equation (for example equation (40) of [47]) reads

$$G = \frac{1}{G_0^{-1} + \tilde{\sigma}(G,D_0)} \tag{10.180}$$

where $\tilde{\sigma}$ is the sum of all two-legged diagrams without two-legged particle insertions, with propagators G and D_0. Thus this equation does not resum two-legged interaction subgraphs (one wavy line goes in, one wavy line goes out). However resummation of these diagrams corresponds to a partial resummation of four-legged particle subgraphs (for example the second equation in (10.183) below resums bubble diagrams), and is necessary in order to get the right behavior, in particular for the many-electron system in the BCS case.

Another popular way of eliminating two-legged subdiagrams (instead of using integral equations) is the use of counterterms. The underlying combinatorial identity is the following one. Let

$$S(\psi, \bar{\psi}) = \int dk \, \bar{\psi}_k G_0^{-1}(k) \psi_k + S_{\text{int}}(\psi, \bar{\psi}) \qquad (10.181)$$

be some action of a field theoretical model and let $T(k) = T(G_0)(k)$ be the sum of all amputated two-legged particle diagrams without two-legged particle subdiagrams, evaluated with the bare propagator G_0. Let $\delta S(\psi, \bar{\psi}) = \int dk \, \bar{\psi}_k T(k) \psi_k$. Consider the model with action $S - \delta S$. Then a p-point function of that model is given by the sum of all p-legged diagrams which do not contain any two-legged particle subdiagrams, evaluated with the bare propagator G_0. In particular, by construction, the two-point function of that model is exactly given by G_0. Now, since the quadratic part of the model under consideration (given by the action $S - \delta S$) should be given by the bare Green function G_0^{-1} and the interacting Green function is G, one is led to the equation $G^{-1} - T(G) = G_0^{-1}$ which coincides with (10.180).

Since the quantities σ and π in (10.179) are not explicitly given but merely are given by a sum of diagrams, we have to make an approximation in order to get a concrete set of integral equations which we can deal with. That is, we substitute σ and π by their lowest order contributions which leads to the system

$$G(k) = \frac{1}{G_0(k)^{-1} + \int dp \, D(p) G(k-p)}, \qquad (10.182)$$

$$D(q) = \frac{1}{D_0(q)^{-1} + \int dp \, G(p) G(p+q)} \qquad (10.183)$$

This corresponds to the use of (10.7) and (10.8) retaining only the $r = 2$ term. Thus we assume that the expansions for σ and π are asymptotic. Roughly one may expect this if each diagram contributing to σ and π allows a $const^n$ bound (no $n!$ and of course no divergent contributions). The equations (10.182) and (10.183) can be found in the literature. Usually they are derived from the Schwinger-Dyson equations which is the following nonclosed set of two equations for the three unknown functions G, D and Γ, Γ being the vertex function (see, for example, [1]):

$$G(k) = G_0(k) + G_0(k) \int dp\, G(p)\, D(k-p)\, \Gamma(p, k-p)\, G(k)$$

$$(10.184)$$

$$D(q) = D_0(q) + D_0(q) \int dp\, G(p)\, G(p+q)\, \Gamma(p+q, -q)\, D(q)$$

The function $\Gamma(p, q)$ corresponds to an off-diagonal inverse matrix element as it shows up for example in (10.103). Then application of (10.8) transforms (10.184) into (10.179).

One may say that although the equations (10.182) and (10.183) are known, usually they are not really taken seriously. For our opinion this is due to two reasons. First of all these equations, being highly nonlinear, are not easy to solve. In particular, for models involving condensation phenomena like superconductivity or Bose-Einstein condensation, it seems to be appropriate to write them down in finite volume since some quantities may become macroscopic. And second, since they are usually derived from (10.184) by putting Γ equal to 1 (or actually -1, by the choice of signs in (10.184)), one may feel pretty suspicious about the validity of that approximation. The equations (10.179) tell us that this is a good approximation if the expansions for σ and π are asymptotic. A rigorous proof of that, if it is true, is of course a very difficult problem.

The 'Many-Electron Millennium Problems' (€10.000)[1]

Since the author spent a nontrivial amount of time on the following questions, he would be happy if anybody could solve this. The formulation is in the grand canonical ensemble, in finite volume $[0, L]^d$ and at some small but positive temperature $T = 1/\beta > 0$.

Let

$$\mathcal{M} := \left\{ \mathbf{k} \in \left(\tfrac{2\pi}{L}\mathbb{Z}\right)^d \,\middle|\, e(\mathbf{k}) := \mathbf{k}^2 - 1 \le 10 \right\}$$

For $(\mathbf{k}, \sigma) \in \mathcal{M} \times \{\uparrow, \downarrow\}$, let $a_{\mathbf{k}\sigma}$ and $a_{\mathbf{k}\sigma}^+$ be the fermionic annihilation and creation operators obeying the canonical anticommutation relations

$$\{a_{\mathbf{k}\sigma}, a_{\mathbf{k}'\sigma'}^+\} = a_{\mathbf{k}\sigma} a_{\mathbf{k}'\sigma'}^+ + a_{\mathbf{k}'\sigma'}^+ a_{\mathbf{k}\sigma} = \delta_{\sigma,\sigma'} \delta_{\mathbf{k},\mathbf{k}'}$$

and $\{a_{\mathbf{k}\sigma}, a_{\mathbf{k}'\sigma'}\} = \{a_{\mathbf{k}\sigma}^+, a_{\mathbf{k}'\sigma'}^+\} = 0$. Let

$$\mathcal{F} := \oplus_{N=0}^{\infty} \mathcal{F}_N = \oplus_{N=0}^{2|\mathcal{M}|} \mathcal{F}_N$$

be the antisymmetric fermionic Fock space,

$$\mathcal{F}_N := \left\{ \sum_{\mathbf{k}_1\sigma_1 \cdots \mathbf{k}_N\sigma_N} \alpha_{\mathbf{k}_1\sigma_1\cdots\mathbf{k}_N\sigma_N} \, a_{\mathbf{k}_1\sigma_1}^+ \cdots a_{\mathbf{k}_N\sigma_N}^+ |1\rangle \,\middle|\, \right.$$
$$\left. \alpha_{\mathbf{k}_1\sigma_1\cdots\mathbf{k}_N\sigma_N} \in \mathbb{C}, \ \mathbf{k}_i\sigma_i \in \mathcal{M} \times \{\uparrow, \downarrow\} \right\}$$

and let H be the grand canonical Hamiltonian for the many-electron system with attractive ($\lambda < 0$) delta interaction,

$$H = \sum_{\mathbf{k}\sigma} e(\mathbf{k}) \, a_{\mathbf{k}\sigma}^+ a_{\mathbf{k}\sigma} + \tfrac{\lambda}{L^d} \sum_{\mathbf{kpq}} a_{\mathbf{k}\uparrow}^+ a_{\mathbf{q}-\mathbf{k}\downarrow}^+ a_{\mathbf{q}-\mathbf{p}\downarrow} a_{\mathbf{p}\uparrow}$$

The momentum sums range over $\mathbf{k}, \mathbf{p} \in \mathcal{M}$ and $\mathbf{q} - \mathbf{k}, \mathbf{q} - \mathbf{p} \in \mathcal{M}$. Consider the partition function Z and the momentum distribution $n_{\mathbf{k}}$,

$$Z(\lambda) := Tr_{\mathcal{F}} \, e^{-\beta H}$$
$$n_{\mathbf{k}\sigma} := Tr_{\mathcal{F}} \left[e^{-\beta H} a_{\mathbf{k}\sigma}^+ a_{\mathbf{k}\sigma} \right] \Big/ Tr_{\mathcal{F}} \, e^{-\beta H}$$

Then the problems read as follows (the first problem should just remind you of what the ultimate goal is, namely the computation of correlation and partition functions):

---

[1] ..since I'm not the Clay Math Institute, you have to be satisfied with that...

Problem 1) Compute explicitly $Z(\lambda)$ and $n_{\mathbf{k}\sigma}$ (*most likely not possible...*).

€10.000

Problem 2) Find a closed set of r integral equations for r unknown functions, $n_{\mathbf{k}\sigma}$ being one of these functions, where all other quantities except the r functions are explicitly given (*likely not possible...*).

€10.000-1.000r

Problem 3) Let $E_0 = E_0(\lambda, L)$ be the lowest eigenvalue of H and E_1 be the second lowest. Let

$$\Delta(\lambda) := \lim_{L \to \infty} \left(E_1(\lambda, L) - E_0(\lambda, L) \right)$$

Prove that in dimension $d \geq 2$ there is a gap $\Delta(\lambda) > 0$ for sufficiently large β. Let p be the number of pages of your proof (LaTeX, 12pt), then the award is

€10.000-100p

Problem 4) Write down an expansion

$$n_{\mathbf{k}\sigma} = \text{leading order term} + \text{error}$$

where the leading order term is explicitly given and the error is bounded by a term independent of volume and temperature and which goes to 0 for $\lambda \to 0$. Let p be the number of pages of your proof (LaTeX, 12pt; a good Ph.D. student should be able to follow it..), then the award is

€10.000-100p

Total award not more than €10.000, good luck..

References

[1] A. A. Abrikosov, L. P. Gorkov, I. E. Dzyaloshinski, Methods of Quantum Field Theory in Statistical Physics, Prentice-Hall, New York, 1963.

[2] M. Aizenman, G.M. Graf, Localization Bounds for an Electron Gas, *Journal of Physics A*, vol. 31, p. 6783, 1998.

[3] P.W. Anderson, W.F. Brinkman, Theory of Anisotropic Superfluidity, in Basic Notions of Condensed Matter Physics, P.W. Anderson (ed.), New York, Benjamin/Cummings, 1984.

[4] V. Bach, J. Fröhlich, I.M. Sigal, Renormalization Group Analysis of Spectral Problems in Quantum Field Theory, *Advances in Mathematics* 137, 205-298, 1998.

[5] R. Balian, N.R. Werthamer, Superconductivity with Pairs in a Relative p-Wave, *Physical Review* 131, p. 1553-1564, 1963.

[6] J. Bardeen, L.N. Cooper, J.R. Schrieffer, Theory of Superconductivity, *Physical Review* 108, p. 1175, 1957.

[7] J. Bardeen, G. Rickayzen, Ground State Energy and Green's Function for Reduced Hamiltonian for Superconductivity, *Physical Review* 118, p. 936, 1960.

[8] G. Benfatto, G. Gallavotti, Perturbation Theory of the Fermi Surface in a Quantum Liquid, *Journal of Statistical Physics* 59, p. 541, 1990.

[9] F. A. Berezin, The Method of Second Quantization, Academic Press, New York, 1966.

[10] N. N. Bogoliubov, On Some Problems of the Theory of Superconductivity, *Physica* 26, Supplement, p. 1-16, 1960.

[11] N. N. Bogoliubov, Lectures on Quantum Statistics, vol. 2, Gordon and Breach, 1970.

[12] N. N. Bogoliubov, D.B. Zubarev, Iu.A. Tserkovnikov, On the Theory of Phase Transitions, *Soviet Physics Doklady* 2, p. 535, 1957.

[13] T. Chakraborty, P. Pietiläinen, The Fractional Quantum Hall Effect: Properties of an Incompressible Quantum Fluid, *Springer Series in Solid State Sciences* 85, 1988.

[14] T. Chen, J. Fröhlich, M. Seifert, Renormalization Group Methods: Landau-Fermi Liquid and BCS Superconductor, Proceedings of the Les Houches session Fluctuating Geometries in Statistical Mechanics and Field Theory, eds. F. David, P. Ginsparg, J. Zinn-Justin, 1994.

[15] J. M. Cornwall, R. Jackiw, E. Tromboulis, Effective Action for Composite Operators, *Physical Review D*, vol. 10, no. 8, p. 2428-2445, 1974.

[16] M. Disertori, V. Rivasseau, Interacting Fermi Liquid at Finite Temperature, Part I and II, *Communications in Mathematical Physics* 215, p. 251-290 and p. 291-341, 2000.

[17] J. Feldman, T. Hurd, L. Rosen, J. Wright, QED: A Proof of Renormalizability, *Springer Lecture Notes in Physics* vol. 312, 1988.

[18] J. Feldman, H. Knörrer, D. Lehmann, E. Trubowitz, Fermi Liquids in Two Space Dimensions, in: Constructive Physics, *Springer Lecture Notes in Physics*, vol. 446, ed. V. Rivasseau, Springer 1994.

[19] J. Feldman, H. Knörrer, D. Lehmann, E. Trubowitz, A Class of Fermi Liquids, in Particles and Fields, *CRM Series in Math. Phys.*, eds. G. Semenoff, L. Vinet, Springer 1998.

[20] J. Feldman, H. Knörrer, E. Trubowitz, The Fermi Liquid Construction, 10 papers, available under http://www.math.ubc.ca/~feldman/fl.html

[21] J. Feldman, H. Knörrer, E. Trubowitz, A Remark on Anisotropic Superconducting States, *Helvetia Physica Acta* 64, p. 695-699, 1991.

[22] J. Feldman, H. Knörrer, E. Trubowitz, A Representation for Fermionic Correlation Functions, *Communications in Mathematical Physics* 195, p. 465-493, 1998.

[23] J. Feldman, H. Knörrer, E. Trubowitz, Mathematical Methods of Many Body Quantum Field Theory, Lecture Notes, ETH Zürich, 1992.

[24] J. Feldman, J. Magnen, V. Rivasseau, E. Trubowitz, An Infinite Volume Expansion for Many Fermion Greens Functions, *Helvetia Physica Acta* 65, p. 679-721, 1992.

[25] J. Feldman, J. Magnen, V. Rivasseau, E. Trubowitz, Ward Identities and a Perturbative Analysis of a U(1) Goldstone Boson in a Many Fermion System, *Helvetia Physica Acta* 66, p. 498-550, 1993.

[26] J. Feldman, J. Magnen, V. Rivasseau, E. Trubowitz, An Intrinsic $1/N$ Expansion for Many-Fermion Systems, *Europhysics Letters*, vol. 24 (6), p. 437-442, 1993; Two-Dimensional Many-Fermion Systems as Vector Models, *Europhysics Letters*, vol. 24 (7), p. 521-526, 1993.

[27] J. Feldman, J. Magnen, V. Rivasseau, E. Trubowitz, Fermionic Many Body Models, in *CRM Proceedings and Lecture Notes* vol. 7, Mathemat-

ical Quantum Theory I: Field Theory and Many Body Theory, eds. J. Feldman, R. Froese, L. Rosen, 1994.

[28] J. Feldman, M. Salmhofer, E. Trubowitz, Perturbation Theory around Non-nested Fermi Surfaces I. Keeping the Fermi Surface Fixed, *Journal of Statistical Physics* 84, p. 1209-1336, 1996; Perturbation Theory around Non-nested Fermi Surfaces II. Regularity of the Moving Fermi Surface: RPA Contributions, *Communications in Pure and Applied Mathematics* 51, p. 1133-1246, 1998; Regularity of the Moving Fermi Surface: The Full Selfenergy, *Communications in Pure and Applied Mathematics* 52, p. 273-324, 1999; An Inversion Theorem in Fermi Surface Theory, *Communications in Pure and Applied Mathematics* 53, p. 1350-1384, 2000.

[29] J. Feldman, E. Trubowitz, Perturbation Theory for Many Fermion Systems, *Helvetia Physica Acta* 63, p. 156-260, 1990; The Flow of an Electron-Phonon System to the Superconducting State, *Helvetia Physica Acta* 64, p. 214-357, 1991.

[30] R. Fernandez, J. Fröhlich, A. Sokal, Random Walks, Critical Phenomena and Triviality in Quantum Field Theory, *Texts and Monographs in Physics*, Springer, 1992.

[31] A.L. Fetter, J.D. Walecka, Quantum Theory of Many Particle Systems, McGraw-Hill, 1971.

[32] J. Fröhlich, On the Triviality of $\lambda \varphi_d^4$ Theories and the Approach to the Critical Point, *Nuclear Physics B* 200, p. 281-296, 1982.

[33] G. Gallavotti, F. Nicolo, Renormalization Theory in Four Dimensional Scalar Fields, I and II, *Communications in Mathematical Physics* 100, p. 545, *Communications in Mathematical Physics* 101, p. 247, 1985.

[34] R. E. Prange, S. M. Girvin (eds.), The Quantum Hall Effect, second edition, *Springer Graduate Texts in Contemporary Physics*, 1990.

[35] I.S. Gradshteyn, I.M. Ryzhik, Table of Integrals, Series and Products, Academic Press, 1965.

[36] R. Haag, The Mathematical Structure of the Bardeen Cooper Schrieffer Model, *Nuovo Cimento* 25, p. 287-299, 1962.

[37] Hanson, E.R.: A Table of Series and Products, Prentice-Hall, New York, 1975, Sect. 89.5.

[38] R. Haussmann, Self Consistent Quantum Field Theory and Bosonization for Strongly Correlated Electron Systems, *Lecture Notes in Physics*, m56, Springer, 1999.

[39] O. Heinonen (ed.), Composite Fermions: A Unified View of the Quantum Hall Regime, World Scientific, 1998.

[40] K. Huang, Statistical Mechanics, 2nd ed., John Wiley & Sons, 1987.

[41] A. Klein, The Supersymmetric Replica Trick and Smoothness of the Density of States for Random Schrödinger Operators, *Proceedings of Symposia in Pure Mathematics*, vol. 51, part 1, 1990.

[42] T. Koma, Spectral Gaps of Quantum Hall Systems with Interactions, *Journal of Statistical Physics* vol. 90, p. 313-381, 2000.

[43] D. Lehmann, A Microscopic Derivation of the Critical Magnetic Field in a Superconductor, PhD.-Thesis, ETH Zürich 1994, and: *Communications in Mathematical Physics* 173, p. 155-174, 1995.

[44] D. Lehmann, The Many-Electron System in the Forward, Exchange and BCS Approximation, *Communications in Mathematical Physics* 198, p. 427-468, 1998.

[45] D. Lehmann, The Global Minimum of the Effective Potential of the Many-Electron System with Delta-Interaction, *Reviews in Mathematical Physics* vol. 12, no. 9, p. 1259-1278, 2000.

[46] D. Lehmann, Resummation of Feynman Diagrams and the Inversion of Matrices, *Journal of Physics A*, vol. 34, p. 281-304, 2001.

[47] J. M. Luttinger, J. C. Ward, Ground State Energy of a Many-Fermion System II, *Physical Review* vol. 118, no. 5, p. 1417-1427, 1960.

[48] J. Magnen, G. Poirot, V. Rivasseau, Ward Type Identities for the 2D Anderson Model at Weak Disorder, cond-mat/9801217; The Anderson Model as a Matrix Model, *Nuclear Physics B* (Proc. Suppl.) 58, p. 149, 1997.

[49] G.D. Mahan, Many Particle Physics, Plenum Press, 1981.

[50] N. Nagaosa, Quantum Field Theory in Condensed Matter Physics, *Texts and Monographs in Physics*, Springer, 1999.

[51] M.E. Peskin, D.V. Schroeder, An Introduction to Quantum Field Theory, Addison-Wesley, 1995.

[52] G. Poirot, Mean Green's Function of the Anderson Model at Weak Disorder with an Infrared Cutoff, cond-mat/9702111.

[53] A.P. Prudnikov, Yu.A. Brychkov, O.I. Marichev, Integrals and Series, Gordon and Breach, 1986.

[54] V. Rivasseau, From Perturbative to Constructive Renormalization, Princeton University Press, 1991.

[55] M. Salmhofer, Improved Power Counting and Fermi Surface Renormalization, *Reviews in Mathematical Physics* 10, p. 553-578, 1998.

[56] M. Salmhofer, Renormalization. An Introduction, *Texts and Monographs in Physics*, Springer 1999.

[57] J. R. Schrieffer, Theory of Superconductivity, Addison-Wesley, 1964.

[58] A. Schütte, The Symmetry of the Gap in the BCS Model for Higher ℓ-Wave Interactions, Diploma thesis, ETH Zürich, 1997.

[59] T. Spencer, The Schrödinger Equation with a Random Potential: A Mathematical Review, Proceedings of the 1984 Les Houches Summer School, eds. K. Osterwalder, R. Stora, North Holland Physics Pub., New York, 1986.

[60] Wei-Min Wang, Supersymmetry, Witten Complex and Asymptotics for Directional Lyapunov Exponents in $\mathbb{Z}^d$, mp-arc/99-355; Localization and Universality of Poisson Statistics Near Anderson Transition, mp-arc/99-473.

[61] D. Yoshioka, B. I. Halperin, P. A. Lee, Ground State of Two-Dimensional Electrons in Strong Magnetic Fields and 1/3 Quantized Hall Effect, *Physical Review Letters* vol. 50, p. 1219, 1983; *Surface Science* vol. 142, p. 155, 1984.

Index